GOING TO MARKET

Going to Market

· · · · · · · · · · · · ·

DISTRIBUTION SYSTEMS FOR INDUSTRIAL PRODUCTS

·

E. Raymond Corey
Frank V. Cespedes
V. Kasturi Rangan
HARVARD BUSINESS SCHOOL

Research Associates:
Bobbi Carrey
Judith Gurney
Constance Kinnear
John E. P. Morrison

Harvard Business School Press
Boston, Massachusetts

The paper used in this publication meets the requirements of the American National Standard for Permanence of Paper for Printed Library Materials Z39.49-1984.

Library of Congress Cataloging-in-Publication Data

Corey, E. Raymond.
    Going to market : distribution systems for industrial products /
E. Raymond Corey, Frank V. Cespedes, V. Kasturi Rangan ; research
associates, Bobbi Carrey [et al.].
        p.   cm.
    Includes index.
    ISBN 0-87584-202-X
    1. Marketing channels. 2. Industrial marketing. 3. Marketing
channels—United States. 4. Industrial marketing—United States.
I. Cespedes, Frank V., 1950– . II. Rangan, V. Kasturi.
III. Title.
HF5415.129.C67   1989
658.8'02—dc19   89-1769
                                                              CIP

This book is dedicated to Dean John H.
McArthur of the Harvard Business School
in appreciation for his strong support
of research devoted to the advancement
of management practice.

CONTENTS

Preface

THE SEARCH FOR PATTERNS

The authors came into this study with backgrounds in teaching and writing about industrial marketing and working with industrial marketing companies. The stimulus to start on what became a five-year undertaking was the simple recognition that industrial distribution systems formed a critical element of marketing strategy and that business practitioners were expressing concern about channels structure and channels management issues. While there was an academic marketing literature, informed by disciplines such as economics, political science, and social psychology, it did not always or consistently reflect a deep acquaintance with practitioner experience and did not deal systematically with the distribution issues identified by managers as being important. Further, the methodologies employed have usually limited the theoreticians' abilities to capture the complexities of the "real-world" behavioral settings.

Like most empirical field-based research, our study began with only a sketchy sense of the issues and the topics to be covered. We started in 1982–1983 by writing four case studies, each describing the distribution program of an industrial business in the Boston area. Our interest, initially, was to learn how marketing managers thought about distribution, what issues were important to them, and what implicit concepts and attitudes seemed to guide their actions.

The willingness of managers to contribute their time and knowledge to helping us develop case studies came largely out of their own concerns about managing channels of distribution. They spoke to us openly with the understanding that we would not use the information without their approval and formal release. Managers also understood at the outset that we would not disguise the names of their companies or of the products they made and sold, although we were willing to disguise proprietary quantified data as well as the names of personnel and of other companies mentioned in the case.

Each of the four cases focused on one or more issues having to do with the design and/or management of industrial channels networks. In describing the dilemmas and options that managers confronted, we found that we could understand more clearly the reasons for their pursuing one course of action or another. The four studies became even more meaningful to us at a later stage when we were able to compare and contrast these situations with the experiences of other companies selling in different markets. The initial selection of case subjects was random except for location (the Boston area) and subject (industrial distribution). We were therefore fortunate in having been able to develop four studies that introduced us to managers' perceptions about industrial distribution and one case, Barry Wright, that turned out to be key in the area of options in organizing multichannel systems, the subject of Chapter 6.

We also searched Harvard Business School's files for other cases in industrial distribution and found ten. They were very useful to us in broadening our understanding of the issues and of the various modes of distribution exhibited in a wide range of settings.

This early work helped us to understand that the mix in a channel system—of direct selling, independent distributors, agents, and brokers—varied considerably from one company to another for reasons having to do with the nature of the product, market demographics, and the resources that companies had available for marketing investments. We were able to identify, as well, the basic functions that distribution channels performed, such as generating demand, carrying stock, modifying the product, shipping, extending sales credit, and providing after-sale repair and maintenance service. We observed that channels mix and the locus of distribution functions in the system varied both by product-market and within product-market. We made the simple observation as well that distribution was not only a means of access to markets but a competitive battleground; and that often the key to market share supremacy was distribution strength.

The next phase of our field research, then, was to study the distribution strategies of competing producers in one particular product-market. For this purpose, we chose load centers and circuit breakers, commonly known as "fuse boxes," a technically mature product used primarily in the electrical systems of residential and light commercial construction. It was described to us as a "distribution-intensive" product-market in which eight electrical manufacturers accounted for more than 90% of the total supply in the United States. These are firms such as General Electric, Westinghouse, Siemens-ITE, and Square D. We interviewed managers in all eight firms and then conducted questionnaire surveys of the firms and of the electrical distributors through which they sold to user-customers. This phase of the study confirmed that distribution strategy did indeed vary significantly among competing firms. It prompted two other observations: first, that the long-term market leadership position held by one firm, Square D, hinged critically on its superior distribution system; second, that each of the four major suppliers tended to position itself in a different market segment with its distribution network, its product line, and its pricing policies particularly suited to the characteristics of that segment.

The next stage was to extend the study to other product-markets. Recognizing the magnitude of such efforts, we were limited in the number of industry studies we could undertake and opted for a mix that would include a high-technology product (computer disk drives), capital equipment (stationary air compressors), and an industrial supply product (oil country tubular goods), in addition to our low-tech component (load centers). One selection criterion was that 60% or more of the total U.S. supply be accounted for by six to eight producers. This limited the number of firms to be interviewed. All four specific product-market sectors on which we focused were suggested to us by business managers who knew the scope and purposes of our work and noted that one or another "might be what you are looking for." Other product-markets were suggested as well but fell short of meeting our criteria or seemed unlikely to add sufficiently to our understanding.

Having interviewed managers in the six to eight largest firms in each of the four product-markets, we concentrated on the leading firm, one or two followers, and a market share challenger. Interestingly, the last of these in each industry was a foreign multinational company: two were Japanese, one a German firm, and the fourth a Swedish firm. This array gave us the opportunity to compare and contrast firms with varying market shares within a product-market and to compare firms with similar market share positions across widely different product-markets. How, we asked, do distribution strategies differ because of differences in the product-market context and/or differences in competitive positioning within a product-market? What are the similarities in strategy among firms in comparable competitive positions across sectors? The research design for this part of our study is shown below:

	Load Centers and Circuit Breakers	Disk Drives	Stationary Air Compressors	Oil Country Tubular Goods
Market Leader	Square D	Control Data	Ingersoll-Rand	U.S. Steel
Follower(s)	General Electric; Westinghouse	Seagate	Joy; Sullair	Lone Star Steel
Market Share Challenger	Siemens-ITE	Fujitsu	Atlas-Copco	Sumitomo

Our interviewees were selected for their specialized knowledge gained through decision-making responsibility and experience in managing industrial distribution systems. Further, within each company we spoke with a variety of people about a given topic. In this respect, our interview data-gathering technique was analogous to those traditionally associated with ethnographic research used by many anthropologists and sociologists, and the aim was an in-depth qualitative understanding of the decision-making contexts and processes. By contrast, our survey data gathering focused almost exclusively on quantitative information (sales and market share data over time, number of distributors or sales personnel, distribution costs, and so on) rather than on personal attitudes, or behavioral and organizational characteristics. In this way, we attempted to assemble a relevant cross-sectional base including qualitative and quantitative data.

The four product-market sectors represent a variety of channels mixes. In contrast to the distributor-intensive mode of marketing load centers and circuit breakers was the predominantly direct selling pattern for disk drives. Oil country tubular goods managers used a balanced mix of direct and reseller channels, and air compressor firms marketed through the broadest channels mix of the four sectors, including direct selling, independent distribution, captive distribution, and agents. In addition, we could observe that in different sectors and among competing producers in the same sector, distribution functions often tended to be positioned at different points in the distribution chain. In some instances, producers took on major functions even though utilizing resellers; in others, the functions were delegated primarily to external channel members. Finally, we found significant and dynamic shifts in distribution patterns occurring in three of the four product-market groups and had the opportunity to study why these systems were evolving as they were.

Our concern, of course, was what made for these differences. What we found was that the distribution systems we studied had evolved as a result of actions taken incrementally over a long period of time. These were moves such as franchising new distributors, buying a bankrupt distributor and thereby launching—often unintentionally—the development of a captive distribution operation, selling to a discount chain, acquiring a competing firm and consolidating two distribution systems, or revising a distributor price discount schedule. In this sense many distribution systems, like Topsy, "just growed." But this does not mean that the path of growth was arbitrary or illogical.

In time we were able to see that the broad parameters of these systems had been strongly influenced by certain fundamental factors having to do with the market environment and with firm-specific circumstances. Early on, we developed the construct shown in Exhibit P.1, in which we hypothesized that the major external determinants of distribution channel design were product, market, competition, and distribution costs; the major internal factor, the business unit context—that is, its resource availability, the breadth and nature of its product line, and its performance measurement systems. At a later stage in our work we split our "market" box into two domains: market segment demographics and buyer behavior. This later construct is developed in Part I and provides the framework for the topical outline of Chapters 3, 4, and 5.

As we developed the model represented in Exhibit P.1, we

Exhibit P.1 Determinants of Distribution Strategy

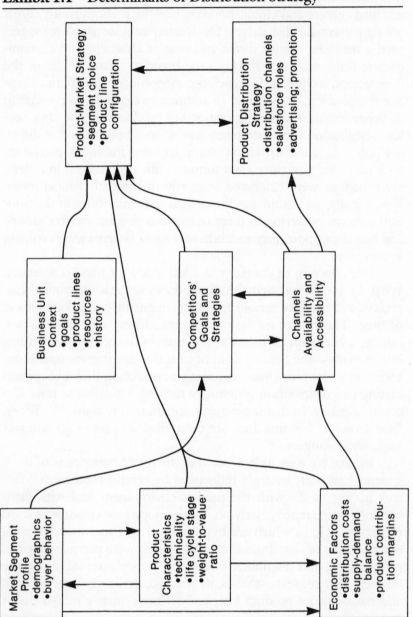

were beginning to find patterns in the ways that distribution strategies in different firms and various product-market sectors had evolved through a long series of apparently temporizing and opportunistic actions. We could see as well that changes in the product-market and business unit environments of each firm created the need for constant reexamination of modes of distribution, often raising agonizingly difficult issues for marketing managers.

At about this stage, we also developed the construct shown in Exhibit P.2. It is a typology of channels functions and of the various channels institutions by which they are performed. This construct provided a framework for understanding the individual producer's channels mix and the locus of distribution functions in the chain that linked factory and user-customer.

Concurrent with our work in the four selected product-market sectors, we studied other firms in other industry settings. We selectively interviewed managers and wrote cases on companies manufacturing such products as computers, electric motors, abrasives, hospital supplies, arc welding electrodes, computer software and supplies, and industrial chemicals. We selected our case subjects to focus on particular issues; by now we had identified a set of generic issues confronting managers concerned with product distribution and searched for companies that had dealt, or were currently dealing, with the dilemmas they posed. Our leads came from the business press, trade journals, current and former students in our MBA and executive programs, and other members of the Harvard Business School Faculty.

We found that as time went on, we developed an "ear" for our subject. Each bit of evidence gathered through formal interviews, casual conversations, and reading began to have greater meaning for us than in the earlier data-gathering stages, and we could quickly compare and contrast new data with those compiled earlier. We were better able to pick up the nuances of distribution system design and to comprehend issues related to systems management. On one occasion, we played back a year-old interview tape recorded in our first field visits and understood considerably more of the interview content than we had during the meeting itself. (We soon gave up taping interviews, however, because the tape recorder seemed to inhibit the free flow of the discussion. Instead, we found it more useful to have at least two members of our team at each interview, taking notes and comparing observations at the end of each session.)

Exhibit P.2   Distribution Systems: Functions and Components

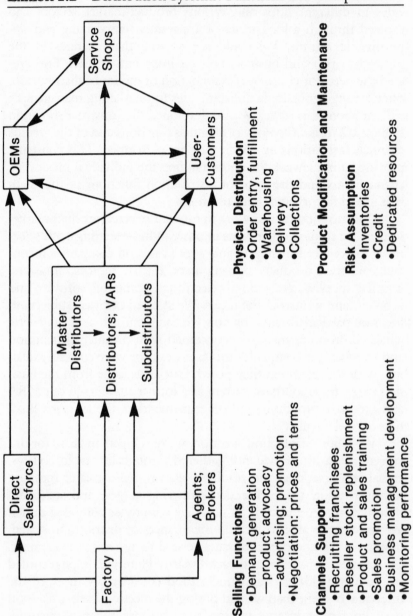

We analyzed our data as the study proceeded, reviewing each case and each set of interview notes to see what topics had been covered and what had been learned. A useful device was to go over the material as though we were preparing to use it for classroom purposes—defining lines of inquiry, analyzing quantitative and qualitative data, and formulating managerial options.

As our field work progressed, we initiated a library search to provide a broader backdrop against which to analyze our field studies. The search encompassed two areas: the history of industrial distribution in the United States and legal aspects of supplier-distributor relationships. We recognized midway in the project that the country context significantly shaped the distribution infrastructure and strongly influenced the options producers had in developing distribution strategies and managing channels. In conversations with business people who came to Harvard Business School from other countries, we learned that distribution in their homelands differed in certain important respects from the U.S. infrastructure. U.S. manufacturers doing business abroad often had to adapt their modes of going to market to different regulatory conditions, different norms with regard to intra- and interchannel competition, different ways of doing business, and different, perhaps fewer, intermediaries through which to reach markets. In some parts of the world, we were told, the legal framework strongly protected the interests of resellers, requiring, for example, that producers using indirect distribution must sell to all dealers who wished to carry their lines. Also, in discontinuing franchisees, producers incurred heavy termination costs in some countries. Often, distributor associations outside the United States limit the entry of new resale businesses and are powerful in setting the conditions of producer-reseller relationships. While it was well beyond the scope of our work to study distribution in other countries, we could, at least, attempt to understand the factors that shaped our own system.

In preparing our chapters on the history of industrial distribution in the United States, we relied extensively on the histories of individual companies and on studies that tracked U.S. economic development. There is, in fact, nothing written on the history of industrial distribution per se; we could only trace the evolving patterns and trends through the use of relatively fragmentary data extending from the War of 1812 to the present. The result, however, is the only description there is of how the industrial distribution infrastructure in the United States has evolved over more than 160 years.

The chapter on legal aspects of distribution is based on the facts and judicial opinion relative to landmark cases, most of which center on distributor complaints about the allegedly illegal actions of their suppliers. While the law in this area has evolved considerably over the past half-century, we concern ourselves with a description of the legal parameters that currently govern channels relationships. The attitudes of the courts toward producer-reseller relationships are relevant as producers establish distributor franchise terms and conditions and seek to ensure conformance.

With the great bulk of our data in hand, we prepared a first-cut topic outline and identified the cases and notes relevant for each subject covered. We found that most pieces of field data touched on several topics, at the same time being especially relevant for one. The outline itself went through several iterations as we wrote and rewrote chapter drafts and elaborated the basic constructs in our minds and on paper.

Incorporating field data in chapters as illustrative material is not easy; the difficulty comes in presenting succinctly the essential points and issues and at the same time providing enough background information so that the reader may understand the product-market/business-unit settings. Our solution has been to include in our text the most broadly relevant data on which our generalizations are based, both to give these observations practical meaning and to allow readers to evaluate for themselves the validity of our conclusions. The details of the four product-market studies are included in Appendix A, to be perused selectively in accordance with the reader's specific interests.

In the course of the study, we interviewed more than 200 managers in over 50 businesses. We conducted an exhaustive search of relevant reference material in Harvard Business School's Baker Library. We made use, as well, of articles in business, legal, and trade journals; industry data services; internal company memoranda; published corporate reports; other Harvard Business School cases; and previous academic research.

What we gained was an understanding of the factors that shape channels design—the mix both of channel types and of the locus of distribution functions in the channels network. We gained insight, too, into the processes of organizing and administering distribution systems and into the issues that often arise in managing channels relationships and effecting changes in the system. We recognized the critical importance of distribution strength as a factor contributing to market leadership. We developed a broadened understanding of the distribution infrastructure in the United States and of how and why it has evolved.

At one level, therefore, this book is about an important but relatively neglected topic in marketing: the design and management of distribution systems for industrial products. At another level, our topic implicates a far broader story: the establishment, maintenance, and (often) dissolution of interorganizational relationships in a dynamic market economy. We hope marketing managers and students of marketing will find our treatment of the former topic relevant and useful, and we hope that the general business reader will find the book thought-provoking with respect to that sector of our economy—business-to-business transactions—which in fact accounts for the bulk of our gross national product.

Acknowledgments

This study is the product of a great many contributions that we gratefully acknowledge. We are particularly indebted to John McArthur, dean of the Harvard Business School, for his generous support of our work in granting us time from teaching and administrative assignments and in funding, unstintingly, the costs of travel, research and editorial assistance, and secretarial support. These substantial commitments, incurred over five years, were made with the expectation that this study would contribute to knowledge in the field of industrial marketing and to the professional development of the authors. We were fortunate to be conducting this study under the aegis of this school. The substantial costs of the project were covered entirely out of the school's general funds for research and project-specific grants from the George Farkas Fund for Research in Distribution each year for five years. In addition, we drew heavily on the school's extensive alumni network and on corporate members of the Associates of the Harvard Business School in gathering field data.

We are grateful indeed to the more than 200 business managers who shared openly with us their knowledge of, and experience in, building and managing industrial distribution channels, sharing their concerns and problems as well. They did so in the interest of contributing to the improvement of business practice and business education and of themselves gaining broader perspectives on principles that might usefully inform the decisions they are called on daily to make.

The four persons listed on the title page as research associates made very important contributions. Bobbi Carrey and Constance Kinnear were involved in the early stages of the work, concentrating, respectively, on questionnaire survey design and the initial field case studies. Judith Gurney conducted the exhaustive library studies that provided the basis for the two chapters on the history of industrial distribution and the one on the legal framework. John Morrison worked on developing field case materials in the later stages of the project, contributing in particular to the chapter on oil country tubular goods as well as to other cases on individual companies.

Professor Alfred D. Chandler, Harvard Business School's Isidor Straus Professor of Business History, reviewed the two history chapters and gave us very useful suggestions. Professor Philip E. Areeda, Langdell Professor of Law at the Harvard Law School, contributed importantly out of his expertise in antitrust law through his commentary on the legal framework chapter.

The book in draft form was reviewed in its entirety by Professor Richard E. Caves, George Gund Professor of Economics and Business Administration, as well as by Associate Professor Richard S. Tedlow, both members of the Harvard Business School Faculty. Their constructive suggestions with respect to factual accuracy, chapter organization, and clarity and brevity of presentation are gratefully acknowledged.

Jane Barrett provided outstanding secretarial support. The high quality and efficiency of her work as she scheduled our field trips; prepared successive iterations of field notes, cases, and chapter drafts; managed the dissemination and tabulation of questionnaire surveys; and maintained project files contributed importantly to the organization of our efforts and to meeting deadlines. We also had great support from Rose Giacobbe, manager of our Word Processing Center, and from her very capable staff.

Members of our publishing division, the Harvard Business School Press, could not have been more supportive and helpful in managing the peer review process, providing useful editorial suggestions, and planning and implementing production and marketing programs for our book. In particular, we are indebted to Joanne Segal, director of the HBS Press; Barbara Ankeny and William Ellet, acquisitions editors; Paula Duffy, director of marketing and senior editor; Maria Arteta, associate director for market research and planning; Christopher Kellogg, production manager; Natalie Greenberg, production editor, and Nancy Gooden, director of book marketing. We also express our sincere thanks to Marilyn Shepherd for her detailed and constructive editorial work on the manuscript.

Our greatest debt is to our wives, who encouraged us in our work at the same time as they took on much more than their shares of family responsibilities. Five years is a long time for spouses to carry the extra burdens of raising children, managing families and homes, and providing moral support. And so to Charlotte Worrall Corey, Bonnie Costello, and Jayanthi Rangan, we express our love and sincere appreciation.

Finally, it is important to note that we alone take responsibility for the final outcome. The analysis of our data reflects our own perceptions, interpretations, and judgments—not those of our business informants or of our academic colleagues. We have been keenly aware, throughout, of the importance of being objective and constructive in drawing conclusions, knowing that business readers will give our ideas serious consideration and that academicians and students, as well, may integrate these constructs into their own frameworks of knowledge about business. We hope this book will serve them well; that will happen, however, only if the practitioners, students, and academicians who read our chapters do so critically and bring to bear their own special insights in the interpretation of our data.

Boston, Massachusetts E. Raymond Corey
October 1988 Frank V. Cespedes
V. Kasturi Rangan

Introduction

Channels of distribution, those networks through which industrial products flow from point of manufacture to point of use, are basic to an industrial economy. If farms and factories are the heart of industrial America, distribution networks are its circulatory system. Composed of many thousands of manufacturers' sales branches, wholesalers, agents, and brokers, industrial distribution systems generate and fulfill demand; they buy and sell, store and transport goods, provide sales financing, often fill the need for after-sale repair and maintenance services, and make markets for used and reconditioned equipment. Distribution networks are loosely organized federations of independent enterprises that are held together by contractual arrangements, informal understandings, and mutual expectations.

Based extensively on the experiences of industrial marketing managers, this book addresses issues with which they cope in designing and managing distribution channels. The first of these, channels design, involves choices with regard to (1) the relative balance of direct sales and selling through external intermediaries; (2) the selection of intermediary types, such as distributors, agents, and brokers; (3) the relative intensity of reseller representation in the producer's market areas; (4) the division of selling responsibility among elements in the network as defined in terms of products, customer categories, and geographic territories; (5) the locus of responsibility among channels members for such functions as sales negotiation, sales financing, maintenance of field inventories, shipping, and providing after-sale service; and (6) sales organization.

Once the structure is in place, marketing managers must then cope with such matters as the nature of the producer-intermediary relationship as articulated formally in the franchise agreement and informally in day-to-day interactions among representatives of the producer and intermediaries. Channels management also involves dealing with such day-to-day issues as inter- and intrachannel competition for user accounts, monitoring and influencing often incongruent market price structures, gaining the support of intermediaries for the producer's marketing programs, and preserving the integrity of the channels network from competitive intrusion.

In the guide that follows, we lay out the structure of the book itself. Our intent is to help readers, first, to understand how the authors have approached this broad and complex subject and, second, to facilitate their reading selectively as meets their interests. While the book is addressed particularly to business managers, we recognize that it will be relevant as well for academics and students of business administration. Among these several audiences, some may have more interest in particular topics, industry segments, and/ or companies described in these pages than in other elements of the book.

A READER'S GUIDE

Part I outlines the basic structure of the industrial distribution system and the factors that influence it. The section opens with a chapter devoted to understanding distribution's role in the market leader's strategy, and supports this analysis with overviews of competitive distribution systems in each of four selected product-markets. The remaining chapters in the section describe the major factors influencing distribution structure, giving examples from industry to support the model we present.

Chapter 2 presents core concepts on which the three following chapters build. It describes the elements of a distribution strategy and then sets forth a model of the factors that tend to shape producers' modes of going to market. These are the nature of the product, market demographics, buyer behavior or modes of product acquisition, distribution cost factors, and the producer's business unit profile, including internal conditions such as resource availability and external factors, namely, competitors' strategies.

Chapters 3, 4, and 5 are devoted to amplifying these elements in our model. Chapter 3 deals with product and market factors as determinants of distribution system design. It also considers the influence of buyer behavior, seeking to understand why purchasers source through one channel or another and, in fact, why and how large buyers, in negotiating with producers, influence the latter's modes of distribution. Chapter 4 is a discussion of distribution costs. It breaks down the expenditures for taking products from factory to point of use and describes the factors that affect distribution cost structures. In particular, it shows the influence of cost considerations in choosing among optional modes of product distribution and in determining the locus of distribution functions within the channels network. Chapter 5 is written from the perspective of marketing managers who have the responsibility for developing and implementing distribution strategies. It looks at those firm-specific factors that shape distribution strategy, such as the availability of resources for selling and distribution, sales performance measures, the business culture, and current commitments to some existing distribution scheme. It also deals with the constraints imposed and the opportunities presented by competitors' market positioning.

Part II reviews the range of options for organizing multichannel systems and deals with topics in channels management. Distribution systems cutting across the boundaries of channel member units as they do—including direct salesforces and a range of external intermediaries—have their roots in, and are managed through, the producer's marketing organization structure. The different industrial companies we have observed exhibit a wide range of organizational arrangements through which distribution channels network relationships are administered. Further, in keeping with the principle that "structure follows strategy," organizational patterns appropriately change when old distribution strategies give way to new. In Chapter 6, we consider alternatives for structuring the field salesforce and the factors that influence choice. Another topic in this chapter is organizing the channels management function. Both matters are operationally important and strategically important as well, in the respect that while structure follows strategy in principle, a distribution system organization, once in place, tends to influence strategy.

These discussions of structure are followed by ones having to do with channels management issues. Chapter 7 deals broadly with sources and uses of channels power—the power suppliers may have in securing the support and cooperation of their channels and the power that resellers may have to control their own strategies free of the influence of supplier-imposed conditions and pressures. In a positive sense as well, this chapter delineates the ways in which suppliers may be effective in building distribution system support for, and commitment to, the supplier's marketing programs. Because supplier-reseller relationships are fraught with legal implications, a brief overview of the relevant regulatory framework is included at the end of Chapter 7. (A detailed discussion of the laws and the body of court opinion that mediate producer-reseller relationships is presented in Chapter 14.)

Chapter 8, "Managing Channels Conflict," deals with issues of inter- and intrachannel conflict and ways producers have coped with it. Designed to be complementary in theory, the several channels in a multichannel distribution system tend often to compete with one another. Actual competition for customers is augmented as well by both heightened perceptions of inter- and intrachannel conflict and a sense of threat on the part of channels members as producers make changes in their distribution strategies. The direct salesforce is perceived as being in competition with resellers, the independent distributor organization with the firm's captive distributor operation, full-service resellers with discount houses, and within a channel, one intermediary with another, both within and across territorial boundaries. Thus, defining the respective "turfs" of the components of multichannel distribution systems and resolving day-to-day conflict are the subjects of Chapter 8.

A note following Chapter 8 takes up the pros and cons of captive distribution as an element in a multichannel system. Operated as a profit-centered business unit in the producer's organization, the captive branch, like independent distributors, typically functions as a reseller of the products of sister departments and a wide range of other products sourced from outside vendors. Perceived usually to be in direct competition with the producer's independent channels but at the same time serving as a significant source of revenue, captive distribution often plays an ambivalent role in a multichannel distribution system. This discussion is intended to lay out the range of issues and to suggest ways of dealing with them.

Chapter 9 extends the discussion of dealing with channel conflict by examining an especially thorny issue in channels management, "The Gray Market Dilemma." Many producers of industrial products—as well as consumer goods—find their products moving out of their authorized channels and into the possession of unfranchised resellers. Personal computers are a case in point, with a significant percentage of sales to users going through such discount outlets as 47th Street Photo. We consider the factors that give rise to gray markets, the roles that gray marketers play in the system, and how they impact producers' authorized channels. While gray markets pose seemingly intractable dilemmas, we provide some perspectives and suggest possible actions for ameliorating conflict.

Chapter 10 is entitled "Coping with Change." We assert that of all the elements of marketing strategy, none is more difficult to change than the distribution system, that efforts to adopt new modes of going to market almost always fly in the face of a complex of deeply rooted commitments—both explicit and implicit—among producers and intermediaries, perpetuated often in a network of personal relationships. While change may be essential for competitive survival, it seldom if ever comes without disadvantage to members of the distribution system previously in place. This chapter considers the kinds of changes that managers are often called on to make, the factors that create the need for these changes, and the forces of resistance with which managers must cope in responding to the imperatives of the evolving product-market environment.

Part III looks at distribution in the United States from a broad perspective, discussing the overall distribution infrastructure, the history and current trends of industrial distribution in this country, and the relevant legal framework. The book closes with a conceptual discussion of industrial systems, raising issues related to future trends and the likely evolution of channels institutions.

Drawing on Department of Commerce data and other statistical studies, Chapter 11 in Part III profiles the industrial distribution infrastructure in the United States. We describe, for example, the balance of direct and resale sales by major industrial goods categories and suggest reasons for the observed differences among product classifications. We also describe the institutions that make up resale channels networks in this country in terms of their average sales revenues, number of personnel, financial structures, and forms of ownership. In addition, we suggest the factors that in the long run have been at work shaping our distribution infrastructure. This discussion is based on a history of industrial distribution in the United States from 1812 to the present, with eras of growth and change being demarcated by major wars: the War of 1812, the Civil War, and the First and Second World Wars. History is useful because in understanding it, we can see the evolution of industrial distribution in this country and the directions it is likely to take. Chapter 12 gives an historical perspective and may be particularly useful for those who are interested in understanding the historical forces that are still at work shaping industrial distribution in this country. With these perspectives, managers may better anticipate change, foresee its directions, and respond in timely fashion.

Chapter 13 is a summarization of that body of federal legislation and court opinion which deals with relationships among producers and external intermediaries, particularly independent distributors that buy and resell. It describes what producers may and may not do legally in seeking to influence what product lines— theirs and competitors'—its franchised reseller will carry, what resale prices it will charge, and where and to whom it will sell. Not intended to be legally authoritative, Chapter 13 has as its purpose to provide marketing managers with a sense of how their strategies and tactics may possibly be viewed by the courts in this country.

While the discussion to this point is primarily from the perspective of the manager in an industrial products company, our final chapter, on distribution in a market economy, takes a broader view. It begins by offering a concept of competition as a game played by *industrial systems*, rather than as single-level games played among competing producers and contending resellers, respectively. There follows a consideration of how our market-driven system works, the factors that shape it, and how it evolves over time in response to change. Finally, we note that these phenomena and the interactions of producers and intermediaries raise important issues of equity and fairness. We conclude by asserting that the ways in which these issues are handled day-by-day and long-term is an essential quality affecting the health and viability of our economic institutions.

The four studies in Appendix A provide in-depth coverage of firms in the product-markets we studied so that readers might consider how producers of directly competing brands develop and implement their respective distribution strategies in an industry setting and in the context of their individual marketing strategies. Each of the four studies describes competitive interaction in a particular product-market, showing the centrality of distribution strategy in overall business strategy. Each case history tells a markedly different story, adding to an understanding of how product-market setting, product life cycle stage, industry economic circumstances, the market share positioning of individual competitors, and the accidents of history produce a wide array of modes of distribution among industrial firms. For those interested in some of these product categories and not in others, the reading here may accordingly be selective.

Appendix B is a "Glossary of Terms," which the reader may find useful as a reference resource. As in so many fields of work, industrial distribution has its own language, and we use it freely in this book. So that the reader can begin the book with a basic knowledge of the channels entities, we now introduce our "cast of characters."

CHANNELS ENTITIES

Distribution networks, often having long histories of transactional exchange and personal relationships, may include a range of channels institutions. These are direct salesforces, independent distributors, captive distributors, manufacturers' reps (MRs) or agents, brokers, repair shops, and even other manufacturers. Although the links in the chain of product distribution from producer to user are familiar to industrial marketers, the distinction among these elements is not always clearly understood.

As employees of the producing company, *direct sales reps* perform a variety of tasks. A key function is to call on those with ultimate responsibility for making purchasing decisions in user accounts and persuade them to buy. Since industrial buying decisions often implicate a number of personnel in the customer organization, the salesperson is likely to be involved with the customer's procurement managers, production personnel, engineers, and financial managers, as well as with outside influencers such as consulting firms and architects. Salespeople negotiate prices and contract terms and often assume after-sale responsibilities in user accounts, such as expediting delivery, advising on product installation and use, processing warranty claims, and accepting returns.

Sales reps calling on distributors perform these and other functions. They train distributor personnel in product technology and selling techniques; they accompany them on sales calls. They typically inspect distributor inventories and may enter stock replenishment orders on the distributor's behalf. Sales reps are their companies' emissaries, as well, in implementing special distributor promotions, product demonstrations, new product launches, and programs aimed at the improvement of distributor business management practices.

A very important function for sales reps calling on both user-customers and resellers is to develop and maintain personal relationships in these accounts. Often producer-reseller and producer-user relationships are of long standing and are built on trust and personal interdependencies; these relationships become personified for them by the producer's sales rep, on the one side, and by customer and reseller personnel, on the other.

Finally, the direct salesforce is an important source of information on product performance problems, possible new applications, customers' plans and sales forecasts, and competitive activity.

Distributors are resellers; they buy goods from producers and sell them to industrial users and sometimes to other resellers. Typically they are organized as independent businesses in the form of partnerships or publicly or privately held companies. While distributor businesses may operate locally, regionally, or nationwide, by far the greatest number of distributors operate locally through one to three sales locations. On average, they generate annual revenues of $2–25 million, as data on channels institutions in Chapter 11 indicate.

Distributor businesses are defined in product-market terms. For example, they may be known as steel warehouses, chemical distributors, construction equipment dealers, and engine distributors. Or they may be identified with particular customer sets, such as hospital supply houses, office supply dealers, oil field supply houses, marine supply stores, and industrial distributors or mill supply houses.

So-called *captive distributors* function like independents with one major difference. They are owned by producers and typically operate as business units within a larger corporate context. They often have a dual mission: to contribute to the revenues and profits of sister divisions by selling products manufactured by these divisions and to contribute profits to the parent company as a reseller of its goods and the goods of noncompeting outside suppliers. As employees of a corporate parent, the managers of a captive distribution unit are compensated by salary and often bonuses. Unlike their independent distributor counterparts, they neither assume the financial risks nor receive the monetary rewards of owning and managing their own businesses.

If the parent organization operates nationally, the captive usually distributes nationally as well but may have uneven geographic coverage. A reason is that many captive distribution systems were built opportunistically through the acquisition of independent distributors in financial stress and/or in need of leadership following the owner's death or retirement. Branches in a captive system may also be established in market areas where the parent company is unable to find adequate representation among local independents.

Another important distribution channel is the *manufacturers' rep* (MR), or agent selling on commission. The MR typically carries limited lines and may represent only five to ten suppliers. Unlike most distributors, which sell to user-customers, MRs generally sell to other resellers—distributors and retailers—as well as to user-customers. In the latter case, the selling may have a high technical content, require multiple relationships in the customer organization, and involve a long time cycle—conditions that typically constrain most distributors from lengthy account development work. Individual MR sales also tend to be for much higher total dollar amounts than distributor transactions.

In addition, some manufacturers may rely significantly on *brokers*. Like MRs, brokers operate on a commission, not taking title to, or physical possession of, what they sell and often not ever seeing it. Unlike MRs, brokers tend to seek business opportunistically, representing different principals and not having long-term relationships with any one. Their functions as go-betweens include (1) finding and locating buyers for large lots of products that have not gone to market through otherwise established channels, (2) identifying buyers shopping for low price in times of excess supply or locating sources of available supplies in shortage periods, and (3) finding buyers and sellers for used equipment. Brokers are dealmakers; they "make markets" and often represent producers in areas where the latter have no other distribution. Brokers are important channels in the United States for such products as crude oil, agricultural commodities, chemicals, industrial machinery, and computers.

The *repair or service shop* can be a specialized entity in the distribution system, operating independently or as a captive of the parent manufacturing company. A more common arrangement, however, is to franchise distributors for field product service, as, for example, in the case of air compressors, farm equipment, or trucks. Postsale service typically provides a relatively steady stream of income and may help to support the distributorship through periods of low sales revenues.

Finally, the *OEM customer* may itself become a conduit through which a supplier's goods flow to market. General Electric's Component Motor Division, for example, sells motors to the GE Major Appliance Division and to other manufacturers of such appliances and equipment as dishwashers, clothes washers and dryers, and heating and air conditioning systems. It also sells spare parts through its distribution systems to service and repair shops.

As becomes apparent, the scope of this book is indeed wide. That, of course, is by choice and predicated on our observation that there was no comparable work that served to chart the territory. The area was defined broadly as well, in the belief that to study any one aspect of industrial distribution as an element of marketing strategy would be to take it out of a context in which all dimensions of the subject are related to all others. Hopefully, others will follow this initiative to explore in greater depth the many facets of those systems which take industrial products from factory to point of use.

• PART I • Designing Industrial Channels Systems

In this first section of the book, we discuss the major factors that suppliers should consider in determining distribution strategy. Chapter 1 focuses on distribution as a major competitive element, giving examples from four product-markets we have studied in depth—load centers, disk drives, air compressors, and oil country tubular goods.

In Chapter 2, we outline the major distribution choices to be made by producers, illustrating the elements of choice with an industry example, that of Ingersoll-Rand. We then describe the factors that influence those choices, namely, the nature of the product, market demographics, buyer behavior, distribution cost factors, and the producer's business unit profile with particular reference to resource availability and competitive positioning. We note the country environment as another determinant of distribution patterns but reserve it for consideration in Part III.

Chapter 3 relates product factors, market factors, and buyer behavior to prevailing modes of distribution, using examples from the four product-markets, as well as a case history—that of Becton Dickinson, a hospital supplies manufacturer. In Chapter 4, with an example taken from IBM, we focus on issues of distribution cost efficiency and cost effectiveness. Finally, in Chapter 5 we explore the nuances of business unit environment—both internal and external—and their impact on distribution strategy, especially the importance of resources to distribution strength. Our primary case references in Chapter 5 are Lincoln Electric and Alloy Rods, makers of arc welding supplies and equipment.

1

1 Distribution: The Leader's Edge

Effective channels networks distinguish market share leaders from followers. Industrial distribution systems add value to the product. They are the means of access to markets and a conduit for products flowing from factory to point of use. As producers seek to gain market access for themselves and to foreclose access to their rivals, channels are also competitive battlegrounds.

In this first chapter, we position distribution as a key element in marketing strategy and discuss its implications in the context of four different product-markets—selected because they vary widely in terms of the nature of the product, market demographics, and the degree of product technical maturity. We look at the strategies of the leading firm, followers, and the major challenger. We investigate these issues: How and why has each competitor positioned itself in its particular product-market? What factors account for its distribution strategies? In the discussion that follows, we draw on the four studies in Appendix A, which describe in richer detail the competitive interplay within each product-market, and the crucial role that distribution has played in the fight for market share.

FOUR REPRESENTATIVE INDUSTRIES

The four product-markets we selected, each with two or more market segments, are the following:

Load Centers and Circuit Breakers. With the lowest unit value and the most mature technology of the four product categories, load centers and circuit breakers (LC&CBs) may be classified as commodities. Commonly known as "fuse boxes," load centers and circuit breakers are devices for interrupting electrical circuits in the event of line defaults. They are found in virtually

3

every electrified residence and in light commercial construction—shopping malls, office parks, theaters.

Competing with such electrical industry giants as General Electric, Westinghouse, and Siemens, the market leader in this product-market has been a smaller and highly focused firm—the Square D Company.

Disk Drives.   At the other end of the technical maturity scale from LC&CBs, disk drives are essential components in computers, often representing a third or more of the cost of the finished product. In the race among computer manufacturers to gain product advantage by providing greater information-processing capability in increasingly smaller units, advancing disk drive technology has played a key role and continues to evolve rapidly.

In the market for 5.25-inch, 30- to 100-megabyte disk drives, a particularly high-volume product category, the leader has been Control Data Corporation. Its leadership, however, has been under constant challenge from such well-known competitors as Seagate, Quantum, Hitachi, and Fujitsu.

Stationary Air Compressors.   Classified as capital equipment, stationary air compressors supply pressurized air for use in manufacturing operations. They have ranged widely in capacity from ¾ to 6,000 horsepower, with unit prices sometimes in excess of $5 million. The largest product category in terms of both units and dollar sales is the 30- to 100-horsepower class.

The market share leader has been Ingersoll-Rand's Stationary Air Compressor Division. Other major competitors in the U.S. market have been Joy Manufacturing, Sullair, and Atlas-Copco, a large Swedish firm.

Oil Country Tubular Goods.   Representing 7–8% of the cost of completing an oil or gas well, oil country tubular goods (OCTG) include the pipe used to drill the hole, the casing that lines the well hole, and the tubing inside the casing and through which oil and gas are brought to the surface. OCTG producers and users recognize this as a three-tier market. At the high end are tubular goods made of special alloys for use in hostile drilling environments. Second-tier products are used in less demanding but still critical applications, such as deep-well onshore and offshore drilling. Third-tier products, lower in both quality and price per ton, are in demand primarily for drilling shallow land wells.

As of the mid-1980s, foreign imports took approximately 63%

of a rapidly deteriorating U.S. market, up from 44% three years earlier. In Tier 1, Sumitomo—the largest importer of Japanese-produced tubular goods and the one with the longest record of participation in the U.S. market—held a leadership position. In Tier 2, U.S. Steel was recognized as the dominant competitor and is credited with having the largest share of the market held by domestic producers. No single producer was recognized as the Tier 3 market leader.

These four product categories may be differentiated along several dimensions: product type, that is, capital equipment, consumable supplies, or components; unit value—relatively low for load centers and high for large air compressors; stage of technical maturity—with disk drives at one end of the scale, load centers at the other; and demand volatility, with OCTG demand the most volatile—and deteriorating—and the demand for disk drives also volatile, but expanding.

Both across and within the four product-markets, the degree of segment concentration also ranges widely. In all cases, there are market segments in which the buyer population consists of a small number of large purchasers, and others in which the customer set is fragmented in size and geographically dispersed.

Finally, purchasing behavior varies from one market segment to another. The purchase transaction may cover a range of products generally related in use (bundled buying), or the item may be bought separately (unbundled). This difference is roughly related to another: whether the product is typically customized to user specifications or taken "off the shelf." Customized products, such as large air compressors, are usually bought as separate items; standard products may or may not be purchased in a single-item transaction.

Load centers are typically bought as items on a list of electrical components for use on a construction site. Disk drive and air compressor purchases are most often unbundled. As for tubular goods, the 20 major oil companies typically negotiate separately for these supplies, but practice seems to vary among the smaller independent drillers.

These distinctions with regard to product, market, and buyer behavior contributed to significant differences among the four product groups in distribution patterns. Load centers go to market largely through independent distributors. At the other end of the scale, the bulk of disk drive sales are made to original equipment manufacturer (OEM) customers by the manufacturers' direct sales personnel with some use of agents and distributors to reach small OEMs and to serve as back-up sources of supply for the large direct customers. But this pattern is shifting somewhat as disk drive prices come down, as the market grows, as segments proliferate, and as resellers, particularly value-added resellers (VARs), develop applications-specialized systems for particular classes of customers.

For both stationary air compressors and tubular goods, the direct-reseller balance is also changing, but in different directions. Increasingly, suppliers have channeled air compressors through independent distributors, allowing their resellers to take on larger capacity units and introducing small compressors into the tradesperson and consumer do-it-yourself markets. On the other hand, major suppliers of OCTG have decreased the participation of resellers in the past decade and have become increasingly involved in direct contract negotiations with large users. This change followed the precipitous drop in oil and gas drilling activity after 1982.

In the section that follows, we consider the distribution strategies of the market segment leaders and of certain challengers in the four product-market categories and the ways in which distribution has contributed to competitive advantage.

THE MARKET SEGMENT LEADERS

What we found in studying these four product-markets is that the leading firms share certain common characteristics: In all cases they have had a first-mover advantage. That is, each had (1) entered the market early with an innovative and/or technically superior product, (2) established a strong distribution network, and (3) built an installed base of products in use, which served to generate demand for replacement sales, peripheral equipment revenues, and, in some cases, after-sale service and parts.

Major Competitive Factors

Early leadership positions had a tendency to be self-perpetuating as long as the leader did not yield to competitors on product superiority, continued to build and hold distribution, and maintained strong direct relationships with major user-customers. Product superiority tends to generate a large installed base and to

Exhibit 1.1   Elements of Distribution Strength

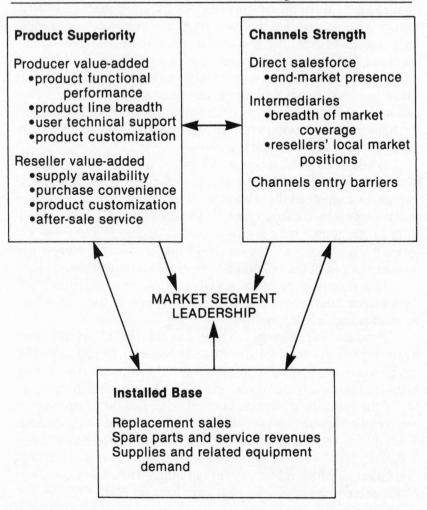

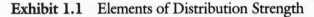

create a strong brand franchise among user-customers, which facilitates building distribution strength. Conversely, effective distribution is essential to the development of the installed base. Together these factors give the producer a competitive edge in maintaining product superiority through access to knowledge about product application technology that influences new product development. Exhibit 1.1 suggests a way of thinking about these relationships.

However, a particular combination of these factors may give a producer a dominant position in some segments but not in others that require a different product-price-distribution strategy. Indeed, competitive superiority in some segments may preclude comparable success in other segments if the distribution policies and practices suitable in one segment are at odds with those needed in another. It may also be true that if taking and holding a high market share in one segment require greater price aggression than in another, competing actively in both would mean yielding potential margins in the less price-sensitive segment. Consequently, competitors may focus on different segments.

While U.S. Steel was a leader in the Tier 2 OCTG market, Sumitomo was in a superior position in the Tier 1 segment. While Square D dominated the residential housing and light construction market segments for load centers, Siemens-ITE had a leading position in the multifamily housing market. Ingersoll-Rand was the market share leader in the above-25-horsepower air compressor market but not in the below-25-horsepower product class.

The following paragraphs elaborate on product superiority, distribution strength, and end-market presence as they contribute to market segment leadership.

Product Superiority.   Three concepts are important here: First, superiority must be measured in terms of a product's value both to end-users and to intermediaries. The value to users may be measured in such terms as cost savings, improvements in the quality of the products they make and sell, increased productive capacity, technical service support, and the expectation of future benefits from the buyer-seller relationship. The product line's value to intermediaries is its potential for generating sales revenues and margins, the extent to which it leads to sales of other product lines, and any "halo effect" the supplier's image may have on that of the reseller. Most important, the value of the brand to the intermediary is measured positively in terms of user satisfaction and negatively by the incidence of product returns for reason of faulty performance.

Second, product values are contributed by both producers and channels intermediaries. Producers add primary value in terms of the product's functional performance qualities, the range of product line offerings, and user technical support related to product application technology. Resellers add value in providing local sources of supply, assortments of products related in use and sourced from multiple producers, and, often, after-sale maintenance and repair service. Both producer and intermediary provide product customization, the intermediary's contribution varying widely—from adding optional features in the case of capital equipment, to preparing formulations in the case of chemical compounds, to linking customized application systems with computer hardware.

Third, the requirements for competitive product superiority usually vary across the product life cycle. In the early stages, the measure may be largely the product's functional superiority and the supplier's technical contribution to the user's business. In later stages, it may rest on full-line availability, prompt delivery, and strong after-sale service support. Industry leaders have typically gained entry and established an early lead through product technical superiority but have then maintained market segment dominance—as competitors have stripped away their technical edge—by developing broad lines and providing other services to both users and channel intermediaries.

Distribution Strength.   Whether the producer sells direct to user-customers, goes through resale channels, or both, having a strong presence at end-market levels is essential to market segment leadership. Such a presence may be fostered by a large installed base of products or may come from the demand-generating activities and technical product-applications services of the direct salesforce at the end-market level. Companies having a strong end-market presence seem often to have developed a base of sales revenue sufficient to support large salesforces deployed near the end-customer's points of purchase and use, thereby permitting continuity in the producer-user relationship.

Another element of channels strength is the system's breadth of market coverage, both geographically and in terms of the range of market segments. Channels strength is also a function of the local market share positions of individual distributors and the size of their user-customer bases.

The ability of a supplier to erect channels entry barriers may contribute to competitive advantage as well. For example, a supplier may impose certain franchise conditions, such as the requirement that the distributor carry the producer's full line or that it not stock competing brands. Further, a channels network that is composed of a large number of small distributors is less vulnerable to competitive intrusion than a network that consists of a smaller number of large reseller firms, each with multiple branches. This is

true for two reasons: First, the relative cost to competitors of court-
ing a large number of small firms, widely dispersed geographically,
may be high. Second, suppliers may be considerably more hesitant
to terminate large distributors for taking on a competitor's line;
even a large supplier may find that disenfranchising a major distributor is a two-edged sword if the distributor accounts for a significant amount of the producer's sales revenues.

New competitors, however, often gain access to particular
markets by bypassing the distribution systems of the established
firms. Especially if the latter have built highly selective channels
networks, the new competitor may not find it difficult to develop
an alternate channels system. In the oil country tubular goods industry, the leading domestic suppliers—U.S. Steel and Lone Star
Steel—tightly limited the number of franchises for their OCTG
product lines. Hence foreign producers, at the time of the boom in
oil well drilling in the late 1970s, found access through other qualified resellers that wanted these lines and could not get them from
U.S. Steel and Lone Star. Similarly, Fujitsu found market access
through regional and local electronics distributors, while the leading American disk drive suppliers opted to go through large national chains.

The Installed Base.   In industrial products, the producer's
cumulative share of products in use often determines future sales.
Given the strong propensity for industrial customers to stay with
brands that have performed satisfactorily, the installed base may
generate continuing replacement demand as well as demand for
new units. The installed base may also be the source of sales of
supplies and ancillary equipment. Further, if the units require postsale service and spare parts, significant follow-on revenues may result for resellers that also function as service points. Typically, resellers do not actively generate original demand but fulfill demand
for a brand that arises from the selling efforts of their suppliers.
Hence, the relative size of the installed base is usually critical in
gaining and holding distribution networks.

These observations are suggested by the factors that have contributed to segment leadership for companies in the four industries
under discussion. We now examine the elements of strategy that
have made for market success in each industry, as well as the strategies of followers and challengers.

Square D

This leading supplier of load centers and circuit breakers established a first-mover advantage with its development of a nondestructing[1] circuit breaker in 1935, and it never yielded product superiority with respect to technology, product quality, or breadth of line. Oriented from the beginning toward resale distribution, moreover, Square D's management quickly built a network of established distributors. Soon after the Second World War, Square D had, in the residential housing and light construction segments, all of the elements of market segment dominance: a large and growing installed base, user brand preference, and a strong reseller network serving these markets.

Its franchise policies specified minimum stock levels, full-line selling, and primary-line conditions[2] to ensure distributor support and at the same time make it difficult for competitors to intrude effectively into its network. Square D's market presence was further enhanced by the active involvement of its large salesforce at both the end-market and resale levels.

Nevertheless, its success in the less price-sensitive residential and light construction market segments constrained Square D from competing aggressively in the price-competitive multifamily and manufactured housing segments. The leading firms in these sectors were Siemens-ITE and Westinghouse, respectively. Square D's attempt to build a position in the do-it-yourself (DIY) market, through manufacturers' reps calling on hardware and home center chains, also ended in retreat because its product line for this market was not sufficiently broad and because Square D was concerned about competing with its electrical distributors. Some of these distributors were already selling to consumer outlets. In addition, some large retail chains were aggressively pursuing the small electrical contractor business in competition with Square D's electrical wholesalers.

The essential elements of Square D's success in the residential housing and light construction market segments were (1) a strong product line in terms of breadth and quality, (2) a strong position

---

[1] Prior to this development, residential circuit-interrupting technology was based on the use of fuses that self-destructed and had to be replaced in the event of a line default.

[2] A requirement that the Square D line be stocked and promoted by the distributor as its major line in this class of products.

at end-market levels by virtue of its large installed base and the active involvement of its field sales personnel in generating demand among user accounts, and (3) an extensive and long-established network of independent electrical distributors operating under franchise terms that encouraged full distributor support and set up effective barriers to competitive entry. Square D's success in these two segments, however, required it to make commitments that left openings for other competitors—such as General Electric, Siemens-ITE, and Westinghouse—in other segments, albeit less attractive ones in terms of revenue and profit potential.

Control Data—Peripheral Products Company

Peripheral Products Company (PPCo) started as a small new venture and was acquired by Control Data in 1962. In 1970, PPCo developed a 14-inch drive that became standard in many non-IBM mainframe systems. The product helped to fuel a nearly 50% per annum growth rate for PPCo through the 1970s.

Unlike Square D, PPCo relied heavily on direct selling as its primary mode of distribution for disk drives. Its salesforce of 70 technically trained representatives was the largest in the disk drive industry. However, PPCo also depended on independent resellers to reach smaller OEMs and to provide back-up stocks in support of orders generated through direct negotiations with large customers. Arrow Electronics and Kierulff, each with national coverage through large numbers of branches, were PPCo's distributors.

PPCo's primary strength was in its product line breadth. It supplied a full line of products in the 5.25-inch, 30- to 100-megabyte range as well as disk drives in the 8-, 9-, and 14-inch sizes. Product line strength enabled it to hold large shares in such accounts as AT&T, Wang, and Digital Equipment. Because of these major account positions, PPCo was in a favored position to participate with these technical standard-setters in developing successive generations of disk drives for their computers.

Product leadership enabled PPCo to build distribution strength. Its large installed base of disk drives provided a replacement and upgrade market. In addition, because the large computer OEMs often set industry standards, PPCo drives were also in demand among smaller OEMs. Further, Arrow and Kierulff benefited as back-up sources of PPCo disk drives for the manufacturer's large

OEM customers. Finally, as a full-line supplier, PPCo offered product line breadth and created the expectation among its distributors and their customers that it would continue to stay abreast of this rapidly changing technology.

Thus, PPCo's product line breadth (1) gave it strength in end-markets, (2) provided the resources for supporting its large and technically qualified direct salesforce, and (3) enabled it to build strong indirect distribution.

Other major competitors faced certain obstacles in attempting to build market share. Fujitsu, for example, was a full-line supplier of disk drives but at a competitive disadvantage in establishing an end-market position. As the largest Japanese manufacturer of computers, it competed with American computer builders, many of which were unwilling to specify Fujitsu components in their equipment. Fujitsu's end-market strength was with the small OEMs. Fujitsu found access to this segment through agents and smaller regional and local distributors that had been passed over by the major disk drive manufacturers seeking relationships with the big national distributors. Nevertheless, this Japanese producer lacked the substantial installed base of a PPCo and was, for that reason, initially handicapped in developing a channels network.

Seagate Technology, another major competitor, was a pioneer in the development of 5.25-inch disk drives. Its first product, the ST506, quickly became the industry standard. As of 1983, Seagate claimed to be supplying over 40% of 5.25-inch rigid drives shipped worldwide. Its largest customers were IBM, Digital Equipment (DEC), and Apple Computer, with these three OEMs accounting for over two-thirds of Seagate's sales in that year. In 1984, however, IBM found itself in the position of having negotiated contracts for disk drives in anticipation of market demand for micro- and mini-computers that was not materializing. IBM placed no new orders and stretched out deliveries on its existing commitments. The resulting decline in Seagate's sales revenues in 1985 forced heavy personnel cutbacks. Seagate then moved to expand its customer base to include smaller OEMs and VARs and to build up its indirect channels accordingly.

Seagate's vulnerability in end-markets was due not only to its heavy dependence on a limited number of large customers. In contrast to PPCo, Seagate had concentrated on the 5.25-inch disk drive, committing its R&D budgets and plant investments to maintaining its leadership in one part of the product category.

To summarize briefly, PPCo held its leading position by virtue of its product line breadth, which tended to have self-perpetuating qualities as long as the company maintained close technical liaisons with the leading computer OEMs. Seagate lost position in the large OEM market and revised its product-market-distribution strategy to develop a new strategy in which it was more heavily dependent on resale channels. Fujitsu had difficulty in gaining distribution strength because it did not have a strong presence at the end-market level and, like Seagate, sought to gain share with smaller OEMs through local distributors.

Ingersoll-Rand

With an estimated 25–30% of the $650–700 million stationary air compressor market in 1985, Ingersoll-Rand (I-R) was the clear leader. Its market share had been built on a broad product line of air compressor types and sizes, the largest network of independent and captive distributor sales locations in the industry, and a direct salesforce negotiating sales of the large, often customized compressors.

Because of the high investments relative to sales revenues that distributors are required to make in inventories of compressors and parts, air compressor producers have typically granted their distributors territorial exclusivity and, in turn, have required brand exclusivity. That is, distributors typically carry one manufacturer's line and are given territories in which each is franchised as the manufacturer's sole representative.

Another salient feature of air compressor distribution is that, unlike the other three product groups, spare parts and after-sale service margins are typically large sources of revenue. Overall, parts and service gross margins tend to make up 30–50% of the distributor's total margins. Further, margins on air compressor parts and service are typically higher (30–35%) than those on new equipment sales (15–25%). This means the size of the installed base of machines in a distributor's trading area is especially significant. The larger the base, the more parts and service revenues it will generate, as well as the more replacement sales.

Given (1) the high level of distribution investments, (2) the role of the distributor's parts and service business in its total operations, (3) the resulting industry norms of territorial and brand exclusivity, and (4) the strategic advantage of a large installed base, the leading producer, in particular, is able to develop strongly protected networks for securing market access. Thus, strong distribution and high market share become cumulatively reinforcing. Against this brief overview of the salient features of air compressor distribution, we now describe the Sullair and Atlas-Copco bids for leadership.

Founded in the mid-1960s, Sullair introduced the so-called rotary screw design, the advantage of which was that the cost of spare parts amounted to only 15–20% of the original equipment cost over the life of the machine. In comparison, the older reciprocating design incurred spare parts requirements amounting to 50–60% of the original equipment price over the machine life.

Sullair's attempts, however, to build a distribution network were handicapped by its narrow product line and by the lower levels of spare parts and service revenues that would be generated by a rotary compressor as compared with other compressor types. Within a decade, it had replaced its relatively few independent distributors with 80 company-owned stores. Initially successful, these channels suffered sharp declines in revenues and earnings in the industry downturn of the early 1980s. Apparently, a major contributing factor was their lack of sufficiently broad product lines to sustain store sales volume in the face of a drop-off in the air compressor business. By 1985, the number of Sullair's company-owned stores had been cut back to 10, and the company once again was attempting to gain market access through independent distributors.

Atlas-Copco of North America (ACNA) entered the U.S. air compressor market as a wholly owned subsidiary of a Swedish company, Atlas-Copco, the world's largest manufacturer of stationary air compressors. Although ACNA supplied a broad line of high-quality air compressors, its initial attempts to develop U.S. distribution in the mid-1970s met with little success. It then undertook to gain entry by persuading independent distributors to take on its "Z series" line of oil-free rotary compressors, a product category in which, at that time, Atlas-Copco enjoyed both product and cost superiority over competing U.S. manufacturers. In particular, it targetted the distribution networks of those producers which did not include oil-free reciprocating compressors (recips) in their lines.

ACNA's next phase of market entry was the acquisition of a U.S. producer, Worthington Corporation, and, with it, a network of independent distributors. With an established distribution base, ACNA then moved strongly to get its distributors to take on other parts of its broad line to replace the competing products of other suppliers.

As of 1985, Atlas-Copco held an estimated 10% of the U.S. market for stationary air compressors. Within the oil-free recip product segment, it was reputed to hold a 30–40% share.

Atlas-Copco's market entry strategy focused initially on penetrating the distribution systems of competitors with a product that (1) had significant user cost and performance advantages and (2) did not compete directly with products in the lines of the distributors' primary suppliers. Next, ACNA acquired a recognized brand name and the distribution network of one of the smaller players in the U.S. market. Having secured a beachhead, it then expanded its market access by inducing distributors to take on more of its product line.

Like Atlas-Copco, Sullair had attempted market entry through product superiority. However, in an industry where the costs of brand switching are exceptionally high, Sullair did not have the requisite product line breadth, as ACNA did.

U.S. Steel

The market supremacy of U.S. Steel (USS) in the Tier 2 OCTG segment is of long standing. USS's leadership was based on its full line of seamless tubing and on the strongest distribution through oil field supply houses of any OCTG supplier. Historically, USS's OCTG line has been limited to distribution through only 20–23 supply houses, a policy that may be explained by the fact that the "oil patch" itself is a limited and well-defined geographic area in the United States and that USS's 20 or more franchisees had multiple locations serving this market. Further, by restricting its representation, USS eased intrabrand competition at the wholesale level, thus preserving distributor margins and making the line very attractive to franchisees. By virtue of its product quality, product line breadth, and distribution strength, the company held strong positions with its major oil company accounts.

In the boom-and-bust of the early 1980s, what had been a stable market—served largely by domestic OCTG producers through highly structured distribution systems—fell apart. Oil and gas well drilling activity had soared to record heights in 1981–1982 and then abruptly dropped off, declining to a 12-year low in 1986. In the meantime, imported tubular goods had made substantial inroads into the U.S. market, reaching a 1982–1984 peak of about 60% of total U.S. consumption. There followed a rash of distributor inventory write-downs and bankruptcies as banks foreclosed on loans used to finance inventory buildups during the boom phase. In addition, OCTG domestic manufacturing capacity was significantly curtailed as marginal mills, unable to continue to operate profitably, were shut down.

The catastrophic events of the 1980s challenged all three dimensions of USS's market leadership—product superiority, distribution strength, and a strong position among large buyers. Many users regarded imported Japanese tubular goods as superior to domestic products. Large buyers, who had traditionally allocated a high percentage of their annual OCTG requirements to USS and its supply houses, began to request competitive bids from other suppliers. Serious inventory losses debilitated oil field distribution networks as OCTG prices plummeted starting in 1982. At the same time, USS's franchisees, like other oil field supply houses, began to take on competing lines of tubular goods often available at prices well below USS price levels.

In the early 1980s, U.S. Steel moved to protect its market share position. Even before the crash, it had broken ground on a new pipe mill in Fairfield, Alabama, the most modern facility of its kind in the United States in producing high-quality seamless OCTG at low manufacturing costs. Increasing its selling efforts among the major oil companies, USS negotiated sales contracts that were then fulfilled by USS distributors. The company also significantly increased its financial commitment in support of distributor and user inventories in the field. The availability of resources for new plant investment and inventory financing made these strategic moves possible for USS. Other major competitors apparently could not or would not make comparable commitments to this market.

Another effective strategy in this industry was that of the leading Japanese competitor, Sumitomo. The exporter focused on a superior product for the Tier 1 market segment because it was handicapped in other segments by its foreign status. Voluntary Restraint

Exhibit 1.2 Distribution Design Determinants and the Market Leaders' Distribution Profiles in Primary Market Segments

	Square D Load Centers and Circuit Breakers	**Control Data** Disk Drives	**Ingersoll-Rand** Stationary Air Compressors	**U.S. Steel** Oil Country Tubular Goods
Primary Market Segment	residential housing	large computer OEMs	industrial plants	major oil companies
Product	consumable supplies in electrical system; mature technology; low unit value	high-tech, high-value computer component; rapidly evolving technology; product design adapted to computer design	capital equipment; large unit prices; relatively mature product technology	tier 2 OCTG; supply item; standardized specifications; mature product technology; significant element in drilling costs
Market Demographics	small residential contractors; geographically dispersed	concentrated; large, highly visible OEM customers and some smaller firms	many large and small users; geographically dispersed	large, highly visible users; geographically concentrated purchase and use locations
Buyer Behavior	bundled buying; KPC[a]: product reliability; prior favorable brand experience; local availability	single-product buying; very large dollar transactions; extensive engineering involvement; KPC: stay on leading edge of computer technology; design compatibility; product reliability; price; local availability to cover fill-in needs	single-product buying; KPC: price; after-sale field service availability; operating costs; operator brand preference	significant shift from bundled buying to single-product buying; KPC: product reliability; low product acquisition costs (product price plus delivery costs); operator brand preference; local supply availability

Distribution Cost Factors	low value per transaction and market dispersion favor two-tier distribution	high product margins and low expense-to-sales transaction ratios favor direct selling to large accounts; two-tier distribution growing because of price-margin declines and proliferation of computer OEMs	varying complexity of product and intensity of information required by customer favor direct selling to large accounts and reseller distribution to smaller customers	large purchasing outlays accompanied by intense price competition favor directly negotiated transactions with large buyers, but stocking and demand fulfillment through local resellers
Business Unit Factors	broad product line; well-recognized brand name; distributor-oriented management; large installed base; ample resources for marketing investment	broad product line; leading-edge computer technology; funding to support large salesforce; technically trained salesforce relative to competitors'	wide line; full technology range; worldwide brand recognition; traditionally direct sales oriented; ample marketing funding	extensive product line; modern low-cost production facilities; strong long-term relations with oil field supply houses
Distribution Profile	all sales through independent distributors; large salesforce calls on large accounts, supports distributor performance, monitors distributor stock levels	90–95% direct sales to large OEMs, 5–10% through two large national distributors, with gradual shift in direction of two-tier distribution to reach smaller OEMs and process fill-in orders from large, direct accounts	direct selling for large machines, through independent resellers and captive distributor branches for smaller horsepower units	emphasis on direct contract negotiations with large oil companies; product stocking and delivery through local resellers for negotiated compensation; limited number of USS franchises, each with multiple stocking/sales locations

aKPC = key purchasing considerations.

Agreements imposed by the federal government greatly restricted the OCTG volumes Sumitomo Corporation of America (SCA)[3] could bring into the U.S. market. In addition, delivery lead times were very long, averaging 4–6 months, and U.S. oil companies had to forecast their needs six months or more in advance because Sumitomo required firm orders in advance of shipment. An even greater deterrent was the fact that both users and oil field supply houses had to make firm price commitments at the time of order, a speculation with little chance of gain in the face of steadily declining prices.

Accordingly, Sumitomo's one viable option was to focus on the Tier 1 segment with special-alloy tubular goods for use in hostile drilling environments and to reach its customers primarily by direct selling. Restricted in import volumes, SCA elected to concentrate on those products which provided the highest revenue per ton and the highest margins. Moreover, the Tier 1 market was not one in which long lead times posed a competitive disadvantage; the most important consideration to buyers was getting a product that would perform under adverse conditions. Unable to compete effectively in the larger Tier 2 market, Sumitomo's strategy gave it a leading edge based upon its particular product superiority.

SUMMARY

Product line superiority in all four industries was a core element in market segment leadership. Companies such as Square D, Control Data, Ingersoll-Rand, and U.S. Steel assumed technical leadership at an early product development stage and held positions of product superiority both through maintaining product quality and by developing broad lines. This product line superiority, carefully protected, was the foundation for each company's large installed base, significant end-market presence, and strong distribution networks.

Concurrent with developing and maintaining product line strength in terms of state-of-the-art technology, breadth of line, and product quality, each leader built strong distribution networks. Their channels systems—driven by the imperatives of product, market demographics, buyer behavior, distribution cost considerations, and business unit factors— were designed to provide effective access to their targeted market segments. Further, they adapted to changes in all of these dimensions as

---

[3] SCA is the U.S. arm of Sumitomo Corporation, a trading company in the Sumitomo group of companies.

product-markets matured. Exhibit 1.2 outlines the considerations that influenced the market leaders' modes of distribution to show the relationship among these factors and their channels strategies.

Companies that challenged the market segment leaders have typically challenged their positions of product superiority. Sumitomo in the tubular goods markets and Atlas-Copco in the air compressor industry are cases in point.

Even product superiority, though, has not ensured easy market access. Strong acceptance among users and resellers takes time to build because of the compelling propensity on the part of both users and resellers to stay with familiar brands and established relationships. Thus, the strategies of new market entrants seem to stress gaining a foothold by offering resellers a product that, in price and/or performance, fills a hole in their product lines and then encouraging distributors to take on an increasingly larger array of the brand.

The relationships among product superiority, distribution strength, and the value of the installed base in achieving market segment leadership are themes throughout the four product-market studies in Appendix A. Each provides detailed stories of competition for major market segments among large industrial producers. Each shows as well how those who have not succeeded in taking leadership in key markets have sought some other viable product-market-distribution combination consistent with buyer needs and the demographic profiles of other segments. A recurring theme, too, is that success in any one market segment often carries with it a set of limiting commitments that may preclude the possibility of comparable leadership positions in other segments.

Having considered the distribution strategies of the leaders in the context of both their overall business strategies and their competitive environments, we turn now to focus on the anatomy of distribution systems per se. We consider the strategic options industrial producers confront in designing channels networks and the basic considerations that lead producers to make those choices in one way or another. In particular, the next chapter develops a construct related to channels design determinants as a basis for the discussion in Chapters 3, 4, and 5.

2 Elements of Distribution Strategy

The major choices industrial goods producers make in developing channels strategies are these:

- relative balance between direct sales and sales through intermediaries
- class of intermediary: independent distributor, captive distributor, agent, broker
- type of intermediary in terms of served markets
- relative intensity of distribution by geographic area
- locus of distribution functions: the allocation of distribution tasks to particular channels in the distribution system
- franchise terms and conditions

This chapter deals with the elements of distribution strategy and suggests those considerations which lead industrial producers to choose particular options in developing distribution channels. We then offer a model to account for both similarities and differences among the distribution strategies across product-markets and among competitors. This construct provides an organizing scheme for the next three chapters.

THE ANATOMY OF A MULTICHANNEL DISTRIBUTION SYSTEM

For illustrative purposes, we begin by profiling Ingersoll-Rand's distribution network for stationary air compressors. As noted in the previous chapter, Ingersoll-Rand (I-R) is the largest U.S. producer of this equipment, a capital good used to supply air power in a wide range of manufacturing processes. We compare I-R's distribution system with those observed in the three other

22

product categories: load centers and circuit breakers, computer disk drives, and oil country tubular goods (OCTG).

Among the distribution systems described in this book, that of Ingersoll-Rand's Stationary Air Compressor Division (SACD)[1] is as complex as any. SACD sold to industrial customers through its direct salesforce, independent distributors, captive—supplier-owned—distributors (known as I-R air centers), and manufacturers' reps (MRs). The air centers and the independents also sold to subdistributors; these were independently owned businesses located mostly in secondary market areas, which sold to small companies such as auto repair shops and individuals such as painting contractors. SACD also marketed small compressors (5 horsepower and under) through manufacturers' rep (MR) organizations, which sold to retail chains such as hardware stores and home service centers. They, in turn, sold to a do-it-yourself (DIY) consumer segment.

Typically, each air center and independent distributor had primary responsibility for a geographic market area; overlapping territories were the exception. The system thus provided for exclusive territorial representation for the air centers and independent distributors. However, these intermediaries sold only small- and medium-sized units.

SACD managers reserved the large compressors for direct sale only. An SACD service unit with 40 trained technicians provided spare parts and after-sale service for large equipment. Distributors and air centers provided parts and service for the products they carried, and they were franchised to sell spares for large compressors, which they did not carry. SACD managers had also established five master distributors, which sold large compressor spare parts to users and to other distributors.

SACD had four separate salesforces selling (1) direct to user accounts, (2) through air centers, (3) to independent distributors, or (4) to the MR network.

The direct sales operation was single-tier distribution. Sales through air centers and independent distributors went through two

---

[1] I-R's Stationary Air Compressor Division made and sold compressors ranging in size from ¾ horsepower to 6,000 horsepower. Three types of air compressors were included in I-R's broad line: rotary, reciprocating, and centrifugal units. While the bulk of SACD's revenues were derived from the sale of new machines, the air compressor aftermarket generated a significant income stream from the sale of spare parts and maintenance and repair services. SACD distributor income, too, came from a mix of new units, spare parts and service sales, and reconditioned compressors.

tiers (i.e., from the direct salesforce to distributors to users). Smaller compressors and spare parts for large machines were sold through three-tier distribution—the former involving sales through distributors to subdistributors, the latter going from master distributor to distributor to user.

Over two decades, as both higher- and lower-horsepower air compressors were added to SACD lines, independent distributors and the I-R air centers had come to play increasingly important roles in its distribution system. Agents, too, gained in significance as SACD sought to establish a position in consumer retail outlets. In contrast, the amount of revenue generated through direct sales declined over that period. In 1981, 45% of SACD's sales were direct. By 1985, about 35% of SACD's sales revenues came through the independent channel, 20% through air centers, 15% from agents selling to consumer channels, and 30% from the direct salesforce. Such a mix compares, on the one hand, with that of load centers, in which 80–90% of sales were generated through independent distributor channels. At the other end of the scale, it compares with disk drive distribution, in which 80–90% of sales were made direct to computer manufacturers, while independent distributors served as back-up sources of supply to large OEMs and as primary sources for small OEMs, computer systems assemblers, and retail stores. What factors account for these differences?

CHANNELS DESIGN OPTIONS

Consider some of the choices SACD management had to make:

- whether to base its distribution system primarily on direct selling or on external intermediaries, such as resellers and MRs
- whether to have a single channel or multiple channels serving different classes of buyers in this diverse market
- whether to have a captive distribution organization competing for many of the same accounts as the independent network
- whether to have multiple distributor representation, or territorial exclusivity, by geographic area

- whether to sell to large retail chains using direct sales reps or MRs (SACD managers attempted for two years to do the former but with no success)
- whether to have one salesforce, or two, or three, or four, to interface with users and with the several classes of intermediaries
- what product categories to commit to each channel
- where in the system the service function would be performed
- where stocks of complete units and spares would be kept
- what pricing structure and price-setting processes would be employed
- what discount levels and commission percentages would be provided for air centers, independent distributors, and MRs, respectively
- how SACD sales reps would be compensated in terms of base salary, bonus, and commission income

Channel Selection

As at Ingersoll-Rand, industrial producers typically develop multichannel distribution systems in which there is some mix of direct salesforces, independent distributors, captive distributors, and MRs with one or more representing the dominant channel(s) in each market segment. The channels mix depends largely on market segment demographics. For all producers, the relative balance among channels elements serving each market segment is a key decision, raising questions such as these for marketing managers: What considerations weigh strongly in favor of direct sales as opposed to distributor sales? Under what conditions are industrial goods producers likely to depend primarily on agents? What are the advantages and disadvantages of going through captive distribution systems?

Direct Sales. The conditions that foster a predominantly direct sales distribution system are (1) a high concentration of buyers, (2) a large dollar amount of individual purchase transactions, and (3) greater needs of buyers for technical product information or for product customization. These three conditions generally held for disk drives and for *large* air compressors, both of which were sold primarily through direct channels. Buyer concentration and transaction size affect sales costs directly. In trading off the costs of supporting a direct salesforce against giving distributors a percentage margin, industrial goods producers often realize higher net returns in selling directly to large accounts as opposed to selling through distributors. Buyers' needs for technical support and for products configured to their particular usage requirements, too, seem often to be best served through direct communication with suppliers.

If sales costs relative to transaction size permit the use of direct selling, there are other advantages in this mode of distribution. Product advocacy is one. Committed to selling their employer's product and measured accordingly, the producer's sales representatives are typically motivated to compete aggressively to win the sale. Independent resellers, on the other hand, with wide and diverse product lines and often carrying competing brands, may have little reason to persuade customers to buy one particular supplier's brand over another. Another advantage is negotiating strength. A direct sales relationship with the user-customer facilitates the negotiation of prices and terms of sale, taking into consideration competitive offerings and the customer's unique requirements. To conduct such negotiations through an intermediary introduces added complexity and issues concerning how to rationalize negotiated prices and terms with published distributor price schedules.

Independent Distributors. The conditions that support a high proportion of reseller distribution are (1) widely dispersed and fragmented markets, (2) low transaction amounts, and (3) bundled purchase behavior, that is, the buyer's propensity to purchase a number of items, often different brands, in one transaction. These three conditions are present, for example, in the residential construction market for load centers and circuit breakers. However, they do not characterize the large industrial construction and prefabricated-housing market segments for these same products. The economics of selling in fragmented markets are likely to favor distribution through resellers—serving local market areas with broad product lines—rather than a direct salesforce.

Industrial distributors range widely in terms of their product lines, the services they offer, the market segments they serve, the breadth of their geographic market coverage, and their modes of selling, that is, outside selling, over-the-counter, or catalog sales. The producer's choice of channel type is largely contingent on its choice of end-market segments.

For example, load centers were marketed through resellers that may specialize in serving either the residential and light commercial construction or the industrial OEM market segments. In other instances, the distribution channel identifies explicitly with a type of product but implicitly with certain end-market segments. We found that electronic parts distributors selling disk drives dealt mainly with OEM producers of electronic equipment, whereas steel warehouses dealt with metals fabricating shops and construction companies.

The essential consideration in selecting a particular type of distribution channel is the channel's ability to serve the needs of the target end-market. Residential contractors want purchase convenience and local availability. Thus, they need reseller sources that are in close proximity and carry broad selections of items used in home building. Users of air compressors need sources of supply that can provide equipment choice, technical applications information, product customization options, installation, spare parts, after-sale service, and trade-in opportunities.

Buyer needs vary considerably by market segment, and the resellers that serve these clusters of buyers tend to shape their own product and service offerings in response to the distinctive requirements of their customers. In the flow of goods to market, the producer's product becomes part of a total package of benefits, tailored to selected classes of user-customers and sold through those channels which address their needs.

Captive Distribution. While a captive distribution network in a multichannel distribution system is a source of marketing strength, it remains the subject of considerable controversy, both within the producer firm and among other elements in the distribution system. This subject is considered in all of its ramifications in an appendix to Chapter 8. To briefly summarize here, captive branches, precluded from carrying competitive product offerings, provide added market access for the products of the parent corporation. These operations can also serve as training grounds for product division managers whose career development would benefit from an experience in end-market selling. Often, captive branches are established in geographic markets where the producer is unable to recruit qualified independent resellers.

On the other hand, captive distribution arms sometimes compete directly with other channels elements, both independent resellers and the company's direct salesforce. Maintaining a captive distribution network may come at some cost in enlisting the full support of other intermediaries.

Manufacturers' Reps. The three conditions that lead toward the use of agents are the same as those which favor direct sales: (1) concentrated markets, (2) large individual transactions, and (3) buyers' needs for technical support and/or product customization. An MR (agent) network tends to be favored over direct selling if resources are not available to cover the fixed overheads associated with a direct salesforce—salesperson salaries and expenses, sales administration, and other overheads. For producers too small to afford these fixed costs or for those which give higher priority to other uses of their capital, agents offer the benefit of minimizing fixed selling expenses.

This consideration has been important for start-up ventures in the disk drive industry, which committed their limited capital primarily to R&D and to developing manufacturing facilities. As these enterprises developed a customer base, direct selling became more cost effective and also facilitated close technical interchange with their customers. Another reason for using MRs is to compensate for a manufacturer's sales organization's lack of knowledge or expertise to serve a particular market segment. In attempting to reach the consumer DIY market, for example, Square D, the leading manufacturer of load centers and circuit breakers, developed a network of 19 MRs in the United States. These agents each carried five to ten complementary lines going to hardware stores, lumber yards, and home centers. They also had considerable expertise in consumer packaged goods marketing, whereas the Square D direct salesforce, calling mainly on distributors and building contractors, had no skills in selling to retail chains.

Still another reason for using agents is that the product line is simply not broad enough to provide the base for supporting a direct sales operation, yet the product requires technical selling. Barry Controls, for example, a manufacturer of materials for reducing noise and vibration in military equipment and industrial installations, sold through ten MR organizations that represented complementary products—power transmission products, electronic components, test and inspection equipment, and shock and vibration controls.

Thus, in recruiting MRs as agents, manufacturers such as Square D may seek a market segment or product application expertise that is not present in their own sales organizations. They may also choose agents over a direct salesforce if, as in the case of Barry Controls, their product line is narrow and can benefit from being part of an agent's broader line. An MR with, say, ten lines might have a product portfolio sufficient to support a small salesforce. Finally, manufacturers may have more urgent uses for their limited capital than investing in the fixed overheads to support a direct salesforce.

Strategic Channels Options

In addition to choosing the types of channels through which to go to market, producers make critical choices on three other dimensions of channels strategy: (1) distribution intensity, that is, the number of intermediaries in a trading area; (2) where in the system different distribution functions will be performed; and (3) the terms and conditions of the distribution franchise.

Distribution Intensity. The level of distribution intensity is an important strategic choice. How many resellers should be franchised in a local market? There are three considerations: The first is the level of channels investments needed—both in inventories required to support a given level of sales and in specialized resources such as equipment and personnel—to service the line. When high distributor investments are required, the supplier's representation must be less intensive in order to attract resellers to take on its line.

User purchasing behavior is the second consideration. If buyers engage in a search process, if the purchase is typically unbundled, and if the buying decision-making process involves the negotiation of product features, prices, and terms, fewer resellers are needed in an area than if convenience is of primary concern to buyers. These two considerations tend to differentiate the channels infrastructures for load centers and for stationary air compressors, as the following data from our distributor surveys indicate:

	Electrical Distributors[a]	Air Compressor Distributors[b]
Approximately how many customers does this location have?	2,250	1,250
Within approximately what distance of your office location are the great majority of your customers located (in miles)?	50	120
About how many product lines do you carry in all?	120	40

	Electrical Distributors[a]	Air Compressor Distributors[b]
What are your annual revenues in this sales location (in millions)?	$5.7	$2.5
How many other distributors represent your prime supplier in this trading area?	6.1	2.3

[a]1986 questionnaire survey of 75 electrical distributors.
[b]1986 questionnaire survey of 79 air compressor distributors.

On average, air compressor distributors' sales territories are much larger, and the number of distributors carrying the same brand are fewer, than those of electrical distributors. At the same time, on average, air compressor distributors have about half the number of customers, and buyers shop over a broader geographic area. Air compressor distributors' sales revenues are less than half those of electrical distributors.

The third consideration in planning distribution intensity is the volume of available business in an area. Often, producers with high market shares and a large installed base of products will have more distributors in a trading area than those with lower market shares. There is some level of potential revenue below which it may be difficult to attract established resellers; thus the lower the producer's market share in a trading area, the fewer are the parts into which it can be divided for purposes of establishing a channels network. Square D, for example, with highest market share, averaged 8.3 distributors per market area, while all other load center suppliers averaged 3.9.

Often the rationale for multiple representation in a geographic area is that different distributor types are needed to reach different market segments—say, residential construction, industrial users, and OEMs in the case of Square D. Distributors may also be franchised because they have loyal customers due to personal relationships, credit extensions, and convenience, or strong positions with certain large accounts.

While multiple representation in a geographic area may address clusters of user-customers, market segments are seldom so neatly compartmentalized as to preclude intense intrabrand competition among distributors. At the same time, a supplier that attempts to mitigate price competition by reducing representation runs the risk of losing sales volume.

Exhibit 2.1   Distribution Systems: Components and Functions

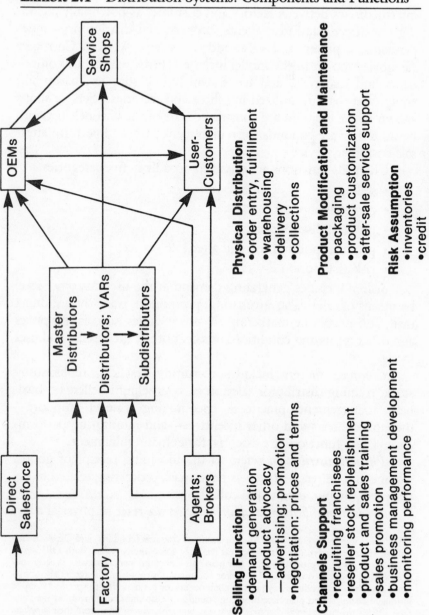

Selling Function
- demand generation
  - product advocacy
  - advertising; promotion
- negotiation: prices and terms

Channels Support
- recruiting franchisees
- reseller stock replenishment
- product and sales training
- sales promotion
- business management development
- monitoring performance

Physical Distribution
- order entry, fulfillment
- warehousing
- delivery
- collections

Product Modification and Maintenance
- packaging
- product customization
- after-sale service support

Risk Assumption
- inventories
- credit
- dedicated resources

Locus of Distribution Functions. Exhibit 2.1 shows the intermediaries between factory and customer as potentially including the direct salesforce, agents, brokers, and distributors—independents, captives, and value-added resellers (VARs). Customer categories shown in the model include OEMs, which buy components and materials, and users, which buy supplies, parts, and equipment that are utilized in offices and factories.[2] Service shops shown in the diagram are considered part of the channels network because they are conduits for replacement parts and perform after-sale customer service.

Channels functions are classified broadly in five categories:

- selling
- channels support
- physical distribution
- product modification and after-sale service
- risk assumption

Selling includes generating demand at the user-customer level by means of sales calls, advertising, promotion, trade shows, direct mail, and telephone marketing. It also includes negotiating prices and other terms and conditions of sale, both at the reseller and user levels.

Channels support includes recruiting resellers, replenishing stock, training distributor salesforces, counseling resellers on business management practices, encouraging reseller support—through contests and other incentives—and monitoring performance. These functions are essential for demand fulfillment.

Physical distribution refers to the incoming receipt of goods, storage and delivery, as well as paperwork processing related to order entry, billing, credit, and collections.

Product modification and after-sale service refer to physical alter-

[2] While distributors, agents, and VARs are classed as channels (conduits) and OEMs as customers, their respective roles (as noted earlier) may sometimes overlap. Both OEMs and VARs are engaged in product design, buy materials and components that they process and/or assemble, and sell to other OEMs and/or end-users. Distributors and agents also sometimes assemble end-products from sets of components, or cut materials to specified sizes, or formulate chemical and pharmaceutical compounds, or customize equipment. At the same time, OEMs may function as channels. They may buy replacement parts from their suppliers and sell them to their franchised repair shops. Often they supplement their own product lines with related items manufactured by other OEMs. The point is that for purposes of modeling distribution channels, while the several categories may be distinguished in terms of their *primary* roles, the functional distinctions among OEMs, distributors, agents, and VARs may not always be cleanly delineated.

ations made in the product as it moves through channels and to its maintenance in use. That includes packaging in accordance with the user's instructions, customizing the product to the purchaser's specifications, and product assembly, for example, combining computer hardware with software programs to produce applications-specialized information systems.

After-sale service—maintenance, repair, and retrofitting—is intended to maintain or improve product performance by replacing worn parts and making physical adjustments throughout the product's lifetime.

Risk assumption is identified as a separate channels function because it may be performed by different channels entities and because it is not necessarily tied to other channels functions. The major risks to be assumed in distribution are inventory carrying costs, customer credit, product liability, and distribution-specific investments. The last includes investments in such resources as warehouses and delivery trucks, office space and equipment, field salesforces, office staffs, specialized demonstration facilities, technically trained sales and service personnel, and information systems—both for internal use and for communicating with customers and suppliers.

The functions of selling, physical handling, product modification, and risk assumption may each be performed at one or more points in the distribution chain. *Channels support* is typically provided through the producer's sales organization, but it, too, could be delegated to intermediaries such as agents and master distributors. What are some of the considerations that influence the locus of the other four basic distribution functions in the channels system?

The responsibility for *selling,* or demand generation at the end-user level, may be retained by the direct salesforce (when this is the only channel), given over entirely to intermediaries, or shared. If shared, the direct salesforce may either call on users along with distributor reps or call alone to negotiate sales that are then passed through distributors for contract fulfillment. Generally, direct sales involvement at the end-use level will be limited to large and strategically important accounts.

Often, responsibility for selling to users is delineated in terms of product categories, specific accounts, and/or market area. At Ingersoll-Rand, for example, distributors selling to industrial markets carry all reciprocating air compressors below 250 horsepower and rotary models under 450 horsepower. Higher-horsepower models are sold exclusively by the direct sales reps, although both distributors and the direct salesforce call on many of the same user accounts.

Physical distribution may be shared among or assumed by either producers or intermediaries. Producers may carry stocks at plant locations, at company-operated warehouses, and sometimes on customers' premises. These stocks may be drawn on to supply the needs of both resellers and user-customers. Typically, distributors supply their customers out of their own warehouses. Often, however, large deliveries in fulfillment of distributor sales may be "drop-shipped," that is, sent directly from the producer's stocking location to the ultimate user, while the billing and credit functions are handled by the distributor.

The locus of physical distribution functions should be viewed as a strategic choice. Relevant considerations include ensuring a high level of on-time order fulfillment and minimizing transportation and inventory costs. Given their magnitude relative to sales, economizing on physical distribution costs may quickly translate into competitive cost advantage.

The locus of *product modification and after-sale service* functions usually reflects (1) the concentration or dispersion of the customer base and (2) the producer's reliance on direct selling or on external intermediaries as the primary channels element. In concentrated market segments served largely through direct selling, it is common practice for the producer's technical personnel to work directly with user-customers in designing customized products and providing postsale repair and maintenance service. When producers assume the responsibility for after-sale service, the choice is often motivated by the desire to ensure high levels of customer support and to develop strong customer relations.

In product-markets where the buyers are smaller and geographically dispersed, markets served largely through distribution, the distributor is often the locus of both product customization and after-sale support services, the latter typically being a significant source of reseller revenue. When the product modification function is passed down the distribution chain, it may be to provide local customization of standard products, for example, optional equipment for heavy machinery. It may be, on the other hand, that the diversity of product modifications needed is so great that small, applications-specialized resellers can handle this function more effectively. An example of the latter situation is that of the so-called VAR (value-added reseller) in the computer industry. These are typically small firms that combine computer hardware with special-purpose software for such uses as medical record-keeping, engineering design work, or legal reference searches.

Risk assumption is often but not always linked with selling and physical distribution functions. Regardless of their source, risks tend to be carried by those who can afford them and by those for whom there is the promise of commensurate rewards. Both characteristics apply, for example, in the case of U.S. Steel's financing OCTG inventories for its distributors and users (see Appendix A, Study IV).

Risks also tend to be carried by those who can best manage them. For example, many producers, not competent to assess specific credit risks at local levels, push sales credit risks onto their distributors.

In summary, the locus of distribution functions may vary across the channels system by type of product, market segment, class of customer, and size of account and/or individual order. From the producer's perspective, the primary considerations in positioning the selling, physical distribution, product modification and maintenance, and risk assumption functions are the following:

- economizing on distribution costs
- maximizing market share, sales revenues, and profits
- optimizing the returns on distribution-specific investment risks
- meeting customer needs for product technical information, product availability, product customization, and after-sale service so as to gain a competitive edge
- maintaining sources of market information

The strategic positioning of distribution functions within the channels system should be guided, then, by a concern for achieving these five objectives to the extent possible, recognizing the need for making some tradeoffs among them. Over the long run, however, distribution strength tends to depend in large part on positioning the responsibilities for channels functions at those points in the chain where they can be performed at the lowest cost commensurate with customer satisfaction.

Franchise Terms and Conditions. The supplier-distributor relationship is typically specified in a formal franchise agreement. This agreement authorizes the reseller to stock and sell the supplier's line, or specific parts of it, in return for the former's agreeing to conform to certain terms and conditions. In specifying terms and conditions, the supplier may seek to (1) ensure that the supplier's line is effectively represented by the reseller, (2) ensure to the extent possible that user-customers are satisfied both by the product itself and by the ancillary services provided by the reseller, (3) discourage intrabrand price competition both to preserve the value of the franchise to distributors and to forestall interbrand price competition, (4) establish distribution entry barriers for competitors that may seek to intrude into the supplier's channels network, and (5) maximize the supplier's revenues and profit margins.[3] Five types of franchise conditions are particularly important:

Full-Line Selling. Suppliers may require resellers to carry the brand's complete line. The purposes for this are three: to ensure that a broad array of the supplier's line is locally available to serve user-customers, to make it difficult for competitors to penetrate the distributor account, and to provide market access for the supplier's new product offerings.

Product Line Exclusivity. Producers may stipulate that resellers not carry directly competing products of other suppliers. A modification of this condition is to require that the supplier's brand be treated by the distributor as its primary line. Like full-line selling, such conditions seek to ensure that the distributor will give the supplier's line full support and that the primary supplier will not lose its share of the distributor's business to competing producers.

Account and Class-of-Trade Constraints. Suppliers may reserve specific accounts and/or types of customers for direct sale and, in effect, declare this business out of bounds for resellers. Usually, the purposes of this condition are to optimize large account revenues and profit margins and to maintain direct relationships with large and important customers for purposes of negotiating prices, providing for technical interchange, and forestalling the penetration of these accounts by competitors. Further, competition between the supplier's direct salesforce and its distributors at the user-customer

---

[3] The legality of franchise terms and conditions is the subject of Chapter 13, "The Legal Framework."

level typically creates confusion and generates interchannel price competition. Account and class-of-trade restrictions minimize such conflicts.

In addition, the terms of the supplier-distributor relationship may provide that the latter sell only to user-customers and not to other dealers, either franchised or not. Transshipments of product at the resale level may be a source of significant price cutting and can subvert the producer's efforts to preserve stable and effective representation in end-markets. (See Chapter 9, "The Gray Market Dilemma.")

Location Franchising. To control the intensity of distribution by trading area and the quality of its channels network, a supplier may specify that each sales branch of multilocation distributors must be franchised individually. Each franchise decision would take into consideration (1) the desirability of adding to the number of the supplier's representatives that may already be in the trading area and (2) the qualifications of the proposed franchisee location to represent the supplier effectively. The alternative would be to franchise at the firm level, thereby permitting all sales locations of a firm to carry the supplier's line. Franchising at the firm level often means that the supplier loses control over the number and quality of resellers in geographic market areas.

Resale Prices. The franchise agreement may require the intermediary to observe resale price schedules set by the supplier. Resale price maintenance typically has three purposes: (1) to preserve margins for all resellers, protecting the value of the franchise and supporting full-service selling; (2) to maintain trading area coverage by minimizing the risk that some franchisees, pricing aggressively, will eliminate others; and (3) to avoid price wars at the resale level from escalating into intensified competition at the manufacturing level. As for the legality of setting and enforcing resale price schedules, producers may specify the prices at which a distributor sells its products and refuse to deal with those who do not conform as long as this action is taken unilaterally and not in conspiracy with the supplier's other distributors.

Other Conditions. There is a range of other conditions the supplier may seek to impose. For example, the terms of the franchise may stipulate that the distributor maintain specified inventory levels, that it make provisions for its personnel to participate in product and product-service training programs, and that it meet certain annual sales quotas. To minimize the producer's manufacturing and shipping costs, it may also stipulate minimum order-size requirements and specify that the reseller submit periodic sales forecasts to facilitate production scheduling.

While the foregoing covers a broad range of terms that producers may specify as conditions of granting distributor franchises, the extent to which the supplier has the power to impose particular terms and conditions in the contract depends on the desirability of the franchise to the reseller. The value of the franchise is likely to be determined by the following:

- superiority of the brand over competing product lines
- installed base, that is, the brand's share of all the units currently in use in the local area
- number of other distributors of the brand in the area
- extent to which the producer actively works to generate demand at the end-market level
- value, as perceived by the prospective franchisee, of the distributor support services offered by the franchisor

A company having the brand preference position and high installed base of an Ingersoll-Rand or a Square D is able to impose terms and conditions such as the above on franchisees that themselves have strong local market shares. Competing suppliers may not have the same clout.

The feasibility of the producer's laying down franchise conditions is also a matter of its willingness and ability to enforce the stipulated terms. Some conditions, such as those which relate to transshipping and price cutting, take considerable resources to police. Others, such as minimum order-size provisions, are inherently simpler to monitor. In general, however, to impose conditions on resellers requires an extensive commitment of personnel to monitor compliance and may not be feasibly undertaken unless the producer has a large sales revenue base to support the required resources. Thus, in a self-reinforcing cycle, firms with large market shares are often better able to impose and monitor franchise terms and conditions.

DETERMINANTS OF DISTRIBUTION STRATEGY

Distribution systems often grow and evolve without benefit of either strategic analysis or a master plan. They are formed over time by moves, often tactical, made in response to competitive actions, market shifts, product line additions, and perceived opportunities in new market segments. Nevertheless, industrial distribution patterns are shaped fundamentally by four factors:

- nature of the product
- market segment demographics
- buyer behavior
- distribution costs

These factors tend to influence distribution strategies of all market segment competitors. Within an industry, however, individual competitors tend to differentiate themselves based on a fifth factor—their respective business unit environments.

Finally, the options available to any one producer are bounded by a sixth factor, the distribution infrastructures of both the industry sector and the nation. Exhibit 2.2 arrays these six factors as they shape the distribution strategies of individual competitors.

Product factors that affect modes of distribution are (1) the product's unit dollar value; (2) technical information exchange needs of buyers and sellers; (3) weight-to-value ratio, a factor affecting the costs of physical distribution; (4) degree of customization required to meet customer requirements; (5) need for postsale field maintenance service; and (6) nature of the production process—whether the product is made to order or mass-produced.

Three *demographic factors* are relevant: market segment concentration and the purchasing power of individual buyers, the diversity of user applications, and the overall level of demand.

Buyers' purchasing behavior and their product usage patterns are clearly reflected in the design of distribution channels. Four aspects of buyer behavior are important: centralized or decentralized procurement modes, single-item or bundled buying, demand predictability, and centralized or dispersed product usage.

Product and market factors frequently translate into *distribution cost factors*. Distribution costs relative to revenues are largely determined by account size, the typical dollar amount of sales trans-

Exhibit 2.2 Determinants of Distribution Strategy

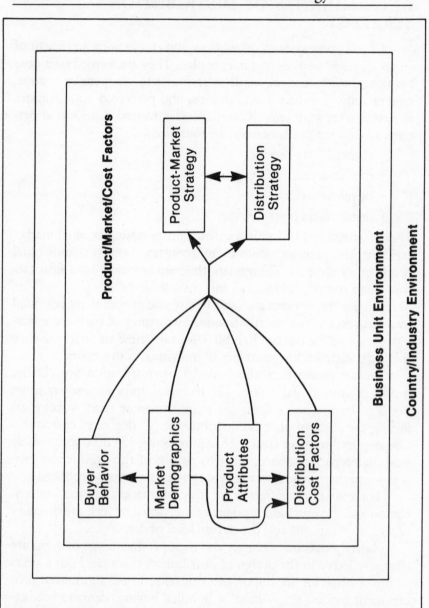

actions, and product gross margins. Other distribution cost factors shape both distribution channels structures and the locus of functions in multichannel systems. For example, the level of inventory risk—as determined by such factors as inventory-to-sales ratios, degree of price stability, and threat of inventory obsolescence—is a critical determinant of whether risk is assumed by producers or intermediaries.

How the distribution strategy of any one firm responds to product-market specifics depends critically on its unique *business unit environment,* in particular, on the availability of resources for marketing investments. A second key element is the individual firm's competitive situation.

The *distribution infrastructure,* both at the country level and at the product-market level, may be delineated in terms of classes of channels institutions, the distributive resources they offer, and distribution demographics—that is, numbers and sizes of firms, product-markets served, and distribution firm concentration by industry sector. An important feature of the infrastructure, too, is the body of law that applies to manufacturer-reseller relationships. The availability of channels institutions and the legal constraints on producers in establishing and managing distribution systems in effect circumscribe the range of options they have in developing distribution strategies in a country and/or a particular product-market.

SUMMARY

This chapter has presented a basic concept of distribution strategy and the considerations that influence producers' choices among a wide array of strategic options. These choices involve the following:

- determining the balance of direct sales and sales through resellers and agents
- choosing among classes of channels—direct salesforces, independent distributors, captive distributors, agents, and brokers—to form a multichannel system
- choosing among types of channels—as defined largely in terms of served-customer categories
- determining relative distribution intensity, that is, the number of locations in a trading area that carry the producer's line
- allocating distribution functions—selling, channels support, phys-

ical distribution, product modification, and risk assumption—among the several classes of channels

- establishing the terms and conditions of the franchise contract, including such provisions as full-line selling, brand exclusivity, account and class-of-trade constraints, location franchising, and resale price maintenance

Having identified these several elements of distribution strategy, we then discussed considerations that influence producers' choices and factors that give producers power to develop effective channels strategies. The producer's base of strength in designing and managing its channels rests primarily on the demand for its brand among end-users and the resulting value of the franchise to intermediaries.

As identified in Exhibit 2.2, the factors that tend to determine distribution strategy are these:

- the nature of the product
- market segment demographics
- buyer behavior
- distribution costs
- the business unit environment
- country/industry environment

We related these factors to producer choices with regard to the direct-reseller balance, the allocation of channels functions, intermediary type, and distribution intensity. Finally, we discussed the impact of country and industry distribution infrastructure.

The construct shown in Exhibit 2.2 provides a framework for understanding how and why the distribution strategies of competing firms in a product-market differ and what explains the differences and similarities in distribution profiles among leading firms in different product-markets. In Chapter 3, we will examine in greater detail the impact of product attributes, market demographics, and buyer behavior on distribution strategy. Chapter 4 deals with distribution costs, while Chapter 5 treats the business unit environment as a determinant of the firm's distribution strategy. The influence of the country context on modes of distribution is the theme of the last section of the book. There, we will describe the history of industrial distribution in the United States from 1812 to the present and the legal framework that mediates buyer-seller relations. Against that backdrop, we show how firms' distribution systems have evolved as the country has grown.

3   Product, Market, and Buyer Behavior

Three key determinants of distribution strategy are product characteristics, market segment demographics, and buyer behavior. These three influence the balance of direct versus indirect distribution and the locus of channels functions among producers, resellers, and other intermediaries. These factors account for many of the differences in distribution modes among the four product-market groups that we have considered, as well as differences among competitors in each of these industries. Other factors also influence modes of distribution—in particular, distribution costs and the business unit context, which we will discuss in the two chapters that follow.

In this chapter, we first relate product, market demographic, and buyer behavior factors to channels design. We then illustrate these relationships with a case history involving contract negotiations between Becton Dickinson and a large hospital buying group. The case introduces a dynamic dimension to product, market demographics, and buyer behavior as determinants of channels design; in this instance, each of the three factors is undergoing dramatic change, with significant implications for Becton Dickinson's distribution strategy.

THREE KEY DETERMINANTS OF DISTRIBUTION STRATEGY

The linkage between product, market, and buyer behavior factors and modes of distribution is well illustrated in the four product-market examples discussed in Chapter 1. Such product-market segments as the residential housing market for load centers, for instance, and the independent drilling contractor market for oil country tubular goods are typically served through multiple channel levels, with independent resellers playing a major role. The products are of relatively low unit value and standardized design, and in these segments buyers tend to be medium- and small-sized businesses widely dispersed geographically. They make purchases intermittently as needs arise, and transactions include other items that are related in use but often made by different suppliers. For example, an electrical contractor may buy several load center assemblies together with wiring, electrical conduits, wall outlets, and electrical switches for installation in a new condominium complex. The contractor will buy familiar brands from a local electrical distributor that carries a wide range of electrical supplies used in home construction. The timing of the purchase will be determined by the building's stage of completion. While convenience and sales credit are important in the purchase transaction, direct communication with the manufacturer is not essential.

In comparison, product-market segments such as disk drives for computers, tubular goods for deep-well drilling, and large stationary air compressors are served primarily through direct selling. The products are of high unit value, often customized to user requirements, and purchased either as single-item buys or as part of an order that includes other items supplied by the same manufacturer. Typically, potential users are large and easily identified by the manufacturer's sales organization. In such transactions, direct producer-buyer communications are often essential in negotiating technical product requirements, scheduling product deliveries to meet usage needs, and providing technical support for the user's operating personnel. In these product-market segments, transaction size and the sales potential of individual accounts are typically large enough to warrant the expense of direct selling. In fact, with large buyers and high transaction values, direct selling is likely to be the more cost-efficient mode of distribution.

Variations on these two basic patterns involve (1) items of relatively low unit value purchased in large transaction quantities, and (2) items of high unit value, such as disk drives, bought as fill-in items by large users. In the former case, the producer may make the sale without the involvement of resellers or may negotiate the purchase directly with the user-consumer but arrange for contract fulfillment (shipping and billing) through local distributors. In the latter case, while the original large quantity order may have been a direct transaction, the user's fill-in requirements may be met through resellers franchised by the producer to carry local stocks.

Product: A Complex of Attributes

The nature of the product essentially determines whether users will rely on producers or resellers as primary sources of technical information and supply availability. Said another way, the nature of the product determines where in the chain of distribution technical communication with user-customers may best be positioned and what may be the most cost-effective supply point relative to buyer needs for quick delivery or scheduled-quantity shipments.

If the product is technically complex, users want a direct relationship with the source of product technology. That may be the manufacturer or other members in the chain of distribution. Usually, the source is the OEM, but for certain types of industrial products, such as custom-built information systems, it may be a value-added reseller. Whether the source of product technology is the OEM or a reseller, users are prone to seek direct ties with the technology source for products that require modification to user requirements, call for significant user education, or are subject to rapid technical change.

The nature of the product and its application also influence users' requirements for supply-source accessibility. Large OEM buyers of materials and components that are used in the products they make and sell may want direct delivery from the manufacturing source. In this way, they can purchase quantities needed according to their own factory production schedules, avoid carrying large in-process inventories and warehouse stocks, and increase the ensurance of on-time delivery. For example, producers of beer and soft drinks often take delivery of metal containers directly from their can manufacturers' local plants; automobile plants source high-volume steel directly from their suppliers' mills.

Items that are of high bulk and/or high weight relative to value are also often delivered direct from the manufacturer's plant or storage area so that buyers can economize on transportation costs and minimize the risk of in-transit damage. Heavy electrical equipment, tank cars of chemicals, and large air compressors are examples. However, shipment from the manufacturer's plant does not preclude a sale being made and billed through a distributor.

On the other hand, small items such as consumable factory

supplies, for which usage is intermittent and often unpredictable, are typically sourced from local resellers. Under these conditions, local distributors may be the most accessible point of supply; in fact, ordering from the manufacturing source may be disadvantageous because many plants are not set up to make quick deliveries to fulfill users' unplanned requirements.

The way in which the product is made is another factor influencing mode of distribution—whether it be mass-produced at relatively constant rates, seasonally produced, or custom produced. When market demand fluctuates but production is relatively stable, or when production is highly seasonal, multilevel distribution systems may be needed to coordinate production and consumption. That is, resellers may perform inventory-carrying functions to buffer variations between production flows and varying levels of consumption. Thus, multitier distribution channels for agricultural products developed in America well before channels for machinery. Because of the highly seasonal nature of crop production, farmers needed intermediaries to find market outlets, to absorb inventories during peak production periods, and to supply market needs. Machinery, however, in the early history of the country, was largely made to order, with inventories of unsold finished units not of sufficient concern to producers to warrant the development of channels networks as stocking institutions.

The Imperative of Market Demographics

Market demographics here refers to the relative degrees of user-industry concentration and of geographic concentration. If a user-industry market segment consists of a limited number of large buyers, it is said to be concentrated; market segments that are composed of a large number of small buyers are described as fragmented. Automobile repair shops would be an example of the latter; auto manufacturers an example of the former.

Buyers may also be concentrated or dispersed geographically. The U.S. market for oil well tubular goods, for example, is concentrated in the Southwest and the Gulf Coast, with smaller markets in California, Appalachia, the Rocky Mountains, and Alaska, where oil well drilling and production are also found. In contrast, buyers of industrial supplies, such as drilling bits and grinding wheels, are widely dispersed throughout the 50 states.

Some market segments are concentrated both in terms of user

industry and geography, for example, tire manufacturing in Akron. Many, such as auto repair shops, consist of thousands of small, geographically dispersed businesses. Others—tile manufacturing in Ohio and New Jersey and furniture manufacturing in the southern states—are geographically concentrated but are composed of relatively small- and medium-sized firms.

Usually, a product-market consists of several user-industry segments that vary in terms of demographic characteristics, ranging from concentrated to fragmented and from geographically dense to dispersed. This is certainly true of the four product-markets described in our first chapter: load centers, disk drives, air compressors, and oil country tubular goods.

Producers try to adapt their modes of distribution to demographic characteristics, tending to reach more concentrated segments directly and generally to use indirect channels to serve relatively fragmented and geographically dispersed segments, as noted earlier.

Some producers elect to concentrate on one market segment; others, such as Ingersoll-Rand, may develop relatively segregated channels systems to serve each of several market segments. Still others opt to compete in multiple segments through one multichannel distribution system with different channel elements—the direct salesforce, distributors, and agents—having different responsibilities in different segments for such functions as selling, the negotiation of transaction terms, customer technical support, delivery, and postsale service.

The Buyer: A Moving Target

Buyers seek to optimize a range of values in their procurement practices, such as low total acquisition costs, purchasing convenience, breadth of brand choice, timely delivery, sales credit, ensurance of future supply, and availability of product services. The relative importance of any one of these values at a given point in time will depend on the nature of the product, its intended use, the value of the purchase, whether it is a first-time buy or a repeat purchase, and often on who is responsible for the purchase decision—an engineer or a production manager, a plant buyer or a procurement manager in a central purchasing operation.

These questions are relevant: Under what conditions will buyers want direct relationships with producers? Under what conditions are they likely to source from intermediaries? When will they turn to producers for contract negotiation but to intermediaries for contract fulfillment (e.g., stocking, delivery, billing, credit, and after-sale service)? Alternatively, under what circumstances might the buyer negotiate a purchase with a reseller but look to the producer for such services as delivery, product warranties, and product maintenance?

Buyers' choices are inevitably limited to their available sourcing options. While a small electrical contractor, for example, may have the choice of buying from an electrical distributor, mill supply house, retail hardware store, or mail-order house, it is not likely to be able to buy directly from Square D, GE, or Westinghouse. Conversely, while a large builder of tract housing may have the option of negotiating directly with manufacturers, it probably cannot source large volume product requirements through the local hardware store. The following discussion therefore assumes in each case that the options being considered are available to the buyer.

The following defines four channels choices from the buyer's perspective and suggests the conditions under which each tends to be preferred. In each case, it is essential not that all conditions apply but that, on balance, the purchase conform to the described parameters.

1. *Direct sourcing relationships with the producer for both contract negotiation and contract fulfillment*

   The scope of the procurement covers a single item or multiple items, related in use and within the individual producer's product line scope.

   The purchase has high salience in terms of
   • dollar amount;
   • level of opportunity and risk for the buyers' production efficiencies and/or end-product performance characteristics.

   Technical interchange is an essential element of buyer-seller negotiations with respect to product specifications and/or product use information.

   Demand is relatively predictable.

   Minimizing product acquisition costs has high priority.

2. *Sourcing relationships with resellers only, for both negotiation and fulfillment*

The purchase is for relatively small amounts.

An individual purchase order may include the products of multiple manufacturers.

The product is standardized or requires minor customization.

Product usage is intermittent and/or unpredictable.

Local availability of supply is important.

Local availability of after-sale service may be important.

3. *Producer-negotiated terms and conditions; reseller contract fulfillment*

Purchases are for large amounts.

The items purchased fall within the scope of the producer's product line.

The purchase decision-making responsibility is centralized within the firm or business unit.

Product requirements are relatively predictable.

The product is used at multiple locations.

Local supplies must be available for fill-in orders when timely deliveries cannot be made from the manufacturer's stocks.

Other important purchase considerations include
• the availability of after-sale field service;
• the continuing availability of supply items to be used with the primary purchased product;
• the availability of sales financing.

4. *Reseller as the primary source; reliance on the producer for acquisition-related services*

Maintaining local source relationships is important to the buyer.

The producer sells only through distributors and agents but not direct, other than to large accounts.

The purchase is for a relatively large physical quantity, and transportation economies may be realized by shipping direct to users from the producer's plants or stocking locations.

The producer maintains its own network of repair shops and makes field service available.

The producer warranties the product.

The producer extends sales financing to reseller customers.

These four types of sourcing relationships, involving different combinations of producers and resellers, reflect buyer preferences under different purchase parameters. Exhibit 3.1 describes channels sourcing options from a buyer's perspective and suggests the considerations that influence choice.

The range of buyer sourcing preferences has important distribution implications for producers. One is that distributors take on different roles with different types of customers. A Norton abrasives distributor, for example, both sells to and services small accounts for grinding wheels and sandpaper and serves as a fulfillment agent and source of back-up supplies for Norton's direct customers. Acting in dual roles, the distributor may be working under two or more sets of prices and discount schedules. It may sell to its small customers at published prices on which it receives a fixed margin; it may sell to large customers at prices negotiated by the producer and receive a margin negotiated by it with the producer and/or the user-customer.

Further, because different sets of customers have different priorities in selecting among channels options, two classes of distributors, full service and minimal service, may be established. In the former category, for example, come Square D's Blue Chip distributors.[1] In return for a range of specialized support services and sometimes for being given classes of products not made available generally to the producer's resellers, these firms agree to observe certain full-service standards of performance. Other resellers in the network, not included in the elite distributor set, are likely to stress low prices and offer minimal services in selling to their clienteles.

In summary, the great variation among buyers with regard to their preferred modes of product acquisition leads producers to develop channels systems and processes that offer customers a range

---

[1] To be classed as Blue Chip distributors, Square D franchisees were required to maintain stocks at specified levels and to reorder to these levels every two to three weeks. In return, they were granted certain stock return privileges. As of 1986, 75% of Square D's distributor locations carried the Blue Chip designation and 95% of Square D's sales through distributors of electrical equipment came through these outlets. In major metropolitan areas, some Square D sales specialists—distributor inventory reps—worked only on implementing and monitoring the Blue Chip programs, programs to which headquarters marketing managers attached considerable importance. The Blue Chip Program had the effect of (1) stabilizing ordering patterns of Square D plants and hence helping to stabilize production levels and (2) ensuring that Square D products were always in stock on distributor shelves. Further, the Blue Chip program was seen as a way of encouraging distributors to stock a broad assortment of Square D electrical equipment, including new product introductions; distributors were more willing to take on new products knowing that they could return the unsold quantities for credit.

Exhibit 3.1 Buyer Behavior Factors and Channels Sourcing Choices

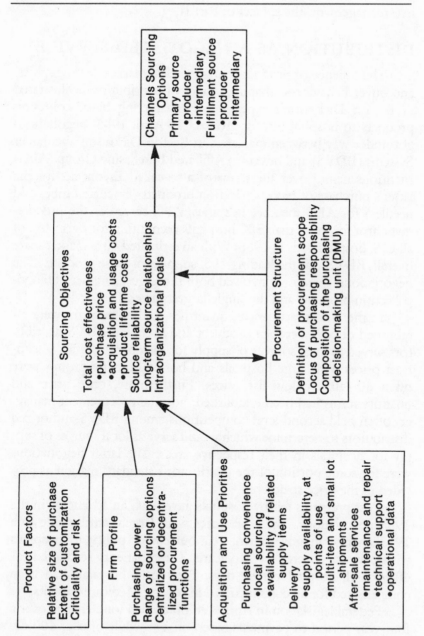

of options. While this may indeed respond to buyer behavior and broaden the producer's market scope, it will also complicate channels management, the subject of Part II.

DISTRIBUTION AS A NEGOTIATED SERVICE

The salience of product characteristics, market demographics, and buyer behavior in shaping modes of distribution is illustrated in Becton Dickinson's experience in marketing blood-collection products to hospital buying groups.[2] In April 1985, negotiations got under way between managers in Becton Dickinson Vacutainer Systems (BDVS) and buyers in Affiliated Purchasing Group (APG) [fictitious name] over the terms of a so-called Z contract for the latter's purchase of blood-collection products (evacuated tubes and needles) for APG member hospitals. APG served as the procurement arm for more than 500 hospitals accounting for over 10% of all U.S. hospital beds in 1985. With an estimated 80% market share overall, BDVS was the leading U.S. supplier of these products. The negotiations, therefore, involved both the largest supplier of blood-collection products and the single largest buyer.

In the Z contract, prices, quantities, and other terms were established directly between producer and user, and BDVS distributors served as local sources of supply for member hospitals. Z contract prices with large hospitals and hospital buying groups were often 30–40% below list prices. Furthermore, once price and quantity terms had been established, hospital procurement managers often held second-level competitions among BDVS-authorized distributors to determine which would serve as local sources of supply for hospitals in their respective areas. The latter negotiations were for some portion of the distributor's standard margin as provided in the price structure.

Cost pressures felt by hospitals gave APG an advantage in the 1985 negotiations. These pressures were largely attributable to a 1983 change in government reimbursement policy to hospitals with patients covered by Medicare. These patients accounted for 40% of all hospital patient-days. Previously, hospitals had been compensated for all costs incurred in serving Medicare patients.

Federal legislation in April 1983 provided for a change over a four-year period to payments based on a schedule of over 400 de-

[2] Frank V. Cespedes and V. Kasturi Rangan, "Becton Dickinson & Company," 9-587-085. Boston: Harvard Business School, 1986.

fined "diagnosis-related groups" (DRGs). Under the new system, the payment to a hospital was based on standard national and regional costs for each DRG, not on the hospital's costs. As a result, hospital administrators came under severe pressure to reduce costs. One response was the acceleration of a trend toward the formation of multihospital buying groups. Buying groups handled purchases centrally, but the individual hospitals they represented were usually free to accept or reject the terms negotiated on a specific item by the central buying group. Some central purchasing organizations were able to gain greater control over member-hospital procurement than others and so had greater bargaining leverage with suppliers. One BDVS manager noted,

> The chains and buying groups structure negotiations on the premise that they can deliver so many thousands of beds to the manufacturer with the best price. But the actual strength of these groups varies. In some, all of their hospitals purchase through the centralized procedure. In others, a large percentage of the member hospitals do not adhere to the centralized procedure. The result is that purchasing leverage differs from one group to the next.

These developments had a profound effect on hospital purchasing practices and, in turn, on the distribution channels serving hospitals. According to a senior executive of one large national distributor of hospital-supply products,

> In the past, our customer was the pathologist, chief technologist, or lab manager. This person's responsibility was to produce quality diagnostic tests on specimens brought into the lab and to do it as fast as possible. A key was to ensure that an adequate supply of products was on hand at all times. It was also the element that these people were least prepared to deal with. Most lab managers and chief technologists had risen to their positions on the basis of their clinical skills, not their purchasing skills. In addition, they didn't particularly enjoy the purchasing part of their jobs.
>
> They tended to do business with a representative they liked and trusted, and who had a product line that encompassed most of their needs. They wanted a company that would manage most of the purchasing job for them.
>
> Also, distributors generally paid little attention to costs, because customers primarily wanted service and were willing to pay for it. After all, hospital reimbursement procedures allowed any increased operating expenses to be passed on to customers.
>
> Those days are gone. The customer is different. Buying influence has moved out of the lab in most hospitals and beyond the

hospital purchasing department to the corporate purchasing department of national multihospital systems. Some distributors probably have over half of their total sales in these national accounts.

In this environment, distributors must lower costs. I believe many distributors will carry only one or two vendor brands in many product categories in exchange for lower prices from those vendors. Moreover, distributors can reduce inventory, transportation, and some administrative costs through consolidation of their product lines.

By standardizing on purchases of certain products, buying groups could negotiate lower prices from the manufacturer *and* from the distributor. The distributor, fearful of losing a big order, often relinquished a significant portion of its margin.

In addition to negotiating favorable prices with manufacturers and bargaining with distributors for further price reductions, hospital buying groups initiated other actions to contain costs. APG, for example, coordinated a program that brought together doctors, nurses, and administrators from different APG-affiliated hospitals to discuss cost-reduction opportunities and to develop specific action plans. The program allowed member hospitals to compare their costs by product line, therapy type, and department. In addition, APG had established a private-label program in which it sought to have its suppliers use the APG trademark on products purchased under APG contracts. By mid-1985, this private-label program encompassed 12 product categories and APG expected to add 30–40 additional products by 1986.

In early 1985, APG also announced its intention of establishing its own distribution network. APG sought to have its suppliers deliver to APG-affiliated distributors that, in return for a larger share of the high-volume APG contracts, would serve APG member hospitals for lower margins than hospital-supply distributors traditionally received. The program would eventually involve a national order-entry system linking these distributors with APG-affiliated hospitals. If the system could achieve sufficient utilization by suppliers and APG member hospitals, hospital costs might be reduced by 3–12% on most supply items.

Negotiations with APG.   In 1982, APG had first sought to standardize its purchase of needles and tubes and had demanded substantial price reductions from BDVS. BDVS had resisted negotiating prices and terms directly with APG headquarters and had continued dealing separately with individual hospitals. When APG established a national purchasing agreement with a BDVS competitor, BDVS's field salespeople had been able to retain most sales of BDVS tubes at individual APG-affiliated hospitals, in part by lowering prices hospital-by-hospital when necessary.

In April 1985, APG announced its intention of establishing a new national purchasing agreement for blood-collection products; this time BDVS management decided to enter into negotiation with APG headquarters. Management felt that the APG system had grown considerably during the past three years, the central purchasing organization had increased its strength with member hospitals, and there was more risk in refusing to negotiate.

BDVS proposed a Z contract with prices it believed to be approximately 20% higher than competitors' proposals. The proposal required APG to "deliver" within 90 days of the initial contract date 95% of its member hospitals' purchases of venous blood-collection tubes and 90% of their purchases of blood-collection needles. If these targets were not achieved within 90 days, prices on BDVS products covered by the contract would automatically increase by 5% during the remaining 21 months of the proposed two-year contract agreement.

APG negotiators rejected this proposal and gave BDVS until August 15 to submit a new proposal. They also announced that they wanted all blood-collection products covered by a national purchasing agreement to be part of the private-label program and thus to carry the APG logo. In addition, they wanted all products covered by the agreement to be supplied through distributors affiliated with APG and provided a list of these distributors. The list did not include most of BDVS's major distributors. According to the APG negotiators, BDVS's competitors had agreed to their original pricing proposals and to both the private-label and distribution demands.

Finally, in late August, BDVS negotiators came to an agreement with APG's representatives that involved making significant price reductions on major items in the procurement on the condition that APG deliver 90% of its member hospital requirements for tubes and needles for the fourth quarter of 1985. In return for these price concessions, APG negotiators conceded that (1) these products would be identified as BDVS brand and not APG private label and (2) contract fulfillment would be through BDVS-franchised distributors and not through those in the APG distributor network.

Commentary: The Buyer's Influence on
Distribution Strategy

Becton Dickinson's negotiations with APG took place in an
environment of dynamically changing market demographics, buyer
behavior, and users' perceptions of the product. Rising costs of
health in this country and legislative response in the form of new
Medicare guidelines for reimbursing hospitals triggered these
changes. While the formation of hospital chains and buying groups
was already a trend, the 1983 DRG legislation greatly accelerated
this development.

Changing Market Demographics. The hospital market
segment, once relatively fragmented, became highly concentrated,
with membership in some buying groups amounting to more than
500 hospitals. Exercising their considerable economic power, buying group purchasing managers sought to negotiate directly with
hospital equipment and supplies manufacturers; previously, individual hospitals had ordered primarily through local distributors
and passed on price increases to patients and insurers.

In effect, the very large and important hospital market split off
from other product-market segments, such as clinical laboratories;
to compete effectively, hospital-supplies producers had to become
directly involved in negotiating the terms and conditions of user
purchase contracts. The change in the basis for Medicare payments
had much less effect on such other hospital-supply market segments
as clinical laboratories, corporate medical departments, and physi-
cians' offices not dependent on Medicare income. These segments
continued to be highly fragmented, with much greater reliance on
local hospital-supply distributors as sources of supply, information
on new products, and broad selections of supplies and equipment.

The Transformation in Buying Behavior. Before 1983,
hospital-supply buying responsibility was in the hands of operating
personnel—lab managers, technologists, and pathologists—whose
performance was measured not in terms of buying acumen but
rather by their professional skills as medical technicians. Motivated
largely by the need simply to avoid out-of-stock crises and to minimize time spent on purchasing tasks, they relied on distributors for
service, with little sensitivity for price.

In sharp contrast, hospital buying groups, especially from
1983 on, concentrated on minimizing total product acquisition
costs, including both the price of the product itself and the costs of

delivery and inventory management. As frequently ordered products used in large quantities, blood-collection items became a logical cost-saving target. While the priority of price and acquisition cost minimization could be explained by the pressure on APG's member hospitals for cost containment, it also reflected another motivation on the part of APG buyers: to control and direct member-hospital purchases. On the one hand, APG control of member-hospital purchases depended on negotiating lower prices than those the individual hospitals could command on their own or through contracts negotiated by other buying groups of which the hospital was also a member. On the other hand, APG's leverage in negotiating for low prices was directly dependent on its ability to deliver purchase volumes under price-quantity agreements with vendors.

As the purchasing decision moved from lab technicians, doctors, and hospital administrators to central buying groups, the locus of responsibility for negotiating prices and contract terms shifted upstream from distributors to producers. In addition, the cost-containment pressures on hospitals were translated into comparable pressures on suppliers and distributors to reduce costs and prices at both levels. Distributors responded in part by reducing the number of brands carried in order to negotiate more favorable prices and terms with their remaining suppliers. In the meantime, brand-name manufacturers stood to lose in some degree their bargaining power with their distributors if hospital buying groups developed private brand lines, and thus undercut a brand preference advantage at the user level.

Further, the strength and indeed viability of the supplier's distribution system may eventually decline if the move to establish buyer-controlled distribution networks is successful. Clearly, such networks would compete with supplier-franchised systems, taking channels development and management out of the supplier's hands.

In sum, this is a case of buyers negotiating not only for lowest-possible product price but for lowest-possible acquisition cost as well. In so doing, they affect the supplier's brand differentiation advantage, the strength of its distribution system, and indeed the supplier's own power vis-à-vis its distributors.

We have seen other instances of large buyers negotiating for product prices and acquisition costs as two phases in the procurement process: for example, the major oil companies engaging the steel companies in competitive bidding for their oil country tubular goods requirements and then negotiating with oil field supply houses for distribution services. While these modes of product acquisition seem limited to large corporations making major unbundled purchases, such customers do represent large portions of the producer's sales revenue and significantly affect its distribution strategy.

Changing Perceptions of Product. Changes in buyer behavior ultimately reflect changes in the way buyers rank purchase criteria. While needle sharpness and lack of defects in blood-collection tubes are important functional product attributes, buying convenience, ensurance of supply, and technical information are also important. As the locus for purchasing responsibility shifted from local hospital operating personnel to centralized buying groups, the changes in market demographics and buyer behavior implicitly reordered buyer priorities among these attributes. Prior to 1983, brand name stood for product functional quality, and distributor relationships stood for ensurance of supply and service. With an emphasis on minimizing total product acquisition costs, buying groups worked hard to negate the influence of brand and of distributor relationships as primary buying criteria for hospitals. For example, by introducing APG labels on high-usage items, the purchasing organization sought to reduce preferences for premium-priced brands on the part of doctors and lab technicians. In addition, by conducting cost studies in member hospitals, APG diminished the effect of product brand preference as a buying consideration, reinforced cost containment as a high priority, and drew attention to the extra and, by implication, unnecessary costs of relying on familiar brand names as a surrogate for product quality and vendor services. The implicit presumption was that all brands were equal with regard to functional quality.

SUMMARY

In summary, the most basic factors influencing the way industrial goods producers go to market are product, market demographics, and buyer behavior. Key aspects of the product are whether it is standard or customized to user specifications, whether its technology is mature or rapidly evolving, and whether it is a relatively major procurement or a minor purchase. Market demography, the relative size and concentration of potential buyers—in terms of both user industry and geography—is a critical factor shaping modes of distribution. Key aspects of buyer behavior are the following: (1) the way the product is used, for example, as a critical component of the production process or perhaps as a supply item; (2) whether the product is purchased as one item on a "shopping list" (bundled buying) or as a major buy, either by itself or with other items related in use and supplied by the same producer; and (3) the nature of the purchasing process—centralized or decentralized, complex or simple buying decision-making units.

Product and buyer behavior factors determine what the user wants from the producer's distribution system in terms of product technical support, supply availability, and service. Market demographics determine the feasibility of the producer's serving those needs directly or through resellers and agents.

Under these conditions—customized and/or highly technical products, high unit price, single-item buying, and large, easily identified potential customers—producers tend to go direct. Under these conditions—low unit product value, mature product technology, bundled-buying behavior, and fragmented markets—producers tend to go through external intermediaries. In buying from the producer—if that option is available—the buyer may have a direct relationship with the source of product technology and may be able to integrate its production schedules with those of the producer to economize on manufacturing costs. On the other hand, buyers sourcing from resellers gain the benefits of purchasing convenience and local availability, important attributes when needs are intermittent and relatively unpredictable. In some instances, the buyer may turn to resellers, as well, as sources of technology. Resellers serve this function, for example, in applications-specialized computer information systems, reseller customized equipment, and special chemical formulations.

All three factors—the nature of the product, market demographics, and buyer needs—influence distribution costs, the subject of our next chapter. Distribution costs, on the one hand, mandate producers' choices between direct and indirect distribution and among such direct channels as personal sales, telemarketing, computerized customer links, and catalog sales. On the other hand, distribution costs are a major cost element covered in the price paid by buyers. The pressures, then, to minimize total acquisition costs strongly influence the buyer's choice of channels sources.

In the next chapter, we identify the elements of distribution costs and show the factors that affect their behavior. We develop the concepts of *cost efficiency* and *cost effectiveness,* with the former relating to percentage cost minimization and the latter concerned with seeking some optimal balance among distribution expenditures and sales revenues and profits. We show, too, how one manufacturer, IBM, sought to gain greater control over its distribution costs for supply items by seeking to alter the balance among direct selling, catalog sales, and the use of computerized ordering systems by its customers.

4 Distribution Costs

Distribution costs are a significant component of producers' total marketing costs. A leading supplier of electrical components estimates its average sales expense-to-revenue (E/R) ratio at over 14% of its sales. A leading producer of expendable supplies calculates its E/R ratio at 20%. Both companies rely heavily on industrial distributors, which realize net margins on their product lines of 18–21%. Thus the total distribution costs included in the prices end-customers pay for these companies' products may then be estimated at roughly 35–40%.

The manufacturer's ability to manage such costs affects its revenues, profits, and market share position directly. Further, the urgency to control costs is heightened by the pressures brought by competitors and customers. Customers seek to get that combination of price and channels services that meets their procurement objectives at the lowest cost. Accordingly, producers' market shares depend heavily on being able to meet buyer needs at the lowest delivered-product cost.

Distribution cost factors—defined here to include both manufacturers' distribution costs and resellers' margins—significantly shape channel structures. For example, the rising costs of sales calls by manufacturers' salesforces have caused suppliers to rely increasingly on resale channels. Resellers' sales costs are often lower as a percentage of sales, in part because they serve many accounts with broad product lines.

For producers, distribution cost management means controlling or influencing two kinds of costs: directly incurred expenses, that is, costs such as personal selling, promotion, and order receipt and processing; and reseller margins. Key issues, then, are the following: Which channels functions will be performed by producers and resellers, respectively? At what costs? How many levels of distribution will the product pass through? And what types of resellers will be included in the channels network—local mill supply houses, large multibranch firms, or industrial catalog-order operations?

Distribution cost management is not limited to factors affecting *cost efficency* as measured in terms of sales expense-to-revenue ratios. It also includes *cost effectiveness,* or the relationships among cost efficiency, sales revenue, and sales margins, with the relevant measures of cost effectiveness being market share and product line profits.

This chapter identifies the elements of distribution costs and suggests ways of thinking about them. We also propose some concepts useful in understanding the implications of distribution cost analysis for distribution strategy.[1]

The quantitative data included in the following discussion are drawn from surveys we conducted in 1986 of distributors and producers. The data are not definitive, since the number of data points is limited. We believe the data base is nonetheless useful in suggesting expense category magnitudes and indicating transaction cost experiences by industry as well as between producers and distributors.

In the following discussion, we (1) identify the elements of distribution cost, consider their relative magnitudes in a total cost context, and suggest the factors that influence cost behavior; (2) discuss sales-cost/sales-revenue relationships; and (3) consider how to optimize costs and revenues through (a) choosing appropriate modes of customer communication and (b) allocating channels functions effectively across the distribution system. To illustrate these issues and choices, we cite the experience of an IBM marketing manager in analyzing rising E/R ratios and considering their relationship to profit margins.

ELEMENTS OF DISTRIBUTION COST AND COST BEHAVIOR

Distribution costs are defined here to include these elements:

Customer communications
- personal selling

---

[1] It is not the purpose of the chapter to present an analysis of the costs of distribution as experienced by industrial goods manufacturers and/or resellers. Such data are not readily available, since many manufacturers and resellers do not systematically gather and analyze them. To our knowledge, those which do so accumulate information according to their own accounting frameworks, making it difficult, if not impossible, either to describe average cost experience by product-market segments or to draw cross-firm and cross-industry comparisons.

- · sales calls
- · travel
- · sales meetings; reporting
- telemarketing
- computer interfaces
- trade shows and demonstrations
- advertising
  - · media
  - · direct mail

Paperwork flows
- order receipt and processing
- billing and collections

Physical distribution
- shipping
- physical handling and storage

Financial risk assumption
- sales credit
- inventory financing

Reseller net margins and commissions

Some of these cost categories are large, while others represent relatively minor elements of distribution costs. The data in Exhibit 4.1, provided by two major suppliers of electrical equipment and a leading manufacturer of industrial supplies, suggest that three cost elements account for the bulk of these companies' total distribution costs. They are (1) salesforce salaries, commissions, and expenses; (2) warehouse and shipping; and (3) reseller margins—the difference between what the reseller pays the supplier and what it charges its customers. These costs are included in the end-product price. We believe that many industrial firms also find these three cost categories to represent major distribution expense factors.

Customer Communications

Communications between buyer and seller take various forms, and in general, the suitability of the medium depends on the nature of the message. The most effective medium for understanding the particular needs of individual buyers and communicating relevant information is personal selling. It provides opportunities for establishing and maintaining personal relationships as an element in selling and for gathering both customer and competitive information.

Exhibit 4.1 Distribution Costs Reported by
Three Leading Industrial Goods Manufacturers for 1985
(% of producers' sales dollars)

	Company A	Company B	Company C
Salesforce salaries, commissions, and expenses	5.7	6.0	6.2
Warehouse and shipping	6.9	7.9	4.6
Order receipt, order entry, expediting, billing, credit, and adjustments	0.7	0.6	0.7
Media advertising, trade shows, other promotional expenditures	1.1	3.8	0.7
Distributor margins (as estimated by the producer)	25.0	25.0	20.0
Totals	39.4	43.3	32.2

Note: Companies A and B are major electrical equipment suppliers; Company C is a leading industrial supplies manufacturer.

In industrial marketing, other media often supplement direct selling. Telemarketing, or telephone communication, may be used to receive incoming inquiries and orders and respond with order fulfillment, product service calls, and lead referrals to sales personnel. Telemarketing can also identify new prospects through "cold calls" or solicit repeat orders from current customers. Computer links between buyer and seller are increasingly a means of providing information on prices, product availability, and estimated delivery times; entering orders; and invoicing with little or no human intervention. Media advertising, direct mail, special promotions, trade shows, and product demonstrations convey information on product performance, prices, and availability. They may also generate sales leads and create a receptive environment for follow-up personal sales calls.

Suppliers and their intermediaries tend to develop a communications mix as an element of sales strategy. Ideally, the choice of media, or media combination, depends on the buyer's information and service needs and on the relative costs of reaching various types of customers through different channels of communication.

The following discussion considers the factors affecting personal selling costs, which typically constitute by far the largest element of customer communication costs in industrial selling.

Exhibit 4.2   Factors Affecting Sales Costs

	Load Centers and Circuit Breakers	Air Compressors	Disk Drives
	(6 firms)	(4 firms)	(2 firms)
Average number of sales calls (monthly)	70	61	40
Approximate percentage of sales time spent			
• calling on user-customers to —sell and service existing products	13%	29%	55%
—introduce new products	8%	12%	27%
• calling on outside sales influencers and specifiers	4%	11%	0
• calling on distributors	59%	34%	10%
• calling on others	5%	5%	0
• completing reports, attending meetings; other administrative duties	11%	8%	8%
Average cost per sales call	$129	$232	$280
Total active user accounts called on	220	110	N/A
Total number of distributor locations called on	1,225	100	N/A
Sales reps' average total annual compensation	$46,500	$37,000	$58,000

Personal Selling. The costs of personal selling, excluding sales administration overheads, are a function of call duration, call frequency, travel requirements, and time taken for such sales-related activities as reporting, attending sales meetings, and training. As Exhibit 4.2 indicates, the use of sales personnel time and cost-per-sales-call data vary considerably among three product-market groups we surveyed.

The differences in cost per call and number of calls per month reflected in these data may be explained by differences in call content and duration. Generally, the more technical the product, the greater the extent of customization; and the larger the purchase, the greater the number of customer personnel involved in the purchase decision, and the longer the sales call. Thus, as we move from the least technical product, load centers, to the most technical, disk drives, a typical salesperson makes fewer calls monthly and the cost per call increases. The agenda for direct sales visits to users of technical products might include discussions of product performance characteristics, cost and performance comparisons with competitive offerings, product modifications to adapt the product to the user's applications, prices, warranties, delivery schedules, operator training, and product defect claims.

In the case of sales visits to distributors, however, such calls are often time-consuming for both technical and nontechnical products. A typical meeting agenda might include distributor stock levels, merchandise returns, new product features, competitive activity, distributor accounts, credit and collections, distributor business practices, and product performance problems. In addition, manufacturers' sales personnel frequently accompany distributor salespersons on customer visits.

Travel costs, a major component of personal selling costs, are a function of distances between the sales office and the sales reps' customers. For example, a sales rep selling oil country tubular goods for Lone Star Steel commented, "If I'm calling on major oil companies in Houston, I can only manage four calls a day at the most. If I'm visiting oil field supply houses in Tulsa, I can get to six. If I'm talking to independent oil companies in Oklahoma City, I can probably see ten a day, all of them on different floors in the same building."

A sales rep may be assigned to call on all of the existing and potential accounts in the region or on some subset of customers. For example, many of IBM's sales reps specialize in particular market segments, such as banking and finance, manufacturing, distribution, transportation, government, education, or medical. Such specialization permits any one salesperson to communicate with his or her customers about specific product applications. It results, however, in greater travel costs per sales rep than for a nonspecialized rep calling on a comparable number of customers in a reduced geographic area.

While a goal in field sales management is to maximize the sales rep's "time over target," a significant percentage of his or her working hours is spent on activities such as completing customer call reports and attending product-training and sales-briefing meetings. Time taken away from customer contact activities is also a major sales expense category. As the data in Exhibit 4.2 suggest, the amount of nonselling time varies among suppliers.

Time spent on sales-related activities will vary with the information needs of the sales rep, the customer, and management. For example, time spent for product training is influenced by the breadth of the product line, product technicality, and the needs of customers for product information. The last will be related to the importance of the purchase and the extent to which product use is expected to create changes in the customer's operations and procedures.

Further, salesforce needs for nonselling time may be affected by changes in product or by personnel reassignments. Product-training costs escalated sharply for the Honeywell Information Systems Division (ISD), for example, when it reorganized its field salesforce in 1986 into two product-specialized groups, one focusing on large systems (the DPS-8 series) and the other on small systems (the DPS-6 series). ISD management had sensed that its salesforce operated at a disadvantage in having to sell both lines. Product line breadth and rapid developments in computer technology made it exceedingly difficult for the sales rep to represent effectively the full range of Honeywell computers and application systems. With the reorganization, the entire salesforce was put through retraining programs to give sales reps in-depth knowledge of their products, product applications, and markets.

While the purpose of product training is to aid sales reps in communicating with customers, completing call reports serves internal communications needs. Typically, sales call reports provide information on individual account activity, personnel changes, competitive developments, customer service needs, and problems having to do with delivery, credit, and product performance. The time required for sales call reporting is related to the needs of the manufacturer for such information in developing and implementing marketing strategies.

Sales Promotion. Compared with consumer goods manufacturers, industrial goods producers spend relatively little on media advertising, direct mail, and trade shows. While advertising-to-sales (A/S) ratios of 10–20% are not uncommon for items such as breakfast cereals, cosmetics, and proprietary drugs, the six manufacturers of load centers and circuit breakers that we surveyed had average A/S ratios of .2%, while their expenditures for trade shows, demonstration, and sales contests amounted to .5% of sales. In comparison, field salesforce salaries amounted to over 3% of sales in 1985.

Nevertheless, industrial trade shows, attended by potential customers, are a suitable vehicle for introducing new products, developing sales leads, and maintaining relationships with existing customers. The National Mine Service Company (NMS), for example, a major producer of underground mining equipment and supplies, reserves about 25% of its promotional budget for 4–6 trade shows annually and a major exposition, the American Mining Congress International Coal Show, held every four years. NMS spent over $1 million on the latter event in 1980.[2]

In addition to demonstrating their product lines at trade shows, manufacturers also contribute funding and personnel time to distributors for demonstrations and displays, as well as paying part of the cost of local advertising. The amount that a manufacturer contributes to a distributor's advertising each year is often set as a percentage of the latter's purchases from the supplier. The amount contributed is typically limited to a certain percentage—say, 50%—of the actual cost of each ad. Other promotional funds provide sales incentives to distributors in the form of awards and prizes for meeting specified sales targets.

Producer expenditures on national and regional advertising are typically the smallest portion of industrial promotion. Ads are placed most often in trade magazines to announce new products, develop brand familiarity, increase customer receptivity to sales calls, build goodwill among distributors, and generate sales leads.

Nevertheless, promotional communication in industrial marketing seems to be supplemental to personal selling. In particular, expenditures on advertising and promotion are usually small. In addition, they seem to be among the first expenses to be cut in economic downturns.

Taking and Processing Orders

The costs of paperwork—for order receipt, shipping, billing, credits and collections, and account adjustments—are not of the magnitude of personal selling costs but are often significant. In fact, the costs of order receipt, collections, and adjustments are univer-

---

[2] Thomas V. Bonoma, "Get More Out of Your Trade Shows," *Harvard Business Review*, January–February 1983, p. 75.

sally understated in the respect that salesforce time spent on these activities is not accounted for separately as order receipt and processing costs but is lumped in with sales call costs.

In any case, order processing and documentation typically involve cadres of clerical personnel, which represent a relatively fixed cost. However, the actual cost per order will depend significantly on the number of items listed, since clerical time spent is likely to vary with the number of lines per order, as is the time spent correcting errors. Thus, as a percentage of the value of an individual order, paperwork costs will vary greatly because processing costs are largely a function of the number of items, not dollar amount.

To reduce clerical overhead for order receipt and processing, some companies have installed computerized links with their distributors and large user-customers. A manager in one such firm commented,

> You [the authors] also asked for information on the relative costs of handling an electronic order versus a mail or phone order. It's difficult to quantify precisely, but we know that, internally, it is 3–4 times easier to handle an electronic order; and that difference will improve as we refine our systems. The other major benefit to us, is that we receive a clean, complete order: the chance for errors, returns, etc., is greatly reduced. From the distributor's viewpoint, the greatest savings should be through a reduction in inventory carrying costs due to the effective reduction of "lead time." Also, the efficiencies of receiving the correct product, on time, because the manufacturer received a clear order.

Sales Expense Analysis: IBM Supplies Distribution. The factors affecting personal selling, order receipt, and processing costs are well illustrated in a case history that describes a study made at IBM's National Distribution Division (NDD) of the costs related to selling IBM supplies. This product line included supply items used with computers, copiers, and word processors. The nature of the product line and of customer ordering behavior made order processing a substantial sales expense item for NDD.

In 1986, concerned that sales expense-to-revenue ratios had risen from 13% in 1968–1972 to 24% in 1986 for the IBM supplies line, William Duffy, manager of Product Planning and Channel Development at NDD, initiated a study that sought to determine the relationship of (1) customer size, (2) order size, and (3)

Exhibit 4.3 IBM Supplies—Direct Sales Account Profile, 1986

Enterprise Size		Firms		Purchasing Locations			Revenue	
		Number	Cum %	Number (000)	Cum %		$ Mil	Cum %
$1 Mil +	15%	29	100	16.4	100	83%	76	100
$500K–1 Mil	of	49	99	9.0	78	of	27	57
$100K–500K	firms	220	96	19.0	66	revenue	42	41
$ 50–100K		195	85	8.7	41		12	17
$ 10K–50K		790	75	16.2	30		14	10
$ 10K		720	36	6.2	8		4	2
Total		2,003		75.5			$175	

Source: IBM, National Distribution Division.
Note: Does not include sales to IBM authorized supplies dealers and wholesalers.

mode of order entry to sales expenses, revenues, and product line margins.

Supplies customers were of three types. First, large corporations, which IBM supplies salespersons called upon directly, ordered either from these representatives or by telephone from a catalog. IBM also made available to large customers a computerized ordering system, although this was just beginning to come into use. Second, user accounts not large enough to warrant direct sales coverage had the catalog/telephone ordering option. Third, IBM supplies distributors ordered by mail and telephone. They could also use an IBM computerized ordering system, but many preferred to use their own computers for generating orders from all their suppliers.

Although the following data, based on Duffy's study, disguise dollar amounts, the relationships among sales volume by size of customer and by size of order are representative of the actual findings. Exhibit 4.3 shows that 15% of all direct user accounts, those firms over $100,000 in size, produced approximately 83% of sales revenue from this class of customer. An analysis of revenue per order from direct accounts (Exhibit 4.4) indicated that 20% of the number of direct orders (120 of 590 orders) accounted for 82% of sales revenue.

Duffy also arrayed the number of orders received, sales revenues, and E/R ratios for the three customer categories as shown in Exhibit 4.5. Exhibit 4.5 data show E/R ratios for orders received from reseller accounts of 10%, from direct accounts through IBM field salespersons of 14%, and through telephone orders, 40%. The

Exhibit 4.4 IBM Supplies—Revenue per Order from Direct Accounts, 1986

			Orders			Revenue	
Order Size	Revenue per Order		Number (000)	Cum %		($ Mil)	Cum %
$500+	$2,785	20% of	{ 55	100	82% of	{163	100
$250–499	340	number	{ 65	91	revenues	{ 25	29
$150–249	195	of orders	90	80		18	18
$100–149	120		80	65		9	10
$ 50– 99	72		140	51		9	6
$ – 49	32		160	27		6	2
Total			590			$230	

Source: IBM, National Distribution Division.
Note: Does not include sales to IBM authorized supplies dealers and wholesalers.

lower E/R ratios for direct sales and for sales to large dealers and wholesalers could be attributed to the large individual order size that typified these two categories. Together these two classes of customers generated 7% of the total number of supplies orders and 60% of IBM supplies sales revenues. The much higher E/R ratio for telephone orders is attributable to the small order size: an average of about $235 per telephone order, with over 70% of these orders for less than $150 each.

Duffy estimated that IBM operators had to complete an average of two telephone calls per order and to prepare invoices for over a million line items annually. In addition, the cost of producing and mailing 250,000 catalogs a year plus special supplements was estimated at 5% of catalog sales.

We come back to Duffy's study and his recommendations to NDD management later in the chapter and again in Chapter 5. At this point, we simply note that his analysis is useful in identifying factors that affect paperwork costs, especially for a product line that includes a large number of small items ordered frequently and in small amounts. These factors are dollar size of order, number of items per invoice, and mode of order receipt. This suggests that managing sales expenses effectively may involve (1) setting bottom limits on order size and (2) developing low-cost modes of order entry and processing and providing customers with the motivation to use them.

Exhibit 4.5 IBM Supplies
1986 Revenue and Expense Analysis

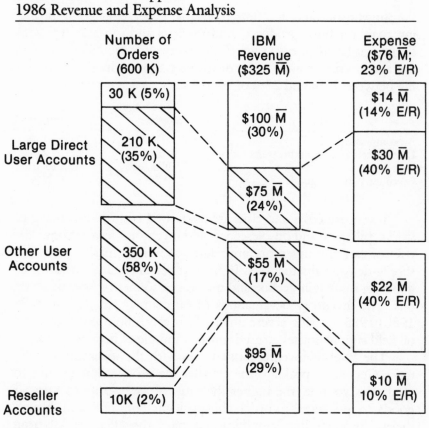

Number of Orders (600 K)	IBM Revenue ($325 $\overline{M}$)	Expense ($76 $\overline{M}$; 23% E/R)

Large Direct User Accounts
- 30 K (5%)
- 210 K (35%)
- $100 $\overline{M}$ (30%)
- $75 $\overline{M}$ (24%)
- $14 $\overline{M}$ (14% E/R)
- $30 $\overline{M}$ (40% E/R)

Other User Accounts
- 350 K (58%)
- $55 $\overline{M}$ (17%)
- $95 $\overline{M}$ (29%)
- $22 $\overline{M}$ (40% E/R)

Reseller Accounts
- 10K (2%)
- $10 $\overline{M}$ 10% E/R

◱ = Telephone orders

Source: IBM, National Distribution Division.

Stocking, Shipping, and Inventory Risk

Inventories and accounts receivable generate ongoing financing costs for both producers and resellers and can also represent significant levels of risk. Our 1986 surveys, for example, indicated average inventory levels at the distributor and producer levels ranging between 17% and 22% of annual sales, as follows:

	Suppliers	*Distributors*
Air compressors	22%	18%
Load centers and circuit breakers	17.5%	22%
Disk drives	N/A	10.1%
Oil country tubular goods	20.5%	N/A

Inventory costs and risks are associated with the financing of stocks, pilferage and obsolescence, and inventory devaluations. The risks of carrying inventory are related as well to market price and supply-demand fluctuations. If product prices decline, the value of goods in stock will decline and may result sooner or later in write-offs. The inventory devaluations of oil country tubular goods in 1980–1985, causing severe financial losses for both producers and oil field supply houses, are a dramatic example of inventory risk.

The level of inventory risk varies with the inventory-to-sales ratio. Generally speaking, the broader the product line relative to total sales volume, the higher the ratio of inventories to sales. If goods are produced to order, this principle does not hold. Nevertheless, product line simplifications have the effect of lowering inventory-to-sales ratios. For example, when Siemens-ITE came out with a load center design that could be mounted either on the wall surface or between studs with the door flush to the wall, it enabled its distributors to cut in half the load center stock levels needed to fill orders from off the shelf. Previously, ITE and other load center suppliers made different models for these two types of installations. This gave ITE an advantage, short-term, in building its distribution network and gaining market share. In its product redesign, ITE had reduced inventory risks for distributors by reducing the amount of stock they had to carry to support a given level of sales.

Another inventory-related risk is product obsolescence, a high risk for products with rapidly changing technologies or specifications. Suppliers may protect their distributors, fully or in part, from this risk by buying back obsolete stock or crediting the distributor's account, but the cost must be absorbed at one level or the other.

Physical handling costs, as a percentage of sales, are a function of weight-to-value ratios and of distances shipped to stocking locations and to customers from the supplier's plant. Oil country tubular goods have a high weight-to-value ratio, and physical handling costs are high relative to total selling costs. Disk drives, in comparison, have a much lower weight-to-value ratio.

Manufacturers and distributors economize on physical handling costs in several ways. First, they typically offer lower prices for carload and truckload quantities than for lesser amounts. Another strategy is to stock fast-moving items at numerous field locations but to stock slow-turnover items at regional centers. In achieving inventory economies through such networking, the manufacturer must balance higher shipping and storage costs that ensure rapid delivery for fast-moving items against lower storage costs and longer delivery times.

Reseller Margins

Not generally accounted for in the producer's revenue stream, resale margins comprise a substantial transaction cost component in end-user prices, often 20% or more. This resale margin covers the distributor's costs of doing business and yields a net profit. The distributor's margin is the difference between what it pays its suppliers and what it receives from its customers. Thus, a distributor's margin depends largely on the supplier's price schedule, which usually stipulates different prices for different purchase volumes. In addition, some suppliers may offer year-end rebates based on annual volumes. The producer's price schedule typically provides the greatest distributor margins on goods purchased in large quantities and sold to user-customers in small quantities. Exhibit 4.6, for example, is a schedule for grinding wheels that stipulates both functional (or class-of-trade) discounts and quantity discounts. The difference between the multipliers in the right-hand column and the one next to it is the functional discount.

Such a schedule indicates the percentage factors applied to the list price of the item to determine, by quantity brackets, the net prices at which the manufacturer will sell to distributors and to user-customers. It *suggests* that the distributor sell to user-customers at these same prices. However, because of local compe-

Exhibit 4.6   User-Customer and Distributor Multipliers

Item Quantity Group	Number of Units	User-Customer Multipliers	Distributor Multipliers
FA	1–9	1.027	0.822
A	10–19	0.689	0.551
B	20–49	0.585	0.476
C	50–99	0.503	0.418
D	100–249	0.447	0.393
E	250–749	0.394	0.355
F	750–1,249	0.379	0.341
G	1,250–	0.365	0.328

Source: Norton Company.

Note: Discounts in this table are expressed as percentages of a list price, not shown here.

tition, resellers may not always—or even often—get the end-user prices their suppliers' price schedules suggest. Thus, resale margins must be calculated with reference to the actual prices distributors charge and not to their suppliers' suggested list prices.

While distributor margins are not directly controlled by producers, they compensate for distribution costs that the producer would incur if it elected to perform the services provided by resellers. A relevant consideration in developing modes of distribution, then, is whether lower total distribution costs would be incurred by selling through distributors or by going direct.

COST-REVENUE RELATIONSHIPS

Sales expense-to-revenue ratios are measures of *cost efficiency,* with low ratios representing good performance. *Cost effectiveness* measures, on the other hand, also take account of revenues and profit margins as reflected in market share and product line profits. Further, it is unlikely in most cases that maximum revenues and profits will be realized at the lowest possible E/R ratios. When William Duffy, for example, went on to compare the E/R ratios by customer order entry mode with the revenues and margins for each channel, it became apparent that high E/R percentages were actually associated with high sales margins and vice versa, as Exhibit 4.7 indicates. Thus, while the average E/R ratio for sales to IBM wholesalers and authorized dealers was the lowest, discounts from list prices for this market segment were the highest. The *net* result was a lower contribution from sales to resellers than to the other two

Exhibit 4.7 IBM Supplies
Average Discounts, E/R Ratios, and Contribution by Channel

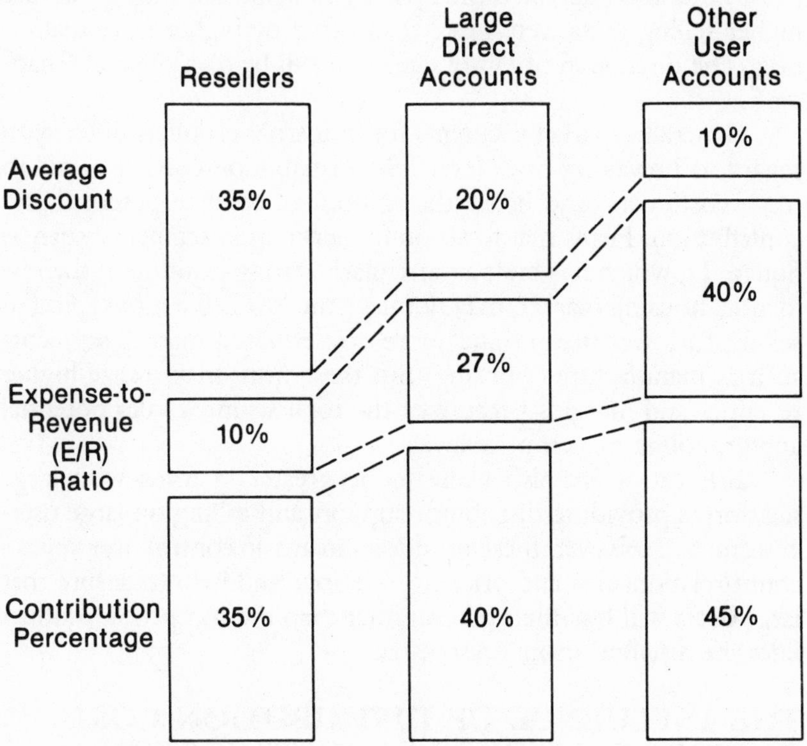

	Resellers	Large Direct Accounts	Other User Accounts
Average Discount	35%	20%	10%
Expense-to-Revenue (E/R) Ratio	10%	27%	40%
Contribution Percentage	35%	40%	45%

account categories. At the other extreme, sales to user accounts through the catalog/telephone channel had the highest E/R ratio (40%) but also generated the greatest contribution (45%) because higher selling costs were more than offset by higher price realization. The discussion of Duffy's analysis will be continued in Chapter 5.

Generally, market segments for industrial products differ with regard to prevailing price levels, the distribution costs required to reach customers, and hence the amount of revenue potential and contribution. Firms that focus on higher-margin segments, such as Square D, which has built a particularly strong position in the residential housing market, may not have the low E/R ratios typical in selling to lower-margin and more concentrated market segments such as manufactured housing. But these firms may realize higher revenues and margins because of the total segment sales potential and prevailing market price levels.

E/R ratios are also likely to be greater in firms with large salesforces providing distributor support and calling on large user-customers. However, these producers maintain control over key account relationships and price negotiations and help to ensure that large users will buy their brand rather than the competing product lines the distributors may also carry.

THE INFLUENCE OF DISTRIBUTION COST FACTORS ON MODES OF DISTRIBUTION

Distribution costs influence both the structure of distribution channels and the locus of distribution functions. Cost effectiveness depends largely on building and maintaining distribution that, on the one hand, is low cost relative to competitors' networks and, on the other, provides a strong end-market presence and brand strength. To achieve effectiveness, producers make choices, first, between direct sales and intermediary channels. This choice represents, in effect, a tradeoff between incurring internal selling costs and "paying" resale margins. Then, in any mixed, or multichannel, system, there are additional choices about the positioning of distribution functions at points in the chain.

Distribution Structure

Specifically, cost factor considerations dramatically influence (1) the balance of direct-indirect selling, (2) the number of levels of distribution, (3) the number and type of reseller firms in a distribution network, and (4) the relative desirability of the producer's granting territorial exclusivity and obtaining brand exclusivity.

The Direct-Indirect Balance. The cost of reaching individual customers relative to the account value is a prime factor in choosing between direct sales and resellers. If, on average, a salesperson representing a disk drive supplier and earning $50,000 in salary can generate $10 million a year in sales from five accounts, the favored option is clearly direct selling. At the other extreme, consider a sales rep working for an electrical distributor, paid $35,000 a year in salary and commissions, making an average of 80 calls a month and accounting for annual sales of $700,000, of which 10–20% may be load centers and circuit breakers.[3] It would be prohibitively expensive for a supplier such as General Electric to reach this sales rep's LC&CB customers directly through salespersons carrying only the GE line.

Number of Distribution Levels. The costs of taking products to market increase with the number of distribution levels through which they pass, since each intermediary requires some margin to cover the costs of providing distribution services. At one extreme is single-tier, or direct, selling; at the other is multitier distribution, with the product moving from the factory through agents, master distributors, distributors, subdistributors, and retail dealers, or some combination thereof. Whether to add or to eliminate distribution levels at different points in the development of markets is often an issue for sales administrators charged with developing and managing distribution channels. For example, General Electric's Component Motor Operation (CMO) served the consumer aftermarket for motors used in major household appliances by selling through three master distributors that sold to electrical wholesalers that in turn sold to appliance repair shops. This arrangement dated back to the early postwar years when the population of consumer appliances was considerably smaller than it is today. In the meantime, other motor manufacturers, bypassing master distributors, were selling to wholesalers. Even though prices charged by the master distributors to electrical wholesalers had declined to the point of yielding the former only marginal returns, competing motor manufacturer prices were even lower on high-volume motors. A key issue confronting CMO marketing managers

[3] These data are based on our 1986 survey of 76 electrical distributors.

as of 1987 was whether to end their master distributor relationships and sell direct to large wholesalers. [This case is considered in Chapter 10 on coping with change.]

In this instance, as the consumer aftermarket for motors matured and became more price competitive, distribution cost and the need to be price competitive became major considerations. By contrast, in the early stages of product-market development, when customers are often less price sensitive and have strong brand preferences, channels cost inefficiencies may be tolerated without major losses in revenues and market share.

Number and Types of Distribution Firms. A significant factor affecting distribution costs may be the salaries and expenses of salespersons required to service the distribution network and monitor its performance. Square D, for example, employed a salesforce of more than 650 people, who spent over 60% of their time calling on 1,300-plus Square D distributors, each distributor firm operating out of 1–3 sales locations.

Control Data's Peripheral Products Company (PPCo), on the other hand, distributed through two large national distributors, working with their headquarters and sales branches through a small, distributor-specialized salesforce. Very little of PPCo's 72-person direct salesforce's time was spent contacting distributors.

Two major considerations affect the choice among selling through a few large national firms, a number of regional firms, or many small local operations: the costs incurred in building and maintaining the channels network and the level of product-specific investments that distributors must make relative to product line sales revenues. In some instances, the absolute dollar amounts of certain product-specific investments are large enough to preclude the use of small, local distributors. For example, large investments are required for technically trained personnel in the disk drive industry. This may be one reason that disk drive manufacturers have often used national and regional distributors, rather than locals. To sell disk drives and other computer-related products, it was essential that distributors develop and support staffs of technically trained computer specialists, representing a fixed cost not likely to be sustainable by small resellers.

Brand and Territorial Exclusivity. Used with reference to distribution systems, the term *exclusivity* has two meanings, *territorial exclusivity* and *brand exclusivity*. A supplier may be represented by a single reseller in a trading area, thus granting the re-seller "territorial exclusivity." A supplier may also seek to develop a network of resellers that do not carry competing lines and thus require "brand exclusivity" of its channels.

Territorial exclusivity is common when the reseller must make large product-specific investments. Especially in the early stages of market development, when the returns are uncertain, these invest-ments would be jeopardized by the reseller's having direct compet-itors in the trading area. For example, in establishing distribution for programmable controllers (PCs),[4] the major manufacturers—Allen-Bradley, General Electric, Westinghouse, and Square D[5]—all opted for territorial exclusivity as an inducement to selected distrib-utors to make the requisite investments. The investments included the cost of demonstration equipment and facilities, the salary of a technical expert, and a minimum inventory of PC units and parts. Square D managers estimated the first year's cost to each franchisee at $150,000. Such an investment was reasonable only if the distrib-utor did not have to compete with other distributors of the same brand for whatever levels of demand could be developed in his or her trading area.

Brand exclusivity, too, is associated with heavy investments in a product line. Investments in inventory that are large relative to the distributor's total assets, as well as significant investments in technical skills or facilities tailored to a specific brand, may prevent the reseller's making comparable investments in a second brand.

Locus of Distribution Functions

Cost factors strongly influence the positioning of such func-tions as selling, physical distribution, and risk assumption along the chain of business units that form a distribution system. Under the pressures of competition, cost efficiency and cost effectiveness be-

[4] PCs are industrial logic controls used in both process and discrete systems to control ma-chining tasks. PCs were initially developed in the mid-1960s to replace electromechanical control devices such as relay switches. PCs are generally a more versatile control system and also provide memory storage. Thus, users can assess data such as the number of parts that pass through a machine, how long a machine has been operating, the time between tool changes, and other useful operating information. They can also change the sequence of op-erations by reprogramming.

[5] Two other large competitors were Modicon, which relied on a direct salesforce, and Texas Instruments, which established a network of systems houses with software and engineering skills for high-tech applications.

come major considerations in who does what in taking products from factory to market.

The demand generation function may be assumed largely by suppliers, assumed largely by resellers, or shared by both. As we discussed in Chapter 2, where the demand-generation function is positioned in the channel depends primarily on customer size and location and the nature of the appropriate mode of communication. The more concentrated the market and the larger the individual account or potential order, the more likely it is that the supplier's salesforce will take an active selling role at the user-customer level. Thus, according to a sales manager in an electrical manufacturing company, "The larger the customer, the more 'out-front' selling we do to help our distributors get the order."

At the same time, suppliers are loath to spend $200–300 per sales call if their distributors can be as effective at $80 per call. That depends on the sales task. The initial sale of a new product to an existing customer or of existing products to a new account may call for multiple messages, high in product application content, to the several members of the buying decision-making unit. In contrast, repeat purchases by satisfied users may be generated through short and simple interchanges of information having to do largely with price and delivery, as calculated from our 1986 surveys.

Compare, for example, the average costs per sales call made by producer and distributor sales representatives in three product-market groups.

Average Cost per Sales Call

	Producers	Distributors
Load centers and circuit breakers	$129	$ 87
Disk drives	280	143
Air compressors	232	106

These differences may be attributed largely to differences in call duration that relate to the number of people visited and the complexity of buyer-seller communications. Typically, a producer salesperson will make longer calls, see more members of the buying decision-making unit, and talk with each about a wide range of subjects. A distributor salesperson may call on one or a few persons in the customer company, and the relevant information exchange may be limited to certain basic essentials, for example, Do you have it in stock? When can I get it? How much does it cost? Often, in that order.

Finally, distribution cost considerations influence the locus of order shipment. Orders received by producers and distributors are not always fulfilled from their respective stocks. Orders received by resellers for large quantities, especially of high weight-to-value products, may be shipped from the producer's inventories to the user-customer. These "drop-shipped" items pass, not through the distributor's warehouse, but through the distributor's books for recording the sale, billing, and collection. For example, McKesson Chemical, the largest distributor of chemicals in the United States, fills 80% of its orders from stock, while another 20% is shipped directly to McKesson's customers from its suppliers' stock. GESCO, the third largest distributor of electrical equipment and a captive distribution arm of the General Electric Company, fills from stock about 95% of its orders from the residential construction market but only 50% of orders received from nonresidential contractors. The remainder is shipped from GE product departments and GESCO's outside suppliers. The difference between the two segments is the size of orders placed and the type of product purchased. Nonresidential contractors buy and install large pieces of electrical equipment not used in residential construction.[6]

Alternatively, orders taken by producers may be delivered from local distributor stocks when immediate availability and delivery in small quantities are important to the customer.

SUMMARY

Distribution cost factors influence the structure of distribution systems and determine where in the channels different functions are performed. The relevant costs may be categorized as follows:

- customer communications
- paperwork flows
- physical distribution
- financial risk assumption
- reseller net margins and commissions

[6] See E. Raymond Corey, "The Role of Information and Communications Technology," in *Marketing in an Electronic Age,* ed. Robert D. Buzzell (Boston: Harvard Business School Press, 1985), p. 35.

Customer communication costs vary significantly among the several modes of communication and even within modes. To a large extent, the form of communication should depend on the information needs of the customer and on where in the channels the particular mode of communication may be performed in a cost-efficient way. However, if sales revenues and margins are considered satisfactory, suppliers may elect to retain control over selling functions even though it may mean higher costs.

Processing costs per order vary with the number of line items ordered and not with the dollar amount of each order. Thus, the leverage points in controlling order-processing costs *as a percentage of sales* are establishing minimum order points and providing customer incentives to reduce order frequency and increase individual order amounts. There is often also significant cost-saving potential in computerized order-entry systems.

Inventory carrying costs, typically a large cost component, depend primarily on the amount of inventory needed for a given level of sales, that is, the inventory-to-sales ratio, and on the risks of inventory devaluation due to technical obsolescence or price deterioration. Freight and storage costs vary with product weight-to-value ratios. Ways of optimizing physical distribution costs include the networking of field stocks, bypassing distributor warehouses with drop shipments direct from producer to user, and providing price incentives for carload and truckload quantities. (Physical distribution cost optimization may be thought of as balancing cost and speed of delivery to satisfy customer requirements.)

How do cost factors affect the design of distribution systems? Distribution functions tend to be performed at different levels in the system, depending on whether they can be supported by available sales revenues and margins. Thus, distribution expense-to-revenue ratios lead suppliers to move often from product-specialized—or customer-specialized—direct selling to full-line direct selling to resellers as they go from highly concentrated, large volume markets to relatively thin ones. A supplier's relative reliance on selling direct versus going through resellers may be explained largely by (1) the different E/R ratios associated with reaching customers of widely varying purchase potentials in markets that range greatly in density and (2) differences in the nature and cost of sales calls made by direct salespersons and reseller sales reps, respectively.

All these cost factors influence the number of levels in the distribution chain and the positioning of such functions as selling and physical distribution. Even in distribution systems that rely heavily on resellers, suppliers may play very active roles in demand generation with large individual user accounts. Distributors may then perform an account-servicing role at a significantly lower cost per call than would be incurred by the supplier's salesperson.

Physical distribution functions also tend to be carried out at that point in the chain where they can be performed at least cost, commensurate with customer satisfaction. As noted above, product weight-to-value ratios affect this choice, with large, bulky items going direct to users from the producer's factory. The need of customers to have local stock availability and to order intermittently in small amounts will also determine shipping strategies. For example, orders negotiated directly by producers may be filled by stocks held in distributor warehouses.

Responsibility for financing sales credit and carrying inventories to serve end-customers may be taken by producers or resellers, depending on the extent of risk. Traditionally, suppliers going through external intermediaries have perceived the costs and risks of carrying inventories and collecting receivables to be in the reseller's domain. However, in some cases (e.g., oil country tubular goods) such risks may not be sustainable at the distributor level and some suppliers have stepped in to assume this burden. In addition, suppliers often finance distributor inventories through providing extended credit, or "dating plans," offering up to six-month terms on receivables. Dating plans not only recognize the financial limits of distributor businesses but also appeal to suppliers as a way of competing for sales by loading up distributor shelves.

Distribution cost factors also influence whether suppliers give resellers and agents territorial exclusivity. Simply stated, if the investments resellers have to make in order to stock and sell a particular supplier's product line are large relative to the sales potential in a trading area, then the supplier is prone to franchise only one agent or distributor in the area to enhance the attractiveness of the investment. As for brand exclusivity, distributors may be more willing to carry only one single brand if heavy investments and costs must be incurred simply to carry the chosen line and if the returns on the investments in additional lines promise only marginal returns.

The need for certain high product-specific investments at the reseller level also limits the number of qualified distributors. Under this condition, suppliers will tend to build their distribution around the larger regional and national distributor chains that have sales volume bases large enough to support investments in specialized sales resources.

5 The Business Unit Environment

Producers of like products tend to focus their marketing strategies on different market segments and to develop different distribution strategies adapted to these target markets. Even when focused on the same market segments, competing producers often develop very different distribution programs. The strategy comparisons among competing producers of load centers, disk drives, stationary air compressors, and oil country tubular goods clearly suggest this observation. Given that producers of comparable product lines have in common the nature of the product, the way the market is segmented, and its demographic structure, why don't they develop similar modes of distribution?

Business Unit Factors

The answer is that each faces, and responds to, its own unique internal business environment and its competitive positioning. For the marketing managers in each competing firm, market segment choice and distribution strategy are strongly influenced by the following factors:

Resource Availability.   The allocation of resources to support distribution investments and expenditures will depend on (1) the priorities established for the use of the funds that the business has available to it and (2) the selling expenses that product line margins can support. Although not of the same order as these two, we also count the product line itself as a resource; its breadth, overall quality, and brand image influence distribution choices.

Performance Measures.   Marketing managers are strongly influenced by their performance measures in developing and implementing distribution programs. These measures are of three sorts: those imposed on the selling function per se, such as expense-to-

84

revenue ratios; business unit sales and profit performance measures; and the business unit's contribution to enterprise goals.

Competitors' Strategies. As the discussion in Part II indicates, market segment positioning is not a random occurrence. Each competitor focuses on the segment opportunities that suit its product line characteristics, brand image, availability of resources for channels networks, and competitive position. The attractiveness of a market opportunity for any one firm will depend largely on how its competitors are arrayed across market segments and on its own relative product line and distribution strength.

Business Unit History. Particularly in the choice between selling directly and selling through resellers, the traditions that have contributed to past success significantly inform future directions. Today's marketing decision makers may have strong biases, arising out of past experience, that either favor reseller-oriented networks or support monolithic direct sales distribution. These attitudes influence future market segment choices and the firm's distribution profile.

Current Modes of Distribution. Once in place, distribution programs have a strong propensity to be self-perpetuating. Marketing managers tend to favor product line developments and distribution strategies that strengthen existing channels and/or the firm's control over its resellers.

In this chapter, we elaborate on these five firm-specific determinants of channels strategy.

RESOURCE AVAILABILITY

Among business unit factors, none is more powerful in shaping how firms go to market than the availability of resources to fund a distribution system. Lack of funding for a salesforce, training and promotional programs, field supply and service facilities, and distribution management functions is likely to leave the least-cost channels strategy as the only option.

Typically, the amount of funding generated internally through product line sales governs the level of selling expenditures. The smaller market share holders make markedly smaller investments in salesforces than do the market leaders, and tend to rely on distributor push to a greater extent. Under conditions of high sales volume and high gross margin realization, firms tend to invest in salesforces and promotion to increase control over resellers and to generate demand at the end-use level. For example, such market leaders as Square D, Ingersoll-Rand, U.S. Steel, and Control Data all have large salesforces working with their distributor networks and calling on end-users. Thus, in defending and growing market shares, the leaders are prone to plow back product line earnings to reinforce marketing channels.

The market leaders have broad product lines, large installed bases, and strong brand images. These factors reinforce distribution channels control by giving segment leaders leverage in establishing franchise terms and conditions that secure distributor support for their product lines and erect channels entry barriers against competitors. Thus, product line strength, product line margins, and the availability of funding for distribution investments—all relative to the comparable resources of competitors—will determine the market segment opportunities open to each supplier of like products. Out of this array, each tends to target those segments in which it may achieve greatest penetration at lowest cost and then to consolidate its market position with distribution investments as sales revenues permit. A comparison of two suppliers of welding products, Alloy Rods and Lincoln Electric Company, reveals some of the effects of resource availability on distribution strategies.

Alloy Rods

Founded in 1940, Alloy Rods (AR), a manufacturer of a line of welding electrodes, was acquired in 1962 by Chemetron, which in turn was acquired in 1978 by Allegheny Ludlum Industries (ALI). In 1985, under a leveraged buyout arrangement, Alloy Rods again became an independent business. With its heavy debt structure and an after-tax profit of between $1 and $2 million, Alloy Rods had few resources to allocate to its distribution program. (As of 1985, AR's debt-to-equity ratio was 16:1.)

Exhibit 5.1 provides an overview of Alloy Rods' distribution channel structure. In 1984, 75% of AR's sales were to independent welding supply distributors; about 13% went direct to these distributors, while 41% went through independent wholesalers and 21% through company-owned wholesalers. Another 13% of sales were made direct to end-users, with the remainder to export,

Exhibit 5.1 Alloy Rods Distribution Channels

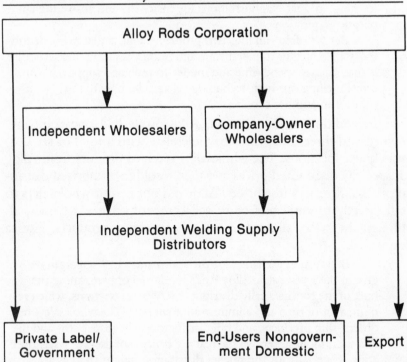

private-label, and government markets. The company's field sales-force called on both distributors and end-users.

About 500 independent welding supply distributors (with a total of about 1,500 locations) represented Alloy Rods. The average AR distributor had about $3 million in total annual sales of welding supplies and equipment. However, electrodes were only one among hundreds of products carried by these distributors, ranging from standard torches, welding guns, gloves, sweatbands, and goggles to special devices for cutting and welding. Further, the average distributor's gross margin was in the 21–28% range on welding materials but in the 9–15% range on electrodes.

AR's managers regularly reviewed their distributors in an effort to build a group with more technical expertise and greater ability to sell AR products. Phil Plotica, AR vice president of sales and marketing, explained,

> We provide distributors' salespeople with training and target technically oriented promotions at them. Our aim is to have the dis-

tributor add or develop sales engineering personnel capable of performing product demonstrations for customers and answering applications questions.

But we often run into two problems. One is that many distributors are reluctant to invest time and resources in this area. Another is that we may spend time and money to train and support a distributor in selling our high-technology electrodes only to lose the distributor to another supplier.

In addition to distributors, Alloy Rods had one company-owned and three independent wholesalers with a total of six U.S. warehouse locations. The wholesalers received product from Alloy at specially discounted prices and then resold to authorized distributors at AR distributor prices. Alloy did not permit wholesalers to sell directly to end-users or to sell competitive product lines, although they were allowed to sell other welding products. Plotica explained,

> Basically, in exchange for the approximately 9% margin we give to the wholesaler, Alloy Rods is relieved of performing the bulk of the routine, daily distributor service transactions, which require a lot of time and administrative support. The wholesalers take orders, ship products, and bill distributors. The distributor and the end-user, though, receive the real net benefits of the wholesale operation. Especially for the smaller distributor, freight charges are lower since the product is shipped in truckload quantities from our manufacturing plants to the regional warehouses of wholesalers. Distributors can also keep lower inventories since delivery time from the wholesalers is typically one or two days instead of nearly a month from our main manufacturing plant.

However, other managers noted that the wholesalers became a communications obstacle between Alloy and end-users in getting feedback on market conditions and customer experience. Speaking of a recently inaugurated telemarketing program, Plotica noted, "The object of this program is not really to sell product, but rather to reestablish communications with our distributors. We want to get feedback, and we also need a better way to send them information. Not too long ago, we discovered that only half of our promotional materials ever reached the appropriate individual at our distributors."

Alloy Rods employed a relatively small salesforce—21 field salespeople responsible for distributor relations, direct accounts, and account development. In 1984, AR spent about $500,000 on advertising and sales promotions, considerably less than its competitors.

Most of AR's limited promotional budget went to product literature, field training seminars, newsletters, and trade shows. The company literature encouraged customers to perform cost comparisons that revealed the value of converting to AR products. In 1981, the company had begun offering a ten-month home study course on welding technology, designed for people with nontechnical backgrounds, especially distributors' salespeople. By 1985, over 4,000 people had enrolled in the course, and more than 1,000 had completed it. However, given the constraints in promotional and training budgets, AR's management could only hope that these initiatives would generate increased interest in AR's product line.

Lincoln Electric

Lincoln Electric Company (LEC) was the largest supplier of welding electrodes and equipment in the United States. It was the perception of one AR manager that,

> Distributors need Lincoln to provide the low-price end of the product spectrum. Also, Lincoln has name recognition among customers and a good reputation in welding equipment. In fact, I believe it's their equipment that drives their electrode sales at distributors. In addition, I think Lincoln's size and aggressiveness make many distributors feel they *must* carry the Lincoln electrode line.

Lincoln sold both welding machines and electrodes to large manufacturing companies. LEC had traditionally used a direct salesforce, but as America's basic industries went into decline in the early 1980s, massive market changes forced management to consider distribution alternatives. Beginning in 1981, according to LEC's president, Don Hastings, who at the time was vice president of sales, "The large companies, who were the backbone of our business for many years, reduced their work forces dramatically, and others just simply closed up shop and walked away. One of our largest industrial-based offices lost 40% of its business in the first three months of 1982."

At the same time, Hastings recognized that small- and medium-sized companies that used welding in production and maintenance operations constituted a strong and growing market that was not being reached effectively by LEC's direct salesforce. Hastings reacted by undertaking a major buildup of LEC's resale distribution system. In January 1982, the field salesforce was instructed to increase the number of distributors of LEC welding machines and electrodes by 50% by the end of April. The drive, called "the 100-days war," was conceived as a blitz-like exercise. In fact, the number of distributors almost doubled; Lincoln added 203 new distributors with outlets in 516 locations. Looking back in 1987, Paul Beddia, Hastings's successor as sales manager, commented, "It made the existing distributors madder than hell to have a lot of new distributors carrying the Lincoln Electric brand and competing with them in almost every city of size, but all existing distributors valued the line and kept selling."

Mike Brooks, distributor sales manager, added,

> Even our own salespeople were skeptical in the beginning. Our old-line "tech reps" took a great deal of pride in having built the arc welding industry. When the welding industry was just getting started, especially during World War II, riveting was our major source of competition. Lincoln's strategy at the time was direct account selling, and I remember having 10 or 11 assigned accounts that gave Lincoln $6–7 million in gross sales in one year. Naturally, we had some resentments about dealers coming along and feeding on what we created, but now that's all changed. Over three-quarters of our salesforce is made up of tech reps who have joined Lincoln in the last decade, and they think of the distributor as their best friend.

Between 1970 and the late 1980s, LEC moved from 75% direct selling and 25% distributor selling to just the reverse. As of 1985, the Lincoln Electric field salesforce included 35 district managers and 126 salespersons, "tech reps," serving customers out of 35 field warehouses. In August 1986, Lincoln purchased the assets of the Airco electrode plant, licensed the brand name, and utilized the Airco distribution system. Lincoln then had approximately 850 industrial distributors[1] with a total of over 3,000 sales locations that carried the LEC equipment and supplies lines. In addition,

---

[1] Typically a third of an industrial distributor's sales were accounted for by industrial gases, 20% by welding equipment, 16% by consumable welding supplies such as electrodes, and the rest by "hardgoods" used by welders on construction sites and in plants, with industrial gases tending to be the most profitable. In carrying the leading line of welding equipment and electrodes, Beddia noted, LEC distributors tended to build business for industrial gases since many industrial customers wanted the convenience of purchasing all their welding supplies from one local source.

LEC had more than 800 rural area LEC distributors with over 2,000 locations.

Over 165 of LEC's industrial resellers, with a total of 450 sales locations, qualified as Flagship Distributors. This group sent their sales personnel to Cleveland for product training at least one week a year and attended product and sales training programs at LEC regional offices. LEC anticipated that at least 75% of the welding equipment and consumables business done by the Flagship Distributors would be LEC-brand business. In return, Lincoln franchised Flagship Distributors to sell certain, more technical items not made available to other distributors. Lincoln also provided them with discounts for mixed carload purchasing.

Each LEC field tech rep served both direct and distributor accounts and in some cases made joint sales calls on user-customers with distributor personnel. In measuring tech reps' performance against sales quotas and in calculating year-end bonuses, all sales were counted equally. According to Hastings, "It doesn't make any difference to us whether the customer wants a direct relationship or buys from a local distributor. That's the customer's choice. We just want the final user buying Lincoln products—not competitive brands. Our tech reps are held responsible for all accounts in the territory, regardless of how they are sold."

Purchasing and reselling in the smaller amounts yielded distributor margins of about 18%, but in the 40,000-pound quantity bracket LEC's prices to distributors and to user-customers were the same. In this way, Lincoln encouraged distributors to buy in the largest supplies quantity bracket, 40,000 pounds (or a carload), and to resell in the smallest, 500 pounds, to realize about a 30% gross margin. As Brooks said, "Our distributors don't always agree with our price policy, but they accept it. They would like a distributor discount in every quantity bracket. Sometimes, though, they'll give away part of their margin in the mid-range quantities to make a sale and maybe get a customer for industrial gases as well."

While LEC's policy was to maintain industry cost leadership through manufacturing efficiencies and to base prices on costs, its policy was also not to discount its published list prices. To cope, then, with low-price competition, LEC field salespeople stressed performance cost savings for the customer. Cost savings through using LEC products were quantified in cost-savings guarantees made under LEC's Guaranteed Cost Reduction Program (GCRP).

This program actually began in 1980 after Lincoln Electric sales management had tried for seven years to sell welding machines and supplies to a major steel mill and was unsuccessful because of lower-priced competition. Beddia, then a district manager, contacted the steel mill's operating people concerned with manufacturing costs and discovered that the difference between LEC list prices and a low competitive offer on a particular order was approximately $15,000. With Hastings's concurrence, Beddia guaranteed to save the steel company $100,000 in welding operating costs if it would place the order with Lincoln Electric. As Beddia remembered,

> We saved them over $100,000 and kept them as a longtime customer, not only for that one plant where we studied their processes and showed them where they could save but in their other plants as well. Actually, electrodes represent only about 15% of the cost of a weldment. So there's a lot of leverage in making process improvements. For this purpose, it really helps to be selling both welding equipment and the consumables because both are part of the welding process.

Since that early sales innovation, LEC had negotiated performance guarantees with over 200 companies as of 1986. According to George Willis, chairman,

> We've had to pay off on a guarantee only once, and even then there's reason to believe that we didn't make the guarantee because the customer's plant people didn't do what they were supposed to do to improve costs. It is important to have a clear understanding regarding what elements of cost will be measured and how and what operating capacity levels are assumed in making the guarantee.

Further, if customers preferred to source through local distributors, Flagship Distributors, too, became involved with LEC tech reps in GCRP.

As of 1985, LEC's sales were $385 million, and the company reported margins of 35%. It had no long-term debt, and its current ratio was 3:1.

A Commentary

In comparison with its smaller competitor, Alloy Rods, LEC had significantly greater end-market presence—a six-times-larger salesforce calling on direct customers and LEC-franchised distributors. In addition, both the desirability of the LEC franchise as a revenue generator and the continuous monitoring of distributor sales performance by LEC tech reps served to secure distributor cooperation and support. LEC's power in the supplier-distributor relationship is evidenced by very low distributor turnover even in the face of doubling the number of LEC franchisees in local trading areas in a very short time span. In addition, LEC could "make stick" a policy that allowed virtually no distributor margin for quantities purchased and resold in the largest quantity price brackets.

By contrast, Alloy Rods—with the bulk of its earnings needed to service debt incurred through recent financing—operated under significant financial constraints in developing its distribution strategy. With only 21 direct sales reps, AR derived over 60% of its sales revenues through wholesalers that dealt directly with AR electrode distributors. The company spent its limited promotional budget on distributor promotions and sales training, but distributor turnover and the failure to reach distributor personnel through direct mail diminished this program's benefit.

As compared with LEC, AR seemed to have markedly less leverage with its distributors. Supplying only electrodes, its share of distributor revenues was less than that of LEC, which manufactured both electrodes and welding machines. AR's direct communications with its distributors had only modest success in conveying relevant information and getting product-market feedback. Further, in contrast to LEC's guaranteed cost performance program, AR merely provided literature encouraging user-customers to make cost comparisons that would demonstrate its products' advantages. In short, for lack of resources, AR did not have a strong end-market presence, whereas LEC did. While LEC, with its superior financial resources and product line breadth, developed an effective "market pull" distribution strategy, AR had to rely largely on "distributor push." In so doing, the latter lost direct communication with both distributors and user-customers. Because of the great dispersion of arc welding customers and the large number of resellers needed to cover such a market, distribution effectiveness would require substantial marketing funding. At lower levels of commitment, it is likely that these investments would be less cost effective in terms of revenue generated per dollar spent.

A firm's overall financial strength is a major determinant of its ability to fund selling expenditures. When funding is ample, as at Lincoln Electric, the tendency is to invest in product distribution by developing the field salesforce and the reseller network as well as by supporting technical and marketing services. When other uses for limited funds, such as for investments in plant and R&D and debt servicing, are more urgent, as at Alloy Rods, significant investments in distribution may be precluded.

Product Line Margins

Since available capital comes largely from sales revenues, firms tend often to move quickly to enhance and/or protect the stream of earnings generated through product distribution. Thus, the long-term investments and the short-term expenditures for product distribution may be determined largely by the contribution of the particular product line, that is, the difference between related out-of-pocket costs and gross revenues.

Not infrequently, distribution system changes are precipitated by declining product line margins. Wright Line, a leading producer of computer accessories, for example, faced with profit deterioration and loss of market share on its line of computer-room supplies, shifted from a monolithic, direct sales mode of selling to a multichannel system that included selling through retail dealers and catalogs as well. Prior to this change, however, although suffering a decline in market share but still enjoying high profit margins, Wright Line's management had responded by adding more direct sales personnel. Only after both market share *and* margins turned down did Wright Line restructure its distribution.

PERFORMANCE MEASURES

In the face of rising E/R ratios, Bill Duffy, manager for product planning and channel development for IBM's line of computer and copier supplies, took the following actions over a period of several years:

- reduced the number of accounts receiving direct sales coverage from 26,000 accounts in 1984 to the 1,800 largest firms in 1986
- reduced the number of IBM supplies field sales reps from

156 in 1984 to 88 in 1986, and the accounts per rep from 165 to 20[2]

- implemented order rules to discourage small orders
- encouraged small direct accounts to source from retail outlets

These moves were motivated by a companywide drive to reduce expense-to-revenue ratios. They would have the effect of shifting IBM supplies business from channels that recorded E/R levels of 40% and 25% to resale channels for which E/R ratios were calculated to be 10%.

It is interesting to speculate that if product line profit margins per se had been Duffy's dominant measure of sales performance instead of E/R ratios, he might have chosen a different strategy, such as strengthening IBM's direct response marketing (DRM) channel (telephone ordering). DRM had the highest E/R ratio (40%), but it also showed the highest contribution because, as noted in Chapter 4, discounts from list prices were significantly lower for DRM customers than for direct accounts and resellers. In fact, the reseller channel showed the lowest contribution. Although at 10%, the E/R ratio for reseller volume was the lowest of the three channels, IBM's price realization from this class of trade was also the lowest, netting the lowest contribution margins. Thus, management's choice of *sales* performance measures strongly influences channels strategy. So do the performance criteria applied to the business unit as a whole.

As our Wright Line example shows, overall business unit performance against revenue and profit goals will also shape distribution strategy. Along with product line margins, *business unit* sales and profit goals at the firm level are major standards against which business performance is monitored. In addition, nonquantified *corporate* objectives shape business unit strategy and, in turn, determine distribution goals and strategies. Following an acquisition, a firm's distribution strategy may change. For example, electrical industry sources have observed that since ITE was acquired by Siemens, the large German electrical manufacturer, it has become considerably more active in building its distribution network and in seeking market share gains through aggressive pricing. The nature

---

[2] The number of locations on which reps called, however, declined only to 740 from 870; user locations per account tended to be greater as firms increased in size.

of ITE's competitive behavior as a business unit within Siemens apparently stands in contrast to its character as a division within Gould, its previous owner, or indeed as an independent company prior to that time. Observers speculate that in supporting and encouraging the development of ITE's salesforce and distributor network, the parent corporation sees it as a channel through which might flow a broad range of other Siemens electrical products. They have also speculated that German firms may set lower profit standards on their business units than do American corporations. If so, ITE's alleged price aggressiveness may have been facilitated by performance measures that give priority to long-term growth over short-term profits.

COMPETITORS' STRATEGIES

Bounded internally by limited resources and sales performance measures, marketing managers responsible for distribution networks may be constrained externally by competitors' market segment positioning and channels strategies. Early market segment entrants may "occupy" certain channels, in effect blocking access to smaller, newer firms. Rodime, for example, a producer of disk drives, had to settle for a second-choice distribution option when its preferred channels choice had been preempted by its already established competitors. Rodime had entered the U.S. disk drive market a few years after other major suppliers had established agreements with such large national electronics distributors as Hamilton Avnet, Arrow Electronics, and Kierulff. Precluded as a practical matter from entry into this channel, Rodime established agreements with three regional electronics distributors. The vice president for sales and marketing at Rodime noted that "we wanted to go through a national distributor, but they already carried product lines from other suppliers, and this channel was inaccessible to us. We therefore put together a national distribution network by establishing distribution through three regionals."

Not only do producers of like products tend to block their channels from direct competitors; indirect competitors may accomplish this for them. For example, because Allen-Bradley (A&B), a producer of electrical controls, is reputed to discourage its strong distribution network from carrying the load center lines of companies that also produce electrical controls, Siemens-ITE has strong representation for its line of load centers through A&B's industrial distributors. According to one source,

> ITE has very strong industrial distribution, but almost by default. The [industrial] distributors need both a switchgear line and a control line. If they carry the Allen-Bradley control line, they are discouraged by Allen-Bradley from carrying a Square D, a General Electric, or a Cutler-Hammer line of switchgear, because all three of these manufacturers have a strong control line and are perceived by Allen-Bradley as major competitors. As a result, Allen-Bradley insists that its distributors carry a switchgear line—such as ITE—that has either a weak control line or no control line at all.

Similarly, Fujitsu confronted barriers when it attempted to establish a foothold in the U.S. disk drive market. Both the unwillingness of the large U.S. computer OEMs to purchase components from an end-product competitor and the resistance of American semiconductor manufacturers effectively precluded Fujitsu from selling through the large national electronics distributors. Instead, this Japanese firm sought distribution through manufacturers' reps and smaller local electronics distributors, which sold to small computer OEMs and information systems assemblers.

The channels blocking actions of market segment leaders notwithstanding, channels access may be difficult for market share challengers because of the high costs to distributors of switching to another brand. Switching costs for a distributor may be especially high when a significant component of its income streams is derived from spare parts and service revenue related to its primary supplier's large installed base, as in the case of Ingersoll-Rand's air compressor dealers.

However, marketing managers who do seek to gain access to channels occupied by competitors have pursued a number of different strategies:

- First, they may choose to gain initial access through offering resellers superior items not in direct competition with any offerings in the distributor's primary supplier line. Once distribution is in place for these products, the market share challenger will steadily push to get distributors to take on other parts of its product line. Atlas-Copco's entry into the U.S. market for stationary air compressors is a case in point. (See Appendix A, Product-Market Study III.)

- A second strategy is to develop new channels, recruiting resellers that have not previously served the target market segment. This is how foreign OCTG producers such as Sumitomo initially gained access to U.S. markets. (See Product-Market Study IV.)

- A third is to do the opposite of what the market leader has done; if the market leader has gone to market primarily through a direct salesforce, for example, the challenger may develop alternate channels. Generally, though, in using a different mode of going to market, the challenger will be focused on a different market "space" from the one occupied by the leader. Examples are Compaq and DEC, both of which developed strong reseller networks at a time when IBM relied almost entirely on its direct salesforce.

- A fourth strategy is to recruit distributors through tie-in offerings, including as part of the franchise other revenue-generating lines for which there is an established demand. A GE or Westinghouse attempting to build distribution for load centers might package this product line franchise with another that covers medium-voltage switchgear, for example.

- Fifth, in some instances producers may gain channels access by offering new distributors immediate supplies in times of shortage. This was a strategy used by Sumitomo in the OCTG market.

- Sixth, suppliers seeking market entry may acquire companies that have established distribution networks. This strategy has been particularly characteristic of foreign companies entering U.S. markets—Siemens, for example, acquiring ITE. (See Product-Market Study I.)

- Seventh, some producers may negotiate large individual end-user sales and then franchise distributors to service these accounts, hoping that they can keep these franchisees, a tactic that Siemens-ITE has reputedly used.

While all such strategies have been used effectively to gain market access, it should be apparent on reflection that none is easy. Each one represents a costly and time-consuming strategic undertaking. Each involves intensive negotiations with whole networks of intermediaries, which are independent both legally and operationally. Each typically encounters, as well, severe competitive reactions, representing as it does a serious challenge to those whose market access may be seriously threatened.

BUSINESS UNIT HISTORY AND THE MANAGERIAL MIND-SET

We were struck by the fact that in a great many of the companies we studied, strong subjective biases seemed to influence the choice between direct selling and indirect selling. If, in the business culture, engineering competence, technical creativity, and technical sales skills have high value, what resellers can contribute to the success of the business is often deprecated.

Speaking in 1985, a vice president at Arrow Electronics, the second largest electronics distributor in the United States, commented,

> Ten to fifteen years ago, manufacturers in the computer and electronics industries viewed distributors as an afterthought, as those people upon whom they could put their excess inventory. The manufacturer almost had a phobia about letting the distributor sell more than 10–15% of that manufacturer's sales volume.
>
> But now many manufacturers we deal with have policies that a certain percentage of their business should go through distributors. One reason is that many manufacturers would rather spend their money on R&D investments than on selling costs. There are also product life cycle issues involved here: the technology has become more familiar and therefore more amenable to selling through distributors. But there have also been significant cultural changes among the manufacturers in their view of distributors.

Given the general shortening of product life cycles in many industries, such "cultural changes" are becoming more common, but belief systems, rooted in past practice, are often inhibiting when it comes to restructuring distribution systems.

Assumptions about the capabilities and place of distributors may derive from the business unit's product line evolution. In the load center market, for example, each competing supplier developed distributor networks and distribution policies from a different product base. Square D started from the "low end" of the load center market—initially selling relatively simple breakers for residential construction purposes—and broadened its line over time to include more highly engineered and sophisticated equipment. Distributors, early on, were an integral part of Square D's marketing strategy and remained important as the company added larger and more technically sophisticated products to its line. By contrast, companies like GE and Westinghouse entered the load center market from a base in large transformers and other highly engineered electrical equipment. Distributors played only a small role in the marketing of such equipment. However, as GE and Westinghouse product lines moved downward in size to include lower-priced items traditionally sold through distributors, the respective roles of direct salesforces, independent distributors, and captive distributors became the subject of constant reexamination and of discussions, because resale distribution was perceived as a relatively foreign element in a traditional distribution strategy.

Despite cost factors that favor indirect channels, suppliers from a high-end product tradition often view intermediaries with initial antipathy. In addition to fearing a loss of "account control" and valuable market feedback, these suppliers may also assume that "distributors get a margin but add no real technical value" or that "distributors can't understand these kinds of products." A marketing manager for a computer products supplier recalled that,

> A few years ago, we were under pressure to improve our marketing expense-to-sales ratios. At the time, our vice president of marketing was being urged by some other managers to consider shaving marketing expenses by using distributors. He asked me to study this possibility and make a report, but he also commented that "a closer look at using distributors will probably kill the idea." Instead, I found that some of our competitors were using distributors extensively and that it made sense for a number of our product lines. But in subsequent discussions, many managers here resisted the idea of paying a margin to somebody who "just stocks the product and takes orders."

Some computer firms have found that their managers resist using distributors even as their products decrease in price and do not generate the margins to support a highly trained direct salesforce. When change must come, moreover, the salesforce often resists the shift to a multichannel system. The experience of Honeywell Information Systems (HIS) provides a case in point. HIS first entered the computer business in 1955 and, throughout most of its history, had focused its product line, service support, and sales organization on selling large mainframe computers directly to user-customers. Recognizing that the market for smaller computers was growing rapidly, however, HIS managers began in 1975 to develop indirect channels of distribution to reach small- and medium-sized accounts for its products. These channels included value-added resellers (VARs) and manufacturers' reps (MRs).

By 1985, 15% of HIS sales revenue was generated by its indirect channels. But HIS managers found that, a decade after the introduction of indirect channels, there continued to be recurrent conflicts between HIS' direct sales reps and the accounts served by HIS' Indirect Sales Operations. Speaking in 1985, a vice president and general manager at HIS explained, "There are 600 people in the direct salesforce, and a high percentage of them really don't believe in indirect channels, because they think it is the competition. We all grew up as a direct selling organization, and there is a problem here of cultural change."

In situations such as those outlined above, the salesforce has typically "grown up" with high-volume or "big-ticket" sales for products that, early in their life cycle, demanded extensive technical assistance and/or applications support as part of the selling process. This sort of selling situation requires certain skills and values on the part of the salesforce that may be the antithesis of the skills and values sales personnel recognize in reseller organizations. A computer firm's direct salesforce, for example, is often composed of highly trained engineers, accustomed to meeting with other engineers in the customer organization, where the sales process often focuses on performance characteristics, technical specifications, applications assistance, or other product features. By contrast, sales representatives for most electronics distributors tend to come from more diverse backgrounds. They are often accustomed to dealing with purchasing personnel at the customer organization, where negotiations may focus on questions of price and delivery for an array of products.

In addition, the producer's sales reps may be wary of working with distributor sales reps who also carry competing products as parts of their lines. The producer's sales rep may perceive that effort spent with the distributor's sales rep is no guarantee that the distributor will in fact sell that supplier's product rather than a competitor's. Finally, accustomed to large volume orders from major accounts, the supplier's salesforce may simply find the smaller volume sales generally available through distributor channels to be "peanuts," not worthy of the effort required to make those sales.

These attitudes play a significant role in channel design decisions. While product, market, and distribution cost factors may shape the limits of what is technically and profitably possible for a given supplier, the business unit's "distribution culture"—a function of its history and product line evolution—often determines how managers respond to specific market developments in formulating or reformulating distribution strategy as needs arise.

THE IMPACT OF CURRENT MODES OF DISTRIBUTION

While at the outset the distribution system is shaped to the imperatives of the product-market environment, once in place the system itself becomes a powerful force, giving direction to product strategy, to selection of new market segments, and also to channels strategy. More than a means of going to market, the system embodies as well a set of explicit and implicit commitments to intermediaries, to customers, and, internally, to programs and strategies. Moreover, these commitments become highly personalized in relationships among managers. Thus, the needs of producers to economize on distribution costs, to maximize product flow through existing channels networks, to enhance their power in the supplier-distributor relationship, and to protect the system from competitive incursion may often drive R&D strategy as well as the choice of distribution programs for new products coming out of the R&D process. An extreme example is Sullair's abortive venture into a wide range of air compressor lines as a means of broadening the base of revenues to support its chain of company stores. These lines not only failed to serve their purpose but were also unprofitable (See Appendix A, Product-Market Study III.)

Competing in the same product-market, Ingersoll-Rand provides a useful example of how new product distribution choices may serve to strengthen the business's existing modes of distribution, although possibly at some sacrifice of potential new product revenues and profits.

In 1985, Ingersoll-Rand's Stationary Air Compressor Division (SACD) was preparing to introduce a new centrifugal air compressor, the Centac 200. At issue was whether this product line would be marketed through the I-R salesforce or through I-R's distributor network. The former option would ensure that purchasers of the new product would have highly qualified technical support; it might mean, as well, greater profit margins for I-R in the short term. On the other hand, I-R would strengthen its distribution network by taking the latter option; it might also blunt other producers' efforts to enter I-R channels with directly competing products. James Clabough, SACD vice president of marketing and sales, explained,

> As of now, we sell all compressors over 450 horsepower only through the direct salesforce, and the smallest centrifugal we've made is 500 horsepower. Now we have a new 200 horsepower centrifugal. While that falls in the size range that our distributors and air centers carry, some of our managers think that all centrifugals should continue to be sold only through the direct salesforce. The distributors haven't had experience with centrifugals, and we're not sure they can provide the technical support needed to maintain these units in the field.

In addition to channels technical competence, marketing managers also considered competition, market potential, and channel relations. The Centac 200 would give Ingersoll-Rand an "oil-free" machine to compete with Atlas-Copco's very successful Z-series compressor, which was sold through distributors. Ingersoll-Rand also eventually planned to expand the Centac line to include other machines at lower horsepower ratings. Finally, Clabough considered the effect of his decision on the relationships among I-R's direct salesforce, distributors, and captive distributors.

In this situation, management's view of the primary role of the new Centac line within I-R's product portfolio would in large part determine the direct versus distributor decision concerning the new product. We can cite a number of roles that the new line might play:

1. Management might seek to maximize profits for the line itself, in which case SACD might want to retain as much as possible of the margins and aftermarket revenues associated with this line. This goal would favor reserving the Centac 200 for direct selling only.

2. By contrast, the line might be viewed as a means of further building I-R's importance and strength with its current distributors, providing them with a new line of air compressors and with access to the aftermarket revenues subsequently generated by this product line. In addition, the line might be viewed as a means of forestalling the entry of competitors' products into SACD's distributor network. In this case, a high priority would be to reinforce I-R's posture as a full-line, exclusive supplier to SACD distributors.

3. Alternatively, the primary role of the Centac line might be to take market share from Atlas-Copco's successful Z-series compressor. In this case, distribution of the Centac line would be strongly influenced by a desire to gain access to those segments and accounts where the competitor's line was currently strong. The line might also be viewed as a means of gaining entry to the competitor's distributor network for SACD's entire product portfolio.

4. Finally, since the division planned to expand the Centac line over time to lower horsepower ratings, this consideration would also influence distribution strategy. The lower-horsepower machines were traditionally sold by I-R's distributors, and building support for these later product introductions might well argue for introducing the initial Centac offering through the established distributor network.

In context, then, establishing a distribution strategy for the line required management to choose among a number of different goals. In particular, management faced a possible tradeoff between (1) the desire to select a distribution channel that would maximize initial technical support and sales of the new Centac line (perhaps more easily done through the direct salesforce) and (2) the desire to use the new line as a means of strengthening, extending, or defending established distribution arrangements for SACD's current product portfolio and planned additions to that product portfolio. Clabough's decision: to add the Centac 200 to the list of distributor-class products. As in this instance, building up the product flow through resellers is often perceived as one way to gain greater distributor support.

SUMMARY

The following aspects of an individual firm's environment all influence the marketing managers responsible for planning and implementing distribution strategies: its resources, its performance measures, the strategies of the competitors it confronts in the marketplace, and the past and present of its distribution programs. These factors condition competing firms' responses to a common product-market environment and account for the wide variety in modes of distribution among suppliers of like products.

Possibly the primary arbiter of distribution strategy for the individual business is the availability of resources, especially those generated by product line margins, to support distribution system investments. The lack of resources—or indeed the more urgent need for limited resources to support other business functions—tends to lead to a greater reliance on intermediaries with a concomitant diminution of control over the flow of product to end-use markets.

On the other hand, the availability of ample funding for distribution—a condition that tends to be characteristic of market share leaders—leads often to increased investments in channels development and support as a means of strengthening the producer's ability to gain channels cooperation, monitor performance at the resale level, and engage in demand-generating activities at the user level.

In addition, we noted the powerful influence of internal performance measures on the business's distribution strategy. The relevant measures include sales revenues, E/R ratios—or selling cost efficiency—and profits.

Corporate purpose, imposed on business unit planning, also shapes distribution strategy. Corporate objectives may be expressed in terms of specific measures, such as E/R ratios, or broader goals, such as cash generation or market share growth. These criteria, imposed often by implication on business units, tend to be reinforced through either increased or diminished funding for distribution systems support.

A key determinant of the individual firm's distribution strategy is its particular competitive environment: What market "spaces" do competitors occupy? With what product lines? And what strength do they have in their relationships with user-customers and distributors? These constraints, plus the firm's own product line, financial resources, and indeed commitments, essentially frame the range of opportunity across target market segments and distribution alternatives. The marketing manager

Exhibit 5.2   Determinants of Distribution Strategy

Product Factors	Channels Design Factors	Buyer Behavior

Product life cycle stage; technicality
• need for user education
• usage feedback needs
Product customization requirements
After-sale service needs
 	
Distribution Cost Factors
Customer communications
Transportation and storage cost
Order receipt and processing costs
Product-specific distribution investments
Financial risks
 	
Market Characteristics
Degree of concentration
Level of market demand
Applications diversity
 	
Country/Industry Environment
Level of demand
Market structures
Transportation/communication infrastructure
Regulatory framework
	Type of intermediary:
	direct salesforce, distributor, agent
	Intermediary profile
	• market segment focus
	• product lines
	• resources
	  financial
	  personnel
	  technical
	• local market position
	  share of market
	  customer base
	Distribution intensity
	Locus of distribution functions
	Number of distribution levels
		Product meaning
		• product essentiality
		• usage predictability
		• price sensitivity
		• time/place utility factors
		Purchasing modes
		• centralized or decentralized buying
		• DMU[a] composition and complexity
		• bundled or unbundled purchasing
		Size of account; size of purchase
		
		Business Unit Environment
		Resource availability
		• availability of capital for marketing investments
		• product line breadth and quality
		• product line margins
		• installed base
		Performance measures
		Business unit and corporate goals
		Distribution history
		Competitors' positioning
		• market segment focus
		• distribution modes

[a]decision-making unit

may seek, and attempt to take positions in, spaces that are either empty or occupied by weaker competitors. The alternative is to take on strong competitors in their areas of strength. We suggested seven different channels entry strategies.

Internal values—for example, a high respect for technical skills, a respect for the traditions that have been the cornerstone of past success—also inform current choices. Firms that have "grown up" relying largely on direct selling to reach markets may tend to eschew adding resale channels. Firms that from the outset have had a tradition of selling through distributors may find it difficult to develop direct channels. In each case, there may be a sense that the new distribution component will tend to build at the expense of the old, competing for common customers and closing in the latter's range of market opportunity. At a basic level, too, engineering values may come into conflict with sales values in any choice of distribution medium.

Finally, successful distribution programs tend to be reinforced by product development programs intended to utilize channels capacity to the fullest extent, to increase distributor income streams, and to foreclose competitive intrusion into the firm's distribution networks. Further, as will be explained in Chapter 10, the reinforcement of existing distribution systems may become an end in itself, making it difficult to respond to fundamental changes in the market environment. Thus, although faced with common product and market factors, competitors have very different internal circumstances and, as a result, exhibit considerable variation in their approaches to markets.

This and the preceding chapters have dealt with the factors that influence patterns of distribution at both the industry and the firm levels. These determinants of distribution strategy are summarized in Exhibit 5.2. This construct, the broad outline of which was first presented in Exhibit 2.2, may serve usefully as a recapitulation of the elements relevant in shaping channels networks and of the parameters of distribution strategy.

• PART II • Channels Organization and Management

In recent years, two broad trends, always present but now accelerating, have influenced the ways in which industrial firms go to market: (1) market demographic changes—the growth in the size of markets for industrial products, their increasing concentration, and market segment proliferation; and (2) the growth of industrial firms in terms of product lines and profit-centered business units.

Supported by growing sales revenues and confronted with a multiplicity of markets, many companies have responded by developing multichannel distribution systems. At the same time, manufacturers themselves are playing an increasingly active role in demand generation at user levels, working through or bypassing distributors. In addition, some manufacturers have integrated forward by developing their own captive distribution systems.

The net effect is that distribution networks and interchannel relationships have become more complex, a complexity that becomes reflected in channels organization and management issues, the subject of Part II. Chapter 6 deals with the structure of the sales and distribution functions. Working from the field back to the business unit and enterprise levels, we look, first, at options for organizing the direct salesforce and, second, at options for organizing the channels management function. Then, at the enterprise level, we take up a particular issue confronting managers in multidivisional companies: Should each product department have its own sales operation? Or should there be so-called pooled sales divisions serving multiple profit-centered business units?

Chapter 7 discusses channels power—the factors that give producers and intermediaries relative power and the ways in which power is used to achieve their respective and often conflicting busi-

108

ness objectives. We focus on what producers may do to build strength in channels relationships. Chapter 7 also sets the stage for the following three chapters, which deal with channels management. It proposes that the issues in managing and adapting distribution systems are more or less tractable, depending on the extent to which the producer is able to persuade or coerce its channels to serve its own evolving marketing strategy.

In a discussion of channels management and concepts of channels power, we are in an area that has important legal implications. Chapter 7 therefore includes an overview of legal issues and of the current status of court opinion that mediates producer-reseller business relationships. This topic is covered in greater detail in Chapter 13, "The Legal Framework."

Chapter 8 considers problems of managing the interchannel conflicts that often develop and the intrachannel rivalry that may erupt as distributors invade one another's territories or attack, through aggressive pricing, one another's customer bases within the same territory. Chapter 9 extends this discussion by taking up what for some producers is a particularly troublesome aspect of channels management: coping with unauthorized channels, or "gray markets."

Finally, Chapter 10 deals with the problems of change as product technology evolves and as markets change. As the competitive milieu shifts, with some market share contenders gaining strength and others dropping back, the need for change in distribution strategies becomes increasingly urgent. At the same time, it is often difficult for managers to see clearly what new modes of distribution are required and even more difficult to implement change, given the ways in which both producers and intermediaries become committed to supporting existing channels relationships.

6 Options in Organizing Multichannel Systems

Distribution networks involve many firms administered by many managements. The structure of a single distribution system thus cuts across the boundaries of large numbers of independently operated businesses, and its organization poses some complex choices.

As we discussed in Part I, the relevant concerns about channels design from the producer's perspective are the mix of channels types, the number of distribution levels from producer to user, and the allocation of channels roles and responsibilities. A related concern is how the producer will organize to deal with its direct customers, on the one hand, and with its intermediaries, on the other. This latter topic is the subject of the present chapter.

How the producer's headquarters sales operation is structured and how salesforce responsibilities are assigned—by product, customer set, market levels, or selling function—are crucial to channels effectiveness. How the producer is organized to interface with its channels is critical, as well, to coping successfully with the day-to-day, tactical issues of channels management.

Specifically, this chapter considers options in organizing a firm's sales operations at the field level. It then deals with how the firm's channels management function is organized. Finally, we consider a special topic in sales organization: pooled sales, that is, having a separate business unit that sells the products of two or more sister divisions. The issue is often whether this is a more or less effective way of going to market than having separate sales organizations for each product department.

110

OPTIONS IN ORGANIZING FIELD SALES

Sales personnel assignments may be defined in terms of

- product sets—categories of products in the product line
- customer sets defined in terms of product application, account type, and/or channel type
- market level, such as distributors or end-customers
- salesperson function, such as customer service or sales training

The extent of salesforce specialization is determined largely by sales expense-to-revenue considerations. Generally, the higher the sales revenue base provided by the product portfolio and the greater the sales potential in a trading area, the greater will be the extent of specialization along one or more of these dimensions. We consider the rationale for each form of specialization in the discussion that follows.

Specialization by Product Category

The rationale for having a field sales organization specialized by type of product relates both to the product and to the customer. One reason to specialize by product is management's desire to focus greater attention on some product categories rather than others. Further, if product technology is complex, individual salespersons can more effectively represent some part of the total line than all of it, which also suggests a product-specialized salesforce. Finally, sales personnel may be specialized by product category if there are different customer sets for different parts of the line or, within the customer environment, different departments and individuals responsible for buying different product groups. Honeywell Information Systems (HIS) is an example in which all three concerns were relevant. In 1986, HIS moved from having one salesforce that sold its full line of small and large computers to user-customers to having sales representatives in each branch office who were specialized in selling part of the line, either small or large systems. Honeywell management believed that large and small systems selling were different and required different kinds of technical expertise on the part of sales personnel. The selling cycle was longer for large systems, often two or three years, as compared with three to six months for small systems. The buying process for large mainframes involved more people and at higher levels in the customer organization. In addition, in any given account, mainframe purchases occurred far less frequently than purchases of small systems. The need for integrating large and small systems applications across the full range of the buyer's requirements was regarded, therefore, as less important in designing a sales organization than providing for a product focus.

HIS management also perceived the market as having three tiers. At the top, or "enterprise," level in large corporations were the mainframe central processing units (CPUs). At a second tier in large companies and in smaller ones, the so-called "departmental" level, were found the mini- and microcomputer systems. At a third tier, "workstations," were the "knowledge workers" in the organization, using personal computers. While minis and micros served to link large CPUs and individual workstations, there were many more sales opportunities for small computers at the departmental/workstation levels.

Given this market segmentation, Information Systems Division (ISD) management believed a full-line salesforce operated at a disadvantage. Product line breadth and rapid developments in computer technology made it exceedingly difficult for an ISD salesperson to represent effectively the full range of HIS computers and applications systems. In the meantime, Honeywell's primary competitors—Wang, Digital Equipment, Data General, and Prime Computer—all concentrated on more narrow product lines.

In addition, with benefit of hindsight, HIS management recognized that in the pre-1986 ISD organization, there had been less sales emphasis on small systems than on large.

Establishing separate salesforces for large and small systems was premised on a fundamental reconception of the market. Whereas Honeywell's marketing managers had previously perceived the customer as wanting to be presented with the full ISD product line at one time, they now distinguished between the large and small computer segments and buying behavior within each and then organized accordingly.

If the large and small computer markets could be strategically uncoupled, then three other considerations supported the logic of separate salesforces for the two product lines. First, ISD's management believed that Honeywell salespersons could be more effective if they were not required to be technically knowledgeable across the full product range, but could build expertise in either small or large systems. Second, competitors' salesforces were organized to provide for such product line specialization. Third, a specialized salesforce devoted to selling small systems would help to achieve one important strategic objective: building ISD's market share in the mini- and microcomputer product segments where significant industry growth was occurring.

Specialization by Customer

Marketing managers can specialize salesforces by customer category in three ways—by application, by account type, and by channel type.

Organization by Product Application. For organizing purposes, industrial markets may be segmented by the ways in which the product is used, that is, applications segmentation. IBM, for example, moved as early as the late 1950s to such a form of sales organization, and by the middle 1960s its Data Processing Division, the marketing arm for its mainframe computers, recognized different industry classifications for which it established specialized sales branches and/or sales personnel in branch offices. These included such industrial sectors as aerospace, manufacturing, process industries, distribution, printing and publishing, consultants and service bureaus, insurance, communications, utilities, finance, education, transportation, and airlines.

A salesforce organization segmented by product application is useful when specialized knowledge of the product use environment gives the seller a competitive edge. That may be especially true when potential customers need product application information; when systems are custom designed to meet user-unique requirements, as in the case of computers; or when applications-specific expertise can support a market position in particular industry sectors or market niches.

Organizing sales to focus on selected applications is, on the one hand, a way for the smaller competitor to utilize limited resources. It is, on the other hand, a way for the larger company to protect against such niche strategies by offering applications expertise across a range of product uses. However, an essential condition of organizing by product-industry application is that the base of revenue in each application segment be large enough to support the overheads associated with this form of specialization. Travel costs and the administrative expense of maintaining and managing applications-dedicated branch offices inevitably increase when the geographic territory is covered by several salesforces rather than one. Further, an applications-trained salesforce may be less flexible than a product-oriented one if investments in applications expertise are not readily transferable among user industries.

Account Type Specialization. A frequently used framework for structuring field sales is type of account—large, medium, or small; business, government, nonprofit—as a way of addressing different patterns of buying behavior. Many industrial companies assign different sales personnel to specific large business, government, medical, or educational accounts because buying modes may differ significantly across these sectors. Similarly, sales reps may be assigned to cover the smaller accounts in a trading area because these accounts may share certain buying behavior attributes as well.

Large account buying is often initiated and carried out by centralized purchasing departments, and the account relationship may involve a sales team representing different technical and selling skills, which deals with the customer's operating and purchasing personnel at multiple locations.

Government agencies and nonprofit accounts that are funded by federal and state sources often have specialized procurement procedures. They may, for example, be legally required to buy through competitive bidding procedures and to source from the low bidder for purchases over a specified dollar amount. In addition, the purchasing decision-making unit in a nonprofit organization, too, may represent a range of constituencies: operating departments, procurement groups, technical consultants, and external funding agencies.

Specialization by Channel Type. Companies that market a high percentage of their output through independent distributors, which serve different market segments, often assign individual salespersons to particular types of resellers, especially in major trading areas. Square D is one example. Square D specialized its sales reps in its major markets in either the industrial/OEM or the residential construction markets. The reps called on distributors that themselves typically addressed one segment or the other. In both cases, sales reps worked extensively with the distributors and their customers, but the nature of the work was different in each case. In calls on industrial customers and OEMs, sales reps became involved in product design and performance specifications and in customizing Square D products for particular applications. By contrast, in selling to the residential market, sales reps worked with contractors to calculate materials requirements by job, and with architects and consulting engineers to ensure that Square D was specified or at least included on the qualified bidders list. All sales went through distributors in the residential housing segment, and therefore Square D's sales reps often accompanied distributors' representatives on calls to contractors to negotiate sales contracts. In the industrial/OEM market, however, sales were sometimes negotiated with large customers without distributor involvement.

By electing to organize field sales at least in part by type of channel, such as oil field supply houses or hospital-supply distributors, manufacturers have essentially adopted a user-industry framework for organizing the sales program. However, channel types may be identified with product categories—as, for example, are steel warehouses, heating and air conditioning wholesalers, or rubber chemicals distributors—rather than with particular industry sectors. In either case, the choice of channels is tantamount to the selection of product-defined and user-industry-defined market segments in which to compete.

Market Level Specialization

In firms selling both direct to user-customers and through distributors, field sales personnel may be assigned to call on one or the other, or may have a mix of direct and reseller accounts to serve. Until 1986, for example, Honeywell's Information Systems Division had specialists for direct sales and for selling to VARs and MRs. Then these two salesforces were merged, and each salesperson had responsibilities for both user and intermediary accounts.

What are the arguments for each arrangement? If there is a need to coordinate the sales program at the distributor level with that at the user-customer level, each salesperson should work at both levels. For example, at Becton Dickinson, field sales personnel negotiated contracts directly with hospital buying groups; the contracts were then fulfilled through local hospital-supply distributors. At Norton Company, large users of grinding wheels bought direct for the bulk of their requirements but often sourced fill-in orders from local mill supply houses. In both cases, salespersons had to deal with both resellers and user-customers for each sale.

When it is not essential to coordinate the selling program across market levels, different salespersons may be focused on each. Ingersoll-Rand, for example, structured field sales in this manner. With distributor and direct sales roles defined according to size ranges of air compressors, distributor-user customer coordination was apparently not important in field operations.

One advantage in having different salespersons selling to distributors and to users is that each type of selling requires a different expertise. Selling to users may call for skills in relating to customer technical and production personnel for purposes of defining requirements, developing product specifications, preparing competitive bids, and dealing with service problems. Selling to distributors has a somewhat different focus of attention: ensuring that the distributor has adequate inventories, that its salespeople are properly trained to sell the product line, that repair and service facilities are properly maintained to service the installed base, that the distributor's financial condition and its business management practices are sound, and that the distributor is actively promoting the line with its customers. In addition, a sales rep assigned to distributor accounts often implements field promotions, such as product demonstrations, media advertising, sales contests, and special price promotions.

Market level specialization also often ensures that the supplier gives full attention to the distributor network. Marketing managers have observed that when their field sales reps call on both distributors and user-customers, the former usually get less attention.

Sales Function Specialization

Depending on what are the key elements of the sales program, it may be useful to allocate dedicated resources to certain sales functions. As noted earlier, Square D assigned field personnel in large industrial areas to work with distributors in defining and maintaining adequate inventories. In other companies, specialized sales personnel are assigned to recruit new distributors or to provide product training to company and distributor sales personnel. In some companies—IBM, for example—field representatives specialize in working with potential customers to design customized products and systems; others concentrate on the field service function by providing maintenance service and by training service technicians.

If the revenue base is sufficient to support such overheads, sales function specialization can ensure the performance of key tasks and allow those responsible for selling per se to concentrate on generating product demand. In fact, sales function specialization may permit a more economic use of field sales resources than if each salesperson were responsible for performing the full range of sales tasks.

THE CHANNELS MANAGEMENT ORGANIZATION

Clearly, distribution systems can be structured in various ways. One option is to establish separate management units for the direct salesforce, for independent distributors, and for agents, each unit having its own resources and administered separately. A second option is to work through a matrix organization in which channels programs are developed by staff managers and executed by an integrated field sales network under the direction of a line manager. Under the latter option, field sales personnel may or may not be identified with particular channels programs. An example of the first option is to be found in Ingersoll-Rand's Stationary Air Compressor Division (SACD). In that organization, seven managers in all, of whom four were channels managers, reported to James Clabough, marketing and sales vice president.[1] (See Exhibit 6.1.)

- A direct sales manager (DSM) was responsible for sales to users of all centrifugal compressors, rotary compressors above 450 horsepower, and reciprocating compressors above 250 horsepower. The DSM's organization included 42 sales personnel, responsible for sales in designated market segments.

- An independent distribution manager (IDM) directed sales activities through independent distributors that sold reciprocating compressors below 250 horsepower and rotary compressors below 450 horsepower. Reporting to the IDM

[1] Other managers in SACD marketing and sales included the following: a director of regional field service supervisors, a manager of SACD sales in Canada, and a director of a telemarketing program, which was primarily a market research and sales support activity.

Exhibit 6.1   Ingersoll-Rand—Organization Structure

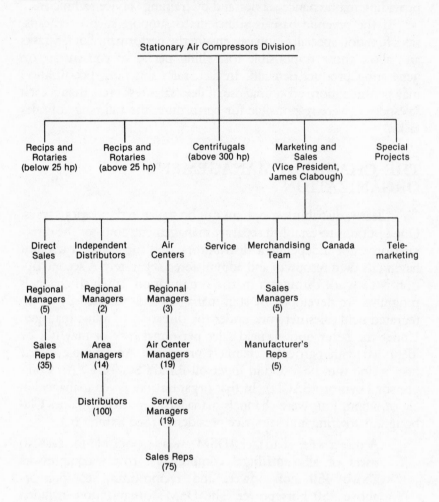

were 2 regional managers, supervising 14 area managers responsible for sales and service to 100 distributors.

An air center manager (ACM) supervised sales through company-owned branches. As in the independent distributor channels, each branch or air center sold and serviced reciprocating compressors below 250 horsepower and rotary compressors below 450 horsepower. Reporting to the ACM were 3 regional managers, each of whom had 6 or 7 sales managers. In all, there were 19 air centers, each having a sales manager, a service manager, 4 or 5 sales representatives, 4 to 8 service personnel, and 3 administrative assistants. Air centers and distributors generally did not overlap one another's territories.

- A merchandising manager (MM) was responsible for sales of do-it-yourself (DIY) products, mainly reciprocating compressors of less than 5 horsepower. The company sold its products through 5 manufacturers' representatives (MRs) to retail chain stores and catalog houses.

By way of comparison, an example of an integrated multichannels structure is Honeywell Information Systems' sales organization, called the Information Systems Division (ISD) as of 1985. As Exhibit 6.2 shows, ISD was organized in the field into 3 areas (sales operations), 8 regions, and 45 branches covering the United States. A branch manager typically supervised the work of a national account manager (NAM) and of 2 resident marketing managers (RMM), each of whom had 3–5 direct salespersons reporting to him or her. In 30 of the 45 branches there were also indirect sales specialists (ISS). The ISD headquarters organization, under Frank Jakubik, included the National Sales Operations, which directed field sales activities, and a National Accounts Program Office (NAPO) to which the national account managers in the field had a dotted-line reporting relationship. There was also the Indirect Sales Operation (ISO), headed by James Murphy, who directed the activities of the indirect sales specialists in the regions. Under Murphy were managers who had charge of a Manufacturers' Rep Program Office and a VAR Program Office, respectively. Also, a member of Jakubik's team, Murray Epstein, served as director of marketing support, a position that gave him responsibility for market research, sales training, and sales compensation systems. Thus, according to Jakubik,

> The sales organization is essentially a matrix organization, with the NSO and ISO headquarters people having goals for their particular channels and with the branch managers also having these same goals and being in control of field organizations that include direct sales reps, indirect sales specialists, and national account reps. I'm not a big fan of matrix management, but when you don't have all the resources to compete across the board, you have to go to matrix management. There's no other way.

Exhibit 6.2 Honeywell Information Systems Sales Organization Chart—1984–1985

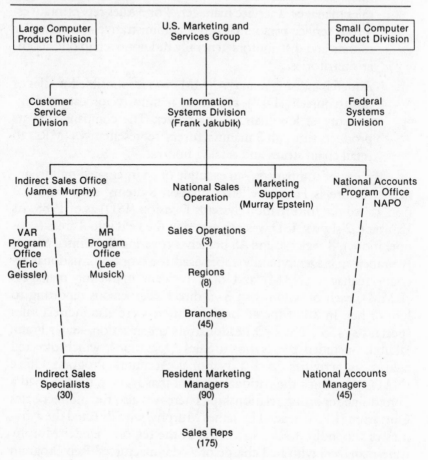

This matrix organization had evolved from a structure much like that of Ingersoll-Rand (I-R). As of 1984, ISD had separate units for national accounts, other direct sales activities, and a reseller network made up of VAR and MR accounts. As Jakubik's comment indicates, the costs of having separate organizations for the three sales programs could not be supported by the revenue stream. In addition, under the earlier structure, field sales managers found it difficult to cope with the conflict that developed between channels.

Under what circumstances is each multichannels system appropriate? Two considerations are pivotal: overhead costs and potential interchannel conflict. The I-R model is likely to incur greater absolute costs because of the overheads associated with staffing four separate channels management organizations, each with its own salesforce. In Honeywell's case, staff supervision of channels programs was the solution because separate line channels management organizations had proved too costly.

Ingersoll-Rand organized its channels system based on an assumption that its several channels could operate independently of one another. The direct salesforce, the independent distributors, and the air centers carried air compressors in different size ranges; the consumer channel sold only the smallest compressors and to an entirely different customer set. At Honeywell, however, ISD's experience with selling to user-customers and resellers had demonstrated repeatedly that the two programs often came into conflict and needed coordination at the field level.

Integration is needed as well when user-customers source from multiple channels. Coordination then preserves the integrity of the customer interface—particularly with regard to product prices—and ensures cost efficiency in servicing accounts. For example, at Wright Line (WL), a computer supplies producer, management integrated three separate channels after recognizing that customers were buying from all three and creating interchannel competition, as well as pricing confusion.

Wright Line, a division of Barry Wright, made and sold products for storing, protecting, transporting, and providing access to magnetic tape, cards, and diskettes used with computers. Its lines also included computer-related furniture: cabinets, terminal stands, workstation desks and chairs, and media protection vaults. Wright Line's 215-person salesforce called directly on approximately

90,000 user accounts, of which about 20,000 yielded 70% of WL's sales.

Until the mid-1970s, Wright Line had faced only three major competitors, all smaller. By 1982, there were 50–75 companies vying for pieces of a rapidly growing and proliferating market.

A 1982 study showed that there had been a marked decline in WL's salesforce productivity, averaging more than 15% for the preceding three-year period, and a significant loss of share in a rapidly growing market. WL management then organized a direct response marketing (DRM) unit to prepare and disseminate WL catalogs and to manage a direct order telephone operation. This unit had its own direct salesforce to call on prospective customers with larger and more complex needs than a telephone salesperson could handle.

Another unit was given the task of securing distribution through other companies' catalogs, those of computer hardware manufacturers, office furniture suppliers, and retail dealers.

The salesforce's mission was redefined: it would sell direct to the 25,000 largest accounts. All three units operated independently of one another.

In the 1982–1984 period, however, WL experienced no growth in commercial sales and sharply declining profits. WL then realigned its marketing organization in January of 1985, with one marketing vice president in charge of managing a single, integrated selling operation.

Looking back in March 1985, one Wright Line manager observed,

> A major reason for the 1982 sales reorganization was the growing financial pressure we were under because sales were growing faster than profits. We tried to bring that in line by reducing sales costs to cover the smaller accounts less expensively than by direct sales calls. We felt that the top 2,500 accounts gave us 80% of our sales volume, so we planned to cover them direct and find other, more cost-efficient ways of reaching the 300,000–400,000 other potential customers. We figured that different channels could be used to reach the 2,500 largest accounts, the 25,000 medium-sized customers, and the 300,000–400,000 smaller ones—this last group by direct mail. We thought we could isolate these segments.
>
> What we found out, in fact, was that we couldn't do that. The first two groups buy through all channels: direct, catalogs, and of-

fice supply stores. The smaller accounts use stores and catalogs and would probably buy direct if they could. Further, the large account list isn't the same from year to year. It keeps changing as last year's small accounts suddenly become this year's major prospects.

Perhaps the biggest problem was the internal fighting among business units. The salesforce (Unit 1) was constantly bickering with the direct response marketing group (Unit 2) over account ownership. Often, salespeople would improperly classify accounts to "hide" them from Unit 2, or worse, tell their customers not to order direct (since the salesperson would not receive credit).

The MediaLink program (Unit 3) was even more controversial. Salespeople genuinely feared the new unit and anticipated that this would be the demise of the salesforce. They would expend considerable effort to document every conflict situation in the hope of persuading the company to reverse the situation and shut down the unit.

Given the fact that Wright Line customers often sourced from all three channels and that the channels tended to be price competitive, it was essential to go to an integrated channels management organization. The Wright Line experiment with managing the three channels independently had proved too costly in terms of market share, profits, and, indeed, managerial personnel; six of ten WL managers resigned in early 1985.

In summary, the Ingersoll-Rand, Honeywell, and Wright Line cases cover three broad options in organizing the channels management function. The I-R model uses separate units for reaching multiple market segments. Such an organization may be effective if (1) the segments are differentiated in terms of the products each buys and of buying behavior and (2) the sales revenue base can support the costs associated with having multiple salesforce and sales management overheads.

The Honeywell model provides product-market segment focus through headquarters staff groups, each responsible for sales program development in a major market segment. In this matrix organization, staff managers give dotted-line direction to field sales branches. At Honeywell, such an organizational structure avoided the costs of having several separate line sales organizations, each serving different product markets. It also facilitated coordination at the field level between the direct sales and the VAR-MR marketing programs.

The Wright Line model integrates direct selling, telemarketing, and catalog sales under a single sales management. Arrived at through slow and painful evolution, this approach recognizes that when many customers purchase through more than one channel, market segment differentiation by product, by application, or by buying behavior is not feasible. What is needed, then, is to avoid interchannel rivalry over customer accounts and pricing incongruities.

POOLED SALES OPERATIONS—A TUG-OF-WAR

In pooled selling, one or more sales operations, defined largely in terms of customer sets, take to market the lines of several product departments. Often, the product department will include product managers who serve as liaison with the sales operation. Often, too, the product department has profit responsibility and ultimate pricing authority, while the pooled sales operation is measured in terms of sales revenue and expense-to-revenue ratios.

At Westinghouse, for example, there are three sales groups, Electric Utility Sales, Industrial Sales, and Construction Sales, each with responsibility for selling both direct and through resellers to their assigned market sectors. They carry the lines of six product departments: Motor Divisions, Transmission Equipment, Distribution Equipment, Electronics Measurement and Control, Industrial Control, and Electronic Components. All are part of the Industries and International Group, one of four Westinghouse business groups.

There are three advantages to pooled selling. First, it enables the firm to cluster through common channels the lines of multiple product departments that are sold to common market segments, thus providing a coordinated customer interface. Second, because of its larger revenue base, pooled sales operations can usually support a more extensive network of field branches and other sales resources. Third, sales organizations representing multiple product departments may be of sufficient scale to specialize their field sales personnel in ways not economically feasible for a single product department's salesforce. Personnel may be specialized, for example, by market level, by class of customer, or by selling function, as the discussion earlier in this chapter suggests.

On the other hand, a consideration in favor of integrating sales operations into the product department may be the need for close ties among sales, R&D, engineering, and production. This is often a requirement of marketing strategy when the sale involves a high degree of technical interchange between buyer and seller with multiple participants on both sides of the transaction. For example, in the early phases of the product life cycle and in the marketing of highly customized products, short lines of communication between the customer and the product department are especially useful. At the same time, the nature of buyer-seller involvement is such that the focus of attention is on the single product.

Even when a compelling case may be made for pooled sales operations, the product department/sales department relationship is often an uneasy one. Illustrative of the strife that may occur are the complaints of product department managers in one electrical manufacturing company.

> This company had profit-centered manufacturing divisions for three product lines: wiring devices, conduit, and wire and cable. All three sold through a sales division.
>
> Sales plans for the company were formalized in November of each year at the sales division's budget review meeting with the president. The three product division vice presidents were invited to attend this meeting. Working within the sales targets established by the product divisions, the sales division presented its program for the following year. The product division vice presidents had an opportunity to comment on the plans and to request changes.
>
> Typically, the sales division's plans were prepared with little consultation with the product departments. In fact, the product department vice presidents seldom discussed anything with the sales vice president, except at bimonthly meetings with the president of the company when deviations in performance from the sales plans were the subject of discussion.
>
> Sales revenue targets were frequently not met, and personnel in the sales and product divisions usually shrugged off responsibility, blaming one another. Although it was hard to "pin down" whose fault it was when sales budgets weren't met, the product division vice presidents believed they were ultimately held responsible since they had profit responsibility.
>
> The comments quoted below were made at a time when the company was in a period of declining sales and profits, and each of

the product department vice presidents was petitioning the president to dissolve the sales division and give each one its own salesforce. They argued that along with profit responsibility should go control over sales operations.

Comments from Various Managers in the Wire and Cable Division

The sales division's main objective is sales dollars. We have to worry about profits. The sales division pushes low-profit, commodity-type products in order to get total sales volume up. We want to push complicated products that are in the development stage so that our profits won't suffer in the long run.

The sales division can shift expenses and we can never track them down! They always make sales-expense ratios look all right. At the end of the year we have to talk assessments nevertheless. We try to watch, but we can't keep our finger on them.

Comments from Various Managers in the Wiring Devices Division

Each sales rep has three budgets to meet. If they get a lucky break on wiring devices and fill their quotas, they then concentrate on the other two products exclusively. We can *never* get any better than budget. We don't realize the full potential of the wiring devices market.

We are very conscious of product planning. For good planning we need good feedback from the field. All we get from the field are a bunch of alibis. They say, "We can't sell this" or "We can't sell that." We ask them what they want, and they can't tell us. We can send out a new product and never hear what happens to it for months!

Comments from Various Managers in the Conduit Products Division

These large construction jobs are field days for purchasing agents. They drive prices down to the extreme. Nevertheless, our salespeople cave in too easily. They sell only on price.

The system concept is out of date. We really have little in common with the other product departments. The industry is sophisticated and doesn't need to buy a packaged system any longer. All three products are purchased at different stages in a construction project.[2]

The contention between the product departments and the sales division seems to hinge primarily on two aspects of the relation-

2 E. Raymond Corey, "Bergman Wire and Cable Company (A-B)," #9-571-059. Boston: Harvard Business School, 1972, pp. 20–23.

ship: (1) incongruous goals and (2) lack of communication. On the one hand, the sales division's goals are these:

- maximize sales revenue
- minimize expense-to-revenue ratios
- achieve sales quotas by product line
- make optimal use of sales time by concentrating on selling established products at competitive prices
- maintain and support the field salesforce

On the other hand, the product department's primary objectives are these:

- achieve sales quotas by product
- maximize product line profitability
- introduce new products
- get feedback on product performance
- gain information from sales personnel and customers that is useful in extending its product lines
- influence the quality and thrust of field sales programs.

In this particular case, the level of contention had escalated largely because of the company's long-run declines in sales revenues and profits and the need to eliminate high-volume but unprofitable product groups. As each party sought to pin responsibility on the other, the profit decline seemed to invite mutual recriminations.

Nevertheless, a tug-of-war between product departments and pooled sales operations seems almost to be inevitable, as is the proverbial contention between manufacturing and sales. It stems largely from differences in goals, the difficulties of maintaining open lines of communication, and the frustrations of product managers charged with making sales quotas for their respective lines but unable to control the selling function. To the extent that product departments and sales departments share common goals, working relations might benefit. In particular, if both could be measured in terms of product line profitability, managers would have a broader context in which to assess expense-to-revenue ratios and to make pricing decisions. A focus on sales expense by itself may in fact lead to lower revenues and profits if it results in scaling down field sales resources, concentrating on high-volume sales opportunities, and reducing new product introductions. Performance goals, too, might usefully cover not only E/R ratios, sales revenues, and product quotas but market penetration for new products, on-time delivery to customers, sales training, and the implementation of promotional programs.

In addition, if there is a generic contention between pooled sales operations and the product departments they represent, integrative communication processes may be useful for dealing day-to-day with conflict issues as they arise. It hardly suffices to have the interchange among personnel in the sales and product divisions limited to bimonthly meetings to discuss deviations from sales goals and an annual meeting to determine sales quotas, to develop budgets, and to set transfer charges for sales services. The more constructive focus might be on the needs of customers, competitive activity, and product line development.

There is also a more fundamental question here: Does defining each product department as a profit center unnecessarily fragment what is basically a single business? If so, that would argue not for dissolving the pooled sales division but for consolidating the three product departments into a single functionally organized business unit. Thus, questions of sales organization may often raise, as in this case, other and more fundamental issues involving the structure of the business itself.

SUMMARY

The effectiveness of a sales organization should be assessed in terms of how it positions the supplier vis-à-vis its customers, its competitors, and its channels. Resolving salesforce specialization issues and delineating the respective product-customer domains of the direct salesforce and resellers must start with the nature of the product, with buyers' modes of product acquisition, and with the meaning of the purchase experience to user-customers. This analysis then becomes the basis for sorting out the dimensions along which the customer interface may be structured, whether it be along lines of product technology, product application, account size and importance, or some other dimension.

Other considerations affect how channels are structured and then managed. Managers should structure multichannel systems to integrate sales activities, especially in situations where customers use more than one channel for sourcing and for product service. Multichannel integration is also needed where there is high potential for interchannel price competition.

Finally, the structuring of a multichannel system must take into consideration the aspect of controllability. Marketing managers dividing sales responsibilities—in terms of products, customers, and sales functions—among potentially competing entities in the channels system must contemplate the relative feasibility of managing the whole system so that it functions in the way that was intended in its design. This, too, is a matter of organization structure. It is also a matter of channels management processes. The latter is the subject of Chapter 8, which deals with coordinating multichannel networks. In Chapter 7, we will discuss channels power, the factors which give producers and resellers, respectively, power in channel relationships and the ways in which power is used.

7 Building Channels Power

Relationships between supplier and distributor are both co-operative and adversarial. Each has a stake in the revenues and profits generated by the marketing system in which they play complementary roles. Contention develops over issues of channels control and ultimately over how revenues and profits will be shared. Suppliers seek to engage intermediaries' full support of their sales programs. Distributors aim at controlling their own product-market strategies without producer interference, maximizing revenues and margins in competition with other resellers and often with their suppliers, and engaging suppliers' resources in support of their own business goals.

From the producer's perspective, the purposes of channels management and control are to

- achieve breadth of market access through widespread distribution
- ensure product flow-through in fulfillment of end-user demand
- gain distributor sales support for its product line with the distributor's customer base
- ensure that its distributors devote their primary efforts to the producer's lines as opposed to competing brands
- maintain orderly end-market price conditions and thus safeguard product line profits and avoid deterioration in the distribution network as a result of price wars
- enable it to make adaptive changes in the distribution system as product-market conditions evolve

The purpose of this chapter is to describe how some of the companies we've seen—for example, Ingersoll-Rand (air compressors), Square D (electrical equipment), Lincoln Electric (arc weld-

ing supplies and equipment), and Control Data (disk drives)—
have successfully managed their channels to achieve these objectives. In particular, we discuss the dimensions of power and the
elements of strength on which suppliers base their relationships
with intermediaries.

This chapter also considers the U.S. regulatory framework,
which mediates producer-reseller relationships. Actions taken by
producers in developing distribution systems and imposing conditions on intermediaries are fraught with legal implications. We deal
briefly with these dimensions of distribution but treat this subject
in greater depth in Chapter 13, "The Legal Framework."

We draw extensively on the experience of the Norton Company as an example of effective channels management. Headquartered in Worcester, Massachusetts, Norton is the world's largest
producer of abrasives, both coated (sandpaper) and bonded (grinding wheels).

ELEMENTS OF SUPPLIER STRENGTH

The terms and conditions that producers seek to impose on
their independent resale representatives are constraints on the latter's freedom in controlling their own business strategies. Conforming to supplier franchise conditions by resellers must be presumed, then, to be in expectation of compensating rewards and/or
penalties from the supplier. What are the conditions that put suppliers in the position of offering sufficiently attractive rewards to
gain reseller support, while at the same time securing reseller acceptance of constraining franchise terms and conditions? We consider this question in the discussion that follows.

Three basic factors give the supplier strength in developing
and managing its channels: (1) the value of the franchise to the
reseller in terms of potential income, (2) the quality of day-to-day
supplier-reseller relationships, and (3) distributor disincentives for
switching to other suppliers or dropping the product line. All three
gave Norton Company considerable strength in attracting and securing the support of industrial distributors that were leaders in
their market areas. With the broadest line of any supplier in the
abrasives industry, Norton held about one-third of the total abrasives market in the United States as of 1986. Its distribution network was acknowledged by both Norton's management and its

competitors to be a fundamental source of Norton's marketing strength.

According to Robert Hamilton, vice president of Norton's Abrasives Marketing Group (AMG),

> About 35–40% of our business goes through our roughly 800 distributors to about 6,000 [direct] accounts. These managed accounts are the ones that Norton's sales reps are assigned to call on. There is another 15–20% of our business that is billed direct to the user and doesn't go through distributors. The remaining 40–50% is handled by distributors, with Norton reps playing a distributor support role by training distributor salespeople and making calls with them on their customers. In all, we market to about 75,000 industrial user-customers. Overall, about 70% of what Norton sold ten years ago went through distributors, and now it is about 80%.

Norton's abrasives line typically accounted for at least 10% of a distributor's sales volume and was usually among the three largest selling lines carried by its distributors. Norton's franchise agreements stipulated that distributors could be disenfranchised for poor performance against annual sales goals or for credit delinquencies. The agreement also stipulated that if the distributor sold its business, the Norton line did not necessarily pass to the new owners. Further, a distributor was required to have a separate franchise agreement for a new branch.

In the case of some large end-user accounts, Norton negotiated direct annual contracts, allowed the account to select certain distributors, and then negotiated a margin percentage with those distributors for servicing the account. Hamilton explained, "You might ask why should we give a distributor a percentage for servicing a large user-account. The reason is that they handle warehousing, delivery, and credit and call on purchasing; also, they may do a better job on the rest of our line if they can get a percentage from servicing our large managed accounts."

Distributor gross margins on Norton products ranged from 15% to 30%, with 22% as the average. During the 1980s, Norton had increased these margins gradually, with the hope that distributors would realize higher profits and not use the added margins to cut prices. From time to time, however, distributors cut prices to capture the accounts of other Norton distributors in their areas or to move into other distributors' territories. According to one Norton manager, "Basically we can't stop them, but we don't give these people sales support if they insist on cutting prices to go after the accounts of our other distributors."

Field Sales Organization—Norton versus Carborundum

Norton's 275 field sales personnel worked closely with distributors, especially the larger ones, on marketing plans and time-and-territory management. The Norton sales rep and the distributor's abrasives specialists periodically reviewed the distributor's business, account by account, and set specific account goals. The Norton rep also monitored the distributor's outstanding receivables and checked distributor stock being held for major customers. In addition, Norton provided distributors with a program designed to help them analyze their profitability by product line. Larger distributors were called on weekly, sometimes daily.

Norton management contrasted its experience with that of Carborundum, also a full-line supplier of both bonded and coated abrasives and historically a strong number-two competitor. Carborundum was acquired by Kennecott Corporation in 1977, and Kennecott was in turn acquired by Standard Oil Company of Ohio in 1981. Norton's sales personnel reported that Carborundum's marketing programs deteriorated during this period. In 1983, Standard Oil announced that Carborundum would leave the bonded abrasives business entirely and only continue with coated abrasives until that portion of the business could be sold. Hamilton commented,

> Going back more than 30 years, the Carbo network and distribution practices looked very much like Norton's. Somewhere along the line, distributor relations gradually eroded. We are guessing that it was because their field sales people were not as good as ours. They were losing the battle in the trenches. Later on, they lost credibility with distributors because in part they were "fishing behind the net." That is, they were saying that it was their policy to go through distributors, but then they would take business direct.

Norton's district sales managers and headquarters managers called frequently on distributor accounts. Norton also had a number of incentive programs for distributors, such as sales contests that stressed getting production orders or getting user-customers to test new Norton products. For a decade, 1972–1982, Norton also sponsored a program in which it entertained as many as 1,400 invited guests each year at the Indianapolis 500 races. Invitees included Norton's own user-customers, distributors, and distributors' guests and spouses. This program was intended, in the words of one Norton manager, to provide a setting where Norton managers, distributors, and their customers could meet informally and establish personal relationships. In addition, Norton had an advisory council of 12 distributors from different parts of the country. Each was chosen for a three-year term. The council met for four days each year to discuss Norton's distributor policies and practices and any changes that Norton's management was considering. Hamilton described the meetings:

> We select a place where we can relax and talk openly together. Then we talk about Norton's sales and marketing policies, strategies, and practices. We really don't want it to be a dog-and-pony show; we're asking for their advice and feedback, but we don't promise, of course, to do everything they say.
>
> Each year the distributors have a three-hour closed session and talk about their concerns. Then they report on their discussions. Last year they told us there was some slippage in the way certain of our policies were being implemented in the field. They also said they liked the program of annual planning meetings our field sales reps have with each distributor but that the execution is very spotty; some sales reps do it well, some poorly, and some not at all. The advisory council members put the spotlight on this inconsistency and urged us to fix it.[1]

Selling primarily through distributors, Norton has gained market share steadily for over two decades. What factors helped to account for this and for the company's distribution strength? We can cite six major determinants:

1. The value of the Norton franchise. With the broadest product line in the abrasives industry, and long-term reliability as a source of supply, Norton has been able to attract and hold distributors that have strong local market share positions and to enforce franchise agreements that require stipulated levels of performance.

2. The Norton line as a significant source of distributor income. Norton's product line is typically among the three best-selling lines carried by its distributors, accounting on

---

[1] E. Raymond Corey, "Norton Company (A): The Carbo Conversion Campaign," #9–585-140. Boston: Harvard Business School, 1985.

average for about 10% of the distributor's sales volume. High distributor margins and a large installed base ensured Norton's distributors of a steady source of revenue and profit. In addition, distributors derived revenue from Norton's direct sales to managed accounts, which were fulfilled out of distributors' stocks.

3. Tactical programs aimed at motivating and monitoring distributors. The time, money, and attention devoted to nurturing good relationships with its distributors ultimately translated into more effective representation of Norton's line by its distributors. Features of Norton's marketing programs—such as the distributor advisory council, the incentive programs sponsored by Norton, and the frequent interactions among individual distributors and Norton's field sales personnel—helped to build these relationships. In addition, the product and sales training provided to distributors by Norton salespeople, as well as joint sales calls and monitoring distributor inventory levels, helped to ensure effective representation at the distributor level.

4. A clear sense of selling strategy and long-term consistency in its implementation. The areas of primary sales coverage for distributors and the direct salesforce and the roles of each are clearly understood. Norton's credibility with its distributors rested largely on the long-term stability of these role relationships.

5. Control of the franchise. By stipulating that successor distributor managements do not automatically take over the prior owner's franchise but must be requalified, and by treating each location of a multibranch distributor as a potential franchisee, Norton has retained control over the quality as well as the intensity of its representation in local market areas.

6. Ongoing information about market segments and distributors' performance. Frequent field visits to distributors by Norton's sales personnel provided Norton with information about the market segments served by distributors.

By contrast, Carborundum, starting from a market position close to Norton's lost distribution strength. According to Hamilton's perceptions, Carborundum had difficulties for these reasons:

- conflict between its direct and distributor selling efforts
- the perception by its resellers that successive managements of Carborundum had incoherent, uncertain channel policies
- erosion of field relationships and the consequent inability to motivate its distributors
- a questioning by distributors of the supplier's commitment to the product lines it sold
- growth of price competition among the company's distributors and the consequent erosion in the value of the Carborundum franchise

Over time, therefore, Norton succeeded and Carborundum ultimately failed in building and maintaining a strong distribution network. The discussion below elaborates on these and other factors that contribute to supplier strength in developing distribution channels.

Value of the Franchise

End-user demand for its product line is, of course, the basic source of strength for the producer in channels relationships. The demand for particular brands is a function of superior product performance, the producer's end-market presence, the breadth of the line—hence the range of customer choice it offers—and the availability of related services, such as user education, product customization, and after-sale repair and maintenance.

In the case of product categories in which producer brand preference has little, if any, influence on the purchase decision, reseller reputation and service will be given greater weight by users, making the reseller less reliant on a particular supplier's brand. To the extent that resellers have viable options in sourcing relatively undifferentiated products from multiple producers, any one supplier has less strength in the supplier-reseller relationship. Under this condition, as well, larger resellers may introduce their own label brands, giving the producer even less control over its distributors.

An important factor that tends to reinforce demand for a particular brand is its installed base, that is, the number of units in use or the volume of the product brand being consumed in a market area. The installed base generates recurrent demand in several ways. If the product has performed satisfactorily over time, there will be a strong replacement or continued-use demand. Satisfactory performance will also give the brand a favored position among competing brands in new applications. An electrical contractor, for example, who has had good experience in using Square D load centers in residential construction projects will be favorably disposed to reorder when undertaking new projects. In addition, sales of a brand tend to generate tie-in sales of products related in use. The purchase of a particular make of load centers is likely, for example, to generate sales of same-brand circuit breakers. Sales of office copiers will stimulate demand for same-brand toner. Finally, demand levels for branded spare parts and repair services are strongly shaped by the number of same-brand units in use. The installed base, then, becomes a considerably important generator of reseller revenues.

In addition to value, the franchise is also enhanced by a supplier's direct salesforce selling to user-customers for order fulfillment through intermediaries. The power to direct this volume through one distributor or another may be used to reward or discipline and thus becomes an important source of supplier strength in dealing with channels. Finally, a brand franchise has value if it adds luster to the distributor's image among its customers and other suppliers, enabling it to expand sales of other products and to take on the desirable lines of other suppliers. In this respect, part of the franchise value may be in its halo effect.

Fundamentally, however, a franchise has value to resellers only in proportion to how much it represents of their total income streams. If it is a minor contributor in the distributor's broad range of products, the supplier's power with distributors will diminish considerably. Similarly, if intensive distribution in a trading area results in carving up market demand for the brand among a large number of resellers, the supplier's ability to enlist distributor support for the line may be limited.

One major contributor to franchise value is the distributor's perception of the supplier's *staying power.* Both end-user product demand and supplier attractiveness to resellers depend on whether the supplier is perceived as being "in the market to stay." Clearly, that consideration affects the willingness of users to commit to lines that may need after-sale service or complementary products and supplies, or products designed or formulated for particular applications. The value of the installed base as a revenue generator for resellers implicitly hinges on some assessment of the certainty of future supply. In turn, the size of an installed base is perceived as evidence of the supplier's commitment to its market. When Carborundum ceased operations as a producer of bonded abrasives in 1983, Norton's emphasis on supplier staying power was a key theme in its campaign to take over major Carborundum user and distributor accounts.

Norton's management urged its salespeople, for instance, to stress what it termed *the $120 million commitment*. Over the previous five years, Norton had spent that amount to modernize and expand its abrasives manufacturing plants. In addition, Norton also stressed to distributors, "Even during the worst business climate, Norton has maintained its salesforce and product engineering department at full strength." Norton's management realized that with Carborundum's departure from the business, both users and distributors would be concerned about the commitment to the abrasives market of any replacement supplier. Accordingly, Norton's management stressed the commitment and staying power of the firm (rather than specific product features) in ads aimed at both distributors and end-user customers. (See Exhibit 7.1.)

The Quality of Day-to-Day Relationships

To a large extent, the quality of supplier-distributor interaction depends on the personal relationships between supplier sales personnel and distributors. The stability and continuity of these relationships is an essential element in building trust, credibility, and a sense of mutual dependency. Hamilton makes this point about Carborundum's relations with distributors as its market position began to crumble:

> Things fell apart after Kennecott bought Carbo. At that time, they recognized they were losing credibility. They had a big meeting in Colorado with all of their present and past members of the Carbo Distributor Advisory Council. The Kennecott CEO addressed the meeting and said, in effect, we love distributors and we're investing in this business; he then announced a big new plant. But they never broke ground.
>
> The whole thing backfired. The distributors were saying, "We don't know what those guys are doing; a lot of new faces are showing up and we don't know if we can trust them." They felt very vulnerable. At that point, price was the only weapon that Carbo had left.

Exhibit 7.1

WHAT TO DO WHEN YOUR ABRASIVES SUPPLIER SUDDENLY DISAPPEARS.

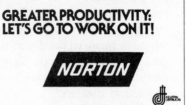

(Norton)   (Carborundum)

Fortunately, it's not every day you pick up the paper – or turn on the radio – only to learn that your abrasives supplier is going out of the business.

Of course others will offer to fill in. But for the thousands of Carborundum customers weighing the alternatives, we have a message:

Norton can match virtually any Carborundum product spec for spec, right down the line.

What else would you expect from the world's largest supplier of abrasive products? We have over 1,000,000 specifications on our active list. Including all types of grinding wheels, coated abrasives and diamond dressing tools.

In the past five years, we've invested more than $119 million to modernize and expand our complete abrasives manufacturing operations. This commitment means added capacity. Even higher quality. Competitive prices. And all our products are backed by the largest, most well-trained sales force and engineering staff in the abrasives industry.

So call your Norton Distributor (he's in the Yellow Pages). He'll find the right abrasive product for your application. Set up a prompt delivery schedule. And provide all the support you need.

Or call 1-617-853-1000 and ask for our Abrasives Hotline at Norton Company, Abrasives Marketing Group, Worcester, MA 01606.

GREATER PRODUCTIVITY: LET'S GO TO WORK ON IT!

NORTON

As appearing in IRON AGE, PURCHASING and AMERICAN METAL MARKET/METAL WORKING NEWS EDITION.

Source: Norton Company, Abrasives Marketing Group.

Supplier and distributor personnel come together around a broad agenda. It includes sales training, communicating new product data, making inventory checks, calling jointly on distributor customers, negotiating prices, conducting distributor promotions, accepting product returns, scheduling distributor payments on overdue accounts, and evaluating the distributors' product service facilities. On the one hand, supplier involvement in distributor operations lends needed support; on the other, it often puts pressure on the reseller to conform to supplier-imposed practices.

Given the range and varied content of such an agenda—sometimes collaborative, sometimes contentious or adversarial—the mutual expectations of each party tend to get defined through day-to-day interchange and the accumulation of informal precedent. Policies, practices, and expectations are established through individual relationships, the quality and stability of which become essential to the effective functioning of the distribution network as a whole. Staff reorganizations can have an impact and may engender great concern on the part of field sales personnel. For example, Arrow Electronics, the second largest distributor in its field, carefully orchestrated the first meeting with a supplier's new president.

In early 1985, one of Arrow's disk drive suppliers had a major reorganization, which included a new president. Shortly after the reorganization, Steven Kaufmann, president of Arrow's Electronics Distribution Division, arranged a meeting between top managers at Arrow and the supplier. One of the Arrow managers present said,

> The primary purpose of that meeting was chemistry: to build more awareness and a better relationship between their management and our management. Steve presented their new president with the history of our company, our sales, how we're financed, and our performance with various product lines. This was basic information, but their new president does not come from an electronics background and was a finance guy; now he knows who we are and what our financial condition is, and those things will make a difference. In addition, if a problem of sufficient magnitude arises, the president of their company can call our president and discuss the problem at that level.

Reseller Switching Costs

Mutual dependency between channel elements and the high costs each side incurs in "changing partners" are sources of strength for both distributors and suppliers. However, the extent of dependency is not likely to be equal. We now consider the factors that would make distributors highly dependent on a supplier, thus giving the supplier greater control in pricing, imposition of franchise terms, and product line support.

Supplier power may rest on the fact that its existing distributors have few, if any, attractive sourcing options and/or that the costs of dropping one supplier's brand to take on another are high. How high depends on the nature of the product, which determines the extent of the reseller's investment in inventories, brand-specific technical knowledge, and other resources. User preference for the existing supplier's lines over those of competitors also supports supplier power. In addition, a distributor does not usually have the option of switching brands if other suppliers already have satisfactory distribution in place in a trading area, another factor enhancing the existing supplier's power in its relationships with its distributors.

In addition, the costs of brand switching for the distributor include liquidating inventories of one line to make place for the other, converting record-keeping systems to new product code numbers, and training sales personnel in the particular features and applications of the new line. Converting a distributor to the Norton line, for example, was a time-consuming and complex operation. Robert Hamilton's description of the process highlights some typical changeover costs for the distributor and its customers:

> Normally, six months are required for a distributor to implement the conversion from a competitor's line to our own. During that period, distributor personnel are trained; records changed over to Norton specifications; announcements, promotions, and advertising developed. Stock orders are prepared and placed to ensure a continuity of supply. Finally, account strategies and plans are developed.
>
> Strategies for converting the end-user accounts are functions of the types of products used by the account (production line items or maintenance items) and the strength of the distributor in the account. MRO [maintenance, repair, and operating supplies] items are the easiest to convert, since a minimum of testing is required, and the Norton brand is widely recognized and accepted. The task is predominantly one of getting the records of the stores and purchasing departments changed over so that replenishment orders will flow to the new Norton distributor. Although mostly clerical, considerable selling and detail work is required.

The conversion of production items is usually more time-consuming. These items, by definition, are critical to the end-user's production process and must perform to the satisfaction of his production managers. If we are lucky, a Norton specification has been tested and approved prior to the conversion, and the distributor can begin accepting orders at once. Otherwise, a testing program must be undertaken to obtain the customer's approval. Since the items involved tend to be custom-made, the testing program can extend to several months.

We also must deal with the issue of the distributor's inventory. The distributor may opt to keep the "old" brand during the testing period mentioned above or to return as much as possible to the previous supplier. What's best to do depends on the situation at the end-user accounts in respect to types of products (MRO versus production) and testing required. In any event, it is likely that the distributor will end up with some dead stock.

The whole process requires a considerable amount of extra time and effort from distributor and Norton personnel. To support our local people, we might send in one or two extra people to work on the conversion for a month or two. They would make customer calls, train distributor personnel, help to convert records to Norton code numbers, and plan special advertising and promotional programs to communicate the change to customers.

A likely pattern is that the distributor's sales of grinding wheels and coated abrasives will dip for the first couple of months, but then recover to a higher level.

Unless the new supplier's installed base is comparable in size or larger than the incumbent's, the distributor's primary cost of brand switching will be the loss of revenue from the incumbent's installed base. Brand-switching costs will be especially high if the reseller is unable to switch its user-customers to the new supplier's lines. Thus, the users' switching costs dovetail with those of the distributor. Deterrents to brand switching for the distributor's customers may include the costs of training operators, liquidating spare parts and supplies inventories, and extensively redrafting manufacturing process and product design specifications. For example, in the market for arc welding electrodes—where users may require a great many different items and no one supplier's set of product specifications matches that of another—changing brands would incur high costs in all three categories. Changeover costs for users may also include obsoleting investments in product use-related items, such as software programs for computers or accessories for capital equipment.

Further, a major deterrent to brand switching for users is often the preference of operators who are responsible for product performance. Operators typically have pronounced brand preferences for such items as construction equipment, plant tooling, oil well drilling bits, electrodes, printing inks, and a wide range of products that (1) may be critical in terms of cost and/or schedule to key business operations and (2) require a certain level of art or technical expertise on the part of line operators. Their influence in the purchase decision is often compelling and usually weighs in on the side of the continued use of products that have performed satisfactorily. Given the choice of (1) staying with the existing brand and finding a new local source of supply or (2) staying with its existing source and changing the product brand, the user will often opt for the former.

Finally, the supplier of the original brand may escalate the distributor's switching costs by calling directly on user-customers to encourage their shifting to another distributor as a source of supply. For example, as Hamilton explained,

> If we can take away a distributor from 3M [the largest supplier of coated abrasives], we're not likely to convert all of its business even though we make sales calls on the end-user accounts with that distributor. The 3M sales rep simply shifts the volume over to another 3M distributor in the same market area. By contrast, if we convert [another competitor's] distributor, we can get almost all of that distributor's business. Other competitors just don't have that much muscle with users.

This ability to shift business from one distributor to another would indeed deter 3M's distributors from dropping the line and would in turn increase 3M's influence with its distributor network. Clearly, a significant source of the supplier's strength is the extent of distributor dependency on its line and the potential cost to the distributor of converting to another supplier's line.

THE LEGAL FRAMEWORK

Channels are managed largely by conditions imposed through franchise agreements and by the actions taken to secure adherence.

The terms and conditions of franchise agreements, as described in Chapter 2, may require that the reseller do the following:

- carry the producer's full line
- either not stock competing brands or treat them strategically as secondary lines
- not solicit business from certain "reserved" accounts, specified classes of trade, and/or beyond the territory in which the reseller is franchised to sell
- observe resale price schedules set by the producer
- maintain specified inventory levels
- meet specified sales quotas

The supplier may secure adherence to these conditions by withholding producer support from nonconforming distributors and by rewarding those which do conform, with resellers resisting such methods of control and sometimes taking legal action against offending suppliers.

Court decisions that have established legal precedent in producer-reseller relationships seem largely to reflect the spirit of the Sherman Act (1890). Relevant legislation also includes the Federal Trade Commission Act (1914), Section 3 of the Clayton Act, and the Robinson-Patman Act (1936) (amending Section 2 of the Clayton Act). In general, the kinds of terms and conditions outlined above and their enforcement are not illegal per se unless they are construed as being "in restraint of trade or commerce." If franchise conditions serve to build and preserve monopoly power as defined by the Sherman Act, and/or if franchise enforcement is carried out through conspiratorial arrangements, both the conditions and the actions to secure adherence are likely to be judged illegal.

Thus, with regard to price restrictions, producers may announce suggested resale prices and refuse to deal with resellers that do not comply *if* this action is taken unilaterally. Questions of legality may arise, however, if in cutting off an offending dealer the producer is alleged to have conspired with other parties, perhaps competing dealers, to take such actions and/or to ensure that the offender does not have indirect sources of supply of the brand.

Restricting distributors from selling to certain accounts or classes of trade or from soliciting business beyond specified territorial boundaries has usually been upheld by the courts. Such restrictions may be suspect, however, if either (1) the supplier is the dominant firm in the market and interbrand competition is weak or (2) the distributor has a local monopoly by virtue of being granted territorial exclusivity by a number of competing suppliers.

Similarly, requiring distributors not to carry competing brands is not deemed to be illegal per se. Such a condition comes into question, however, if it may be demonstrated that the restraint has an adverse effect on competition. It may be, for example, that a brand exclusivity arrangement in effect closes market access to competitors.

Requiring distributors to carry the producer's full line, another condition that may be included in the franchise, is legally permissible but may be challenged if the condition requiring the reseller to purchase and/or stock stipulated quantities effectively precludes it from stocking other lines. Full-line forcing is also a legally questionable tactic if it involves unrelated lines, for example, requiring a distributor to stock both air compressors and air conditioners.

Other court actions have arisen in connection with terminating distributors. Such actions may fall within regulatory bounds if a dealer is terminated by the producer, acting unilaterally, for not conforming to such franchise terms as those described above.

A supplier may also terminate a distributor for failure to meet certain specified performance standards related to sales goals or the adequacy of sales and service facilities, or because the reseller is considered to be a poor credit risk. Further, the producer is not prevented from dropping some distributors in a trading area and taking on others to build its distribution strength in the area.

In general, then, the regulatory framework that applies to supplier-reseller relationships seems to give producers considerable latitude in managing the flow of goods to market. It is important, nevertheless, that franchise terms and conditions, and their implementation be perceived, if challenged, as elements in some legitimate overall business plan. In particular, any actions taken to enforce franchise terms and conditions must be neither conspiratorial nor the means of building a market share position in restraint of trade.

This statement very briefly gives the flavor of legislation and court opinion relevant in supplier-reseller[2] interchange. The more

---

[2] It should be noted that this description of the regulatory framework applies only to producer-*reseller* relationships and not to producer-*agent* relationships, the crucial difference from a legal perspective being that resellers take title to the goods, while agents, selling on

detailed treatment of the subject in Chapter 14 provides the legal references and case synopses on which this overview is based.

SUMMARY

Producers' objectives in channels management are to gain broad market access, enlist distributor support, serve the needs of end-user customers, block competitors from their distribution system, preserve the integrity of the system itself, and maximize product line profitability. What market leaders such as Norton have in common is their ability to manage their distribution networks so as to accomplish these purposes. This ability rests on (1) the value of their franchise to resellers, (2) the quality of their day-to-day relationships with their distributors, and (3) the costs distributors would incur by dropping their lines in favor of competitive offerings.

Franchise value depends largely on user brand preference. To a considerable extent, this relates to the size of the producer's installed base, that is, its percentage share of units in use or the volume of the brand being consumed. Favorable user experience with the product fuels demand for continuing product supply, for new units, for replacement units, and for related supplies, peripherals, spare parts, and service.

The value of the franchise may be augmented by producer-generated demand at end-user levels, through advertising and promotion, and in particular through personal sales calls. In many cases, reseller revenues may be significantly enhanced by fulfilling sales contracts negotiated with end-users by their suppliers. Supplier *staying power,* the perception built through a past record of service to customers, current investments in the business, and known financial strength, is a vital component of franchise value. Especially in industrial markets where producer-distributor-customer relationships tend to develop in a climate of mutual long-term dependency, the producer's reputation for commitment to its markets may considerably strengthen its competitive standing. The reverse is all the more true: a concern on the part of distributors and users that the producer is not a viable long-term source of supply and service, or may be "on the way out," may lead quickly to a deterioration of its channels system and of its user-customer base.

Another factor contributing to franchise value for the distributor is the halo effect of a well-known and respected brand image. The Lincoln Electric name in arc welding equipment and supplies, for example, tends to build sales location traffic for distributors and thus to increase their

---

commission, represent their principals. In the latter case, the producer's right to instruct an agent with regard to prices and accounts to be served does not come into question.

revenues from related lines. The value of the brand franchise, and the relative strength it may give producers in managing their distribution channels, is a function of the percentage of reseller revenues the brand represents. Clearly, the greater the amount, the more brand dependent is the distributor and the greater is the ability of the producer to impose franchise terms and conditions and to secure adherence.

Personal relationships among supplier and distributor representatives, day-to-day, are also exceedingly important in securing reseller support for the producer's marketing programs. Close and continuing contact secures reseller conformance to the terms of the franchise and helps to ensure that the sales location is properly stocked and equipped.

A quality that is critical to establishing strong relationships with resellers is consistency in strategic objectives and distribution policies. Trust, credibility, and fairness are essential conditions, as well, for building channels relationships and for managing intricate distribution processes.

Suppliers gain strength in channels relationships, as well, to the extent that there are strong disincentives for their intermediaries to switch to competing brands. In many cases, the process of moving out old-brand inventories, taking in new stocks, and converting customers is time-consuming, and the cost high in terms of personnel time, lost sales, and inventory write-offs. How high is the cost for distributors will depend largely on user-customer switching propensity. If the brand currently in use by operators in the customer environment is the one they know and prefer, if converting specification numbers in stockrooms and on process documents is onerous, and if the brand producer's field reps have strong direct relationships with users, switching costs may be prohibitive. In such an instance, the distributor taking on a new brand will almost certainly suffer some deterioration in its customer list for the particular product.

Finally, a deterrent to dropping a producer brand may simply be the distributor's lack of more attractive options. If other producers of strong brands already have effective channels in place, they may be hesitant to risk the negative reactions of their existing distributors by taking on new ones and possibly intensifying local market intrabrand competition.

In addition to describing the factors that contribute to supplier strength, we discussed briefly some of the legal implications of imposing franchise terms and conditions on resellers and securing conformity. As long as conditions and actions taken are neither conspiratorial nor deemed to be "in restraint of trade," the U.S. regulatory framework imposes no serious constraints for producers in dealing with resellers.

This discussion of building channels power from the producer's perspective appropriately precedes a discussion of channels conflict. In the three chapters that follow, we consider issues of channels conflict, dealing with unauthorized channels (gray markets) and with bringing about change in the distribution system as product-market conditions change. Effectively coping with channels management problems in these three related areas depends essentially on what power the producer has in its relationships with intermediaries and on how that power is used.

8 Managing Channels Conflict

In theory, direct salesforces, independent distributors, and captive distributors in a multichannels distribution system complement one another. In fact, they are often in conflict. Why? First, it is difficult to define their respective roles and product-market boundaries explicitly. Having defined their roles, it is even more difficult for producers to ensure that intermediaries operate within those boundaries. Both inter- and intrachannel conflicts often result. Accordingly, producers' major concerns are (1) maintaining and delineating product-market boundaries among channels and (2) resolving day-to-day channel conflict issues.

Adversarial channels relationships are of several sorts:

- the direct salesforce versus resellers
- captive versus independent distributor networks
- mass merchandisers versus local full-service distributors
- distributor versus distributor within and across territorial boundaries

In this chapter, we consider the nature of conflict in each of these categories and the ways producers seek to resolve channels frictions.

DIRECT SALES AND RESALE CHANNELS COMPETITION

A simplistic delineation of selling responsibilities is that direct salespersons cover the large accounts, and resellers take the small ones. In theory, carving up the market in this way serves three producer objectives: achieving distribution cost efficiency, maintaining direct sales relationships with large and important customers, and getting broad market coverage. In practice, the large/small dichotomy is an ill-defined and fluid boundary, which intermediaries are seldom prone to observe.

From the producer's point of view, the problem has several facets. Reseller intrusion in accounts served by direct sales often leads to end-market pricing inconsistencies, confusion among user-customers, and salesforce demoralization. A manager for Wright Line, a computer accessories supplier, reported this experience:

> We'd find, for example, that our sales rep would work for six months to cultivate some large account, and then one day would find that an office products dealer would call and tell the customer that he could sell the same Wright Line products out of his MediaLink catalog [WL's catalog] for less than our rep was quoting.
>
> Our sales reps complain that their customers see Wright Line products in these catalogs at lower prices than what they are quoting. Every month there's at least one instance where one of our reps has lost an order to a catalog account or whose credibility is being questioned by a customer who saw lower published prices for the same products in someone else's catalog.

As these occurrences suggest, a major contributing factor in interchannel competition lies in differential resale price structures. Different channels obtain supplies of product at prices that may give one channel a cost advantage over another. Generally, companies that sell to different classes of buyers at different prices try to ensure that product sold at the lowest prices does not flow into channels and market segments in which prevailing prices are higher.

One way that some companies have sought to reduce direct sales reseller conflict is to set quantity discount schedules that enable large user accounts to buy large quantities at prices equal to or lower than the prices charged to distributors for like quantities. For example, the Lotus Development Corporation's price schedule for its software programs included the well-known Lotus 1-2-3 spreadsheet program, which became effective as of 1985. Exhibit 8.1 shows the quantity price schedule for distributors that purchased from Lotus and sold to retailers. Exhibit 8.2 is the schedule of prices direct to large retail chains; to large OEMs such as IBM, DEC, and Wang; and to large software users. As a comparison of the two schedules indicates, "house accounts" buying direct in quantities as low as 2,000 units could purchase at 1% under the distributor price; at 40,000 units and over, they had a 5% price

Exhibit 8.1   Lotus Distributors' Volume-Purchasing Terms

Annual Units Purchased[a]	Discount Off Retail List (%) Price[b]
50,000 and below	36
50,001 to 100,000	37
100,001 to 150,000	38
150,001 to 200,000	39
200,001 and over	40

[a]A unit means one 1-2-3, Symphony, or Jazz program. Other Lotus programs, companion products, or upgrades did not count toward discount volumes but were purchased at the same discount levels achieved through unit purchases.
[b]These discount levels became effective in September 1985. Prior to that, they were each 5% higher.

Exhibit 8.2   Lotus House Account Volume-Purchasing Terms

Annual Units Purchased	Discount Off Retail List (%) Price
2,000 to 12,000[a]	37
12,001 to 20,000	38
20,001 to 30,000	39
30,001 to 40,000	40
40,001 and over	41

[a]All house accounts had to commit to a minimum of 2,000 units. Since a majority of such accounts were large retail chains, there never was a problem getting this commitment.

advantage. Given these price relationships, there would be no room for resellers to underprice Lotus in selling to accounts able to buy in quantities of 2,000 and above.

Prior to September 1985, however, distributor volume discounts had been 5% higher across the board, giving distributors a price advantage over large user-customers in all quantity brackets. The Lotus price revisions implemented an explicit decision to serve large accounts through its direct salesforce.

While some firms have utilized price schedules that discourage distributors from selling to direct accounts, the inherent competition between the direct salesforce and resellers is more often fostered by producer pricing practices, which encourage resellers to compete with the direct salesforce for large volume business. Deeper discounts for distributors than for user-customers in the same quantity brackets allow distributors to quote to large customers prices that are below producer list prices.

In some cases, the large customer may take a price-cutter role. A significant disparity between the producer's list prices and the much lower prices it quotes on competitive bids for large quantities can create arbitrage opportunities for user-customers that buy at lower price and move surplus supplies into market segments where producer prices are higher.

Minimizing direct sales–reseller competition becomes a matter, first, of clearly defining their respective domains and, second, of coordinating pricing actions to reduce the rewards that may be gained by moving goods from one class of buyer to another. However, this may be easier said than done if the producer is to be price competitive in segments that differ in bargaining power and in the pricing conventions (e.g., competitive bid, published list, discounted list) they observe. Pricing coordination, as an arbitrage deterrent, may be supplemented by tracking, to the extent possible, large shipments to the low-price user-customer, both direct and through distributors, to ensure that the quantities purchased are not in excess of those which will be consumed by the purchaser. While tracking large shipments is frequently done by suppliers, this action, too, is never a panacea.

In the discussion that follows, we turn to defining the respective roles of the direct salesforce and resellers.

Establishing Reseller–Direct Sales Boundaries

Why do producers want to draw lines between their direct salesforces and their independent distributors in moving their products to market? There are several reasons. The first and most important is profit maximization. Dealing directly with large accounts may entail lower expense-to-revenue ratios and higher net margins than does selling through a distributor at wholesale prices. A second, and related, reason is that it may be important to control the large customer relationship rather than leave it in the hands of an external intermediary. Third, each customer has needs that may better be served through one channel or another, and each product requires its own level of sales and technical effort. There may be opportunities, then, to match product groups and customer sets to types of channels and to optimize selling effectiveness.

Producers may draw the lines in at least three ways. The first and probably the most common is to reserve certain specifically named accounts or classes of accounts for direct sales. The second is to channel certain product categories through resellers and others through the direct salesforce. The third method is to set some limits expressed in terms of order size or of customer annual purchase volume that delineate the domains of direct sales and resellers.

In choosing a way of drawing the lines, it is relevant to take into account the way user-customers buy. Further, any approach to defining channels turf should, of course, be implementable. In the discussion that follows, we consider customer class, product class, and purchase amount as dimensions along which selling responsibilities may be allocated between a direct salesforce and an independent distributor network.

Demarcation by Customer. Reserving certain customers of classes or accounts for direct selling is a long-established practice. Despite occasional challenge, it has survived the scrutiny of the courts as being a legitimate exercise of the supplier's rights in choosing to whom it will sell—as long as such exercise does not have a detrimental effect on interbrand competition and is not an element of a conspiracy in restraint of trade.[1]

As discussed previously, certain accounts may be reserved for direct selling because direct sales expenditures will be less than the net reduction in account revenues represented by the distributor discount. In addition, suppliers may wish to have direct relationships with their large accounts for purposes of negotiating prices, providing technical support, and meeting the large customers' other service needs. A direct relationship may be important as well if negotiations with large buyers tend to set market price levels.

With a long tradition of selling its lines of stationary air compressors through a direct salesforce only, Ingersoll-Rand's Stationary Air Compressor Division (SACD) continued to reserve certain designated accounts and classes of customers for direct selling after the division began building an independent channels network in 1960. Its policies in this regard were explicitly laid out in its sales manual as follows:

---

[1] For legal reference, see *Donald B. Rice Tire Co.* v. *Michelin Tire Corp.*, 638 F2d 15 (4th Cir) cert denied, 454 US 864 (1981), and *Red Diamond Supply, Inc.* v. *Liquid Carbonic Corp.*, 637 F2d 1001 (5th Cir), cert denied, 454 US 827 (1981).

1. It is the policy of Ingersoll-Rand Company to sell directly to all users those items of equipment not specifically described in the selling agreement. Inasmuch as Ingersoll-Rand calls on these customers direct to sell engineered products, we also reserve the right to sell all products on a direct basis to accounts which are designated as "Special Accounts."

2. It is the marketing policy of Ingersoll-Rand Company to sell our equipment direct to Original Equipment Manufacturers and National Accounts, where this best serves the interest of the customer.

3. It is the policy of Ingersoll-Rand Company to encourage and assist established and officially designated distributors to sell equipment of the type described in the Selling Agreement to state, county and city governments. However, Ingersoll-Rand reserves the right to sell to local governments direct if the distributor is not properly representing us or is unable or unwilling to handle the sale.

4. It is the policy of Ingersoll-Rand Company to sell directly all of its products to the United States Government and all of its agencies, including the Armed Forces.

5. It is the policy of Ingersoll-Rand to sell distributor products through privately owned distributors. However, when the privately owned distributors are not growing or showing sufficient market penetration or where we cannot find private capital or expertise, Ingersoll-Rand reserves the right to operate company-owned Distributor stores.

At least in the early days of multichannel distribution for SACD, one reason for preserving direct sales relationships with "special accounts" and certain classes of customers was the level of technical expertise and service that these accounts required. It is likely that SACD's management regarded their technical resources as being better than their new distributors'.

Equally important, the economics of direct selling to this class of customer would clearly be more attractive than serving it through indirect channels. Beyond the cost effectiveness of chan-

Exhibit 8.3 Allocation of Sales Responsibility for Stationary Air Compressors by Type and Size to Sales Channels for 1960, 1973, and 1984

	1960	1973	1984
Direct Salesforce	Recips 50 hp and over Rotaries 150 hp and over	Recips 150 hp and over Rotaries 450 hp and over All centrifugals	Recips 250 hp and over All centrifugals
Distributors	Recips under 50 hp	Recips under 150 hp Rotaries under 150 hp	Recips under 250 hp Rotaries under 450 hp
Air Centers		Recips under 150 hp Rotaries under 150 hp	Recips under 250 hp Rotaries under 450 hp
Manufacturers' Reps			Recips 5 hp and under

neling product through direct sales, there were significant revenues to be obtained from aftermarket service and spare parts sales.

Delineation by Product Classification. Some companies distinguish between reseller and direct sales roles in terms of the products given to each channel to sell. Again, SACD is a case in point. In addition to class-of-customer boundaries, SACD differentiated between its direct salesforce and resellers in terms of product size and type. However, an increasing number of SACD products were being reclassified to the distributor domain. Exhibit 8.3 indicates the buildup in the list of distributor-class products between 1960 and 1984.

Product-class distinctions between channels implicitly assume that potential customers can be sorted out by the category of product they buy (in this case, large, medium, or small air compressors) and the channels through which they typically source. But as noted in the Wright Line example the same customers often wish to buy through different channels at different points of time, depending on the purchase order size, urgency of need, and desire for purchase convenience. Furthermore, they are typically customers for the full product line range.

Thus, product-class distinctions for purposes of delineating direct sales and reseller domains are useful only when, as in the case of Ingersoll-Rand, broad customer categories can be linked with particular product groups and channels.

Account Potential and Size-of-Order Boundaries. As noted earlier, many manufacturers act on the premise that the role of resellers is to serve the smaller accounts. When accounts grow to some predetermined size, the direct salesforce may then take them over. There are, of course, two problems with moving customers from a reseller to a direct sales relationship. One is that the customer itself may have some preference for buying from local distributors rather than sourcing direct. The other is that such a practice hardly builds goodwill with the resale channels network.

Size-of-account guidelines served, for example, at Control Data's Peripheral Products (PPCo) business unit to distinguish direct customers from those which would be reached through distributors. The following comment from one of PPCo's distributors suggests the reseller's concern about losing large accounts to its supplier:

> PPCo has probably the largest salesforce of any company selling disk drives, and their order quantities for the direct versus distributor cutoff point tend to be somewhat lower than the industry norm. The result is that we may develop a small account into one that now orders at least a thousand units annually, but the fruits of that work go to the manufacturer. That's of course a generic issue in any manufacturer-distributor relationship. But my point is that we can keep an account "our account" longer with some other manufacturers because of their pricing policies.

Bounding resale and direct market segments by size of order may be an even less desirable method of categorization. This method means sorting out, day-to-day, those sales leads to be approached by direct salespersons and those to be turned over to resellers. First, it is often difficult, if not impossible, to know how big a business opportunity might be in terms of immediate potential revenue and/or follow-on sales. Second, serious difficulties often arise in persuading direct sales reps to turn over leads to resellers and vice versa, regardless of revenue potential. Companies attempt to resolve this difficulty by offering lead referral bonuses calculated, usually, as a percentage of the dollar amount of a consummated sale. Such incentives, however, seem often to be ineffective because they fall short of what the referring agent, either reseller or direct sales rep, could earn by making the sale itself.

At Honeywell Information Systems Division (ISD), field sales managers were instructed to channel all sales opportunities under $150,000 through resellers (in this case, VARs) and, above that amount, to the direct sales reps. Going to market through multiple channels, however, sometimes led to questions of which channel would best serve ISD's competitive positioning in any particular instance. A headquarters marketing manager commented,

> Usually a direct sales rep doesn't want to have competition from the indirect sales channel and often will not give a referral to the VAR reseller on his own initiative until it looks like he is not going to be able to get the business anyhow. In one instance recently, in a large account, a piece of business involving 200 systems for large dollars was out for bids. When the Honeywell direct sales rep realized that he was losing the business, he wanted to call in a VAR reseller because it was a specialized application, but it was too late. The winner turned out to be a large computer supplier who had teamed up with a systems house. What we would like to say to the branch manager is "Put Honeywell's best foot forward to make sure that we win the business with the best solution, and we will reward you no matter where the business comes from."

The marketing manager's example suggests another weakness in drawing lines based on size-of-order or account potential: Such a guideline gives no consideration to which channel type in a multichannel system is best equipped to secure the order and to provide account service.

ISD's general manager spoke of two other concerns in turning leads over to resellers:

> Our biggest VAR account also carries a competitor's line. If an ISD rep puts in a lot of time helping it to get an order, it's time wasted if the VAR doesn't supply DPS-6 systems but sells competitive equipment instead.
>
> As a practical matter, too, it may be difficult to know in any individual instance on which side of the line the sale should fall. Let's say, for example, Harvard puts out a bid for a small system, and maybe a VAR dealer who is a specialist and has a lot of experience has a good solution to what Harvard wants. Maybe the business is only potentially half a million dollars, and our direct sales rep shouldn't spin his wheels going after it. But on the other hand,

maybe the job requires a lot of networking where Honeywell is strong, and it will mean a foot in the door at Harvard and we can sell a lot more equipment later on, such as LANs and control systems. But if we don't have the best solution and it's a specific application that doesn't have any follow-on opportunities, then maybe it's best that that piece of business be picked up by one of our VARs. In general, I don't want my direct sales reps going after the small leads and being diverted from the $1 million opportunities.

In summary, producers attempt to minimize direct sales–reseller contentions in several ways. One way is by using different channels for different product sets. If market segments can be differentiated in terms of the products they buy and the channels they use, this strategy may be effective. It is not effective, however, if customers tend to buy from several channels sources, either concurrently or at different points in time, along a wide range of products.

Another approach is to segregate direct and resale market segments by size of account. The drawback here is the inevitable strain in supplier-reseller relationships that results when resale accounts grow in size and are taken over as "house" accounts. Further, shifting a customer from reseller to producer roles may not be in accord with the customer's wishes.

Delineating direct and reseller markets by size of potential order is often impractical to enforce because salespeople are inherently reluctant to turn over leads. In addition, size of order indicates neither future revenue potential nor the relative advantages of producer and reseller reps in serving the account.

Probably the most workable scheme is to specify, at the time of granting the franchise, that the franchisee may sell to all customers and classes of trade except those specifically reserved for direct selling. In establishing account and class-of-trade restrictions, the producer sets explicit boundaries that are relatively simple to administer, are consistent with customer buying behavior, and will serve to maximize sales margins. However, in any case boundaries between direct sales and resellers are difficult to draw, as a practical matter, with any assurance of restricting each to its intended domain.

RESELLER RIVALRY: INDEPENDENT DISTRIBUTORS VERSUS CAPTIVE BRANCHES

"If GESCO puts a captive branch in New Bedford, I'm dropping my GE lines!" exclaimed the owner of one multilocation electrical distributor in the Greater Boston area. Feelings run strong among distributors carrying the lines of suppliers that sell through captive distribution operations[2] in their territories. The complaints of the independents tend to be based on a sometimes ill-founded perception that the producer favors captive branches in its pricing practices, in allocating product supplies, and in providing a range of support services. Ironically, captives often perceive independents as the favored channel. Franchisees' appeals for support in their competition with captive branches go to their suppliers' product department managers and sales representatives, who are dependent on them in making their sales quotas. Product department managers will in turn tend to distance themselves from the captive distribution operation and exert their influence internally to contain its growth and market power. Thus, independent distributors may elevate the interchannel rivalry to another level by working to exacerbate the product department–sales department relationships, and often with considerable effect. Their weight may be felt to an even greater extent if they speak through an organized distributor advisory council.

Two principles may usefully be observed if the two channels are to coexist: sales program stability and market segment differentiation. Program instability—a pattern of opening and closing captive sales branches in different market areas, a changing mix of products going through these branches, or inconsistent price and promotion policies—is almost certain to escalate the level of conflict. To the extent that the role of the captive operation is known and understood in the market and remains relatively constant, independent resellers may develop their own market positions with some degree of long-term security.

In addition, to the extent that the captive operation is focused on different product-market segments than the independents, the sense of unfair competition and local market rivalry may be lessened. We have, in fact, observed a tendency for each channel to differentiate itself in terms of customer groups. For example, captive distributors in the electrical equipment industries, such as GESCO and WESCO, seem to focus on the large construction

[2] As discussed earlier, captive distribution systems are organized as business units in industrial manufacturing companies and serve as wholesale channels through which the producing divisions may move their products to market, in addition, typically, to going through independent resellers. Like any independent chain, the captive network consists of a complex of sales branches that stock and sell a wide range of items sourced not only from sister departments but from outside suppliers as well.

market segment, leaving residential construction to independents. The reason for this focus may be to build on the strength of their engineering competence. Nevertheless, by targeting those classes of accounts not served by small, local independents, the producer-owned distribution network may cool the emotional reactions of the independent reseller, which perceives its single most formidable competitors to be its suppliers. An appendix to this chapter considers, in greater depth, the role of captive distribution in the corporate environment. In many large industrial product firms, both those with and those without a captive distribution system, there seems to be considerable uncertainty about its relative benefits and mission.

THE SPECTER OF THE DISCOUNT CHAIN

Independent resellers perceive themselves as competing, on the one hand, against captive distributor branches and, on the other, against large national chains—chains such as W. W. Grainger in the electrical supplies industry, and retail mass merchandisers, such as Grossman's, True Value and K mart—selling through catalogs and through multiple branch locations. While these channels often do not compete directly for the same customer sets as the independents, they are nonetheless threatening to the small, local industrial distributor. Traditional independents typically react by pressing their suppliers not to franchise the large chains, and may threaten to drop their suppliers' lines if they do.

Producers can respond to this form of interchannel rivalry in one of three ways. The first is to give up one channel as the price of retaining the support and cooperation of the other. For example, Square D entered consumer retail channels in 1981 by franchising large mass-merchandising chains to carry selected lines of electrical products that would appeal to do-it-yourself customers as well as to tradespersons engaged in electrical contracting and small construction. It discontinued this facet of its distribution program five years later, in part because of concerns about interchannel competition. Some Square D distributors were already selling to retail outlets, and Square D's managers did not want to be in the ambivalent position of supplier and competitor. Nor did the managers think it desirable to have consumer retail outlets competing with Square D industrial distributors for the business of the small electrical contractors. There was also some concern that the low prices of the retail chains would serve to undercut the business of the small electrical contractors engaged in home remodeling. In this case, the response was to forgo incremental sales revenues that might have been realized through mass merchandisers in favor of supporting the independent industrial distribution channels.

Interchannel rivalry such as that experienced by Square D would seem difficult to contain without the producer sacrificing the support of one channel or another. As in the case of Square D, then, one option is to be highly selective in the types of channels that make up the distribution system so as to avoid interchannel rivalry.

The second way that the producer may elect to ease interchannel conflict is by developing different brands or by supplying a private brand to the mass merchandiser. Again, the cost may be some loss of revenue as compared with marketing under a single, widely accepted brand name. A third option, of course, is to accept some level of conflict and reseller resentment in the interest of gaining full market access through a range of channels.

The answer clearly depends on an assessment of channel revenue potential, the value among users of the producer's brand name, and the strength of the producer's field salesforce both at the distributor and the user levels. The greater the producer's power in the end-market, the more tolerance it may have for interchannel competition without risking the disaffection of its local distributor network, and the greater its freedom in going to market through a variety of channels.

TERRITORIAL COMPETITION
WITHIN CHANNELS

On the one hand, producers often find it essential to have multiple representation in a trading area; on the other, they may wish to constrain the extent of price competition among distributors both within and across territorial boundaries. Intensive intrabrand price competition at the resale level may discourage resellers from stocking and actively promoting the supplier's line. It will certainly suppress distributor value-added services, which increase the reseller's costs and lower its margins. Finally, intensive intrabrand price competition among resellers may escalate into intensified interbrand price competition at the producer level, as producers lower prices to resellers, enabling the latter to reduce their prices in turn.

Resale price competition may result from the practice of franchising a large number of distributors in a trading area. It may be fueled as well by producers' quantity discount schedules that encourage distributors to order in amounts larger than they need to serve existing customers. Excess inventories may then be moved either by competing for other distributors' user accounts or by shipping to user-customers or resellers in other sales territories. Further, distributors may divert, to other end-users or resellers, parts of shipments purchased at substantial discounts and intended to fulfill producer contracts with large end-users.

Depending on the conditions that lie behind intensive reseller price rivalry, possible ameliorative actions include (1) reducing the intensity of distribution—although this action may decrease sales revenues, (2) adjusting quantity discount schedules if it appears that distributors are "dumping" at low prices, and (3) withholding various forms of producer support from those resellers which are chronic price cutters. As one marketing manager explained, sometimes a producer can discipline a distributor quite effectively:

> Distributor X was hell-bent upon reducing our prices in the market. Repeated requests didn't seem to help. Luckily, Neil, our sales rep, controlled the four largest customers in that territory. He simply routed the following quarter's order through our other distributor, who had a more disciplined approach to the market.
> Frankly, the customer didn't care about prices; Neil's support and relationship were more important.

All of this presumes that tight control of prices at the resale level is in the best interests of the producer. It may be, however, that price competitiveness at the resale level is necessary to preserve the producer's market share, particularly if its suggested list prices are relatively high. Accordingly, any measures that the producer may take to secure conformance to any suggested pricing schedule should be informed by a knowledge of price competition at local market levels.

SUMMARY

Designed to reach a range of market segments in cost-effective ways, multichannel distribution systems nevertheless generate inter- and intra-channel rivalries. Generally, it is in the producer's interest to contain intra-brand competition among different classes of resellers for the business of particular classes of accounts and to discourage intensive intrabrand resale price competition.

Mitigating competitive rivalry among and within channels is considerably facilitated if the producer has widely accepted product lines and a strong end-market presence. Such presence is needed, first, to obtain relevant market information for purposes of formulating tactical plans and, second, to work with distributors in influencing their sales programs. Strong field sales operations are also important in monitoring, to the extent possible, the flow of goods through distributors to users, particularly under producer-negotiated contracts.

The extent of conflict in multichannel systems depends also on the structure of the system. Producers may have to make some choices among channels that are inherently in conflict—such as small, local industrial distributors and large, national discount chains—and between more or less intensive distribution. In addition, effective channels management requires that any boundaries drawn among channels in a multichannel network be enforceable. That is, the delineation of selling responsibilities ought to be consistent with the way user-customers buy, and ought to reflect the natural capabilities, market segment orientations, and strategies of different types of intermediaries. In other words, the strategy should be implementable at the field level.

This discussion has stressed the significance of pricing inconsistencies as a prime source of inter- and intrachannel price rivalries. Clearly, a central element in any program to constrain price contention at the resale level must closely integrate distributor quantity/price schedules, suggested resale prices, and prices arrived at through producer negotiations. In practice, however, inconsistent pricing tactics often result from the needs of manufacturing plants to maintain steady levels of production and to move inventories into the channels.

Finally, constructive channels relationships are built on the clarity and stability of the producer's distribution strategy and practices. The most successful firms have consistent marketing policies, coherent distribution systems structures, and a strong market presence. In addition, they seem carefully to avoid preferential treatment among channels in the interest of enlisting the full cooperation of each.

In the next chapter, we treat still another category of interchannel conflict: authorized versus unauthorized channels competition and the resulting emergence of so-called gray markets, in which goods may be purchased at prices substantially below manufacturers' suggested list prices.

APPENDIX
A NOTE ON CAPTIVE DISTRIBUTION SYSTEMS

The conflicts that develop in the independent distributor–captive distributor rivalry are reflected, as well, inside the producer's organization as different internal constituencies support one channel or the other. In fact, captive distribution business units typically have an unclear mission, with their dual objectives—maximizing sales, profits, and return on investment; and maximizing the sales revenues of their sister product departments—often at odds with strategic and operating decisions.

Given these conflicting performance measures, the captive sales organization's contribution to corporate profits relative to its use of corporate assets is often questioned by product department managers. On the other hand, captive distributor managers often express the view that the strength they derive from the parent-company affiliation may be more than offset by the competitive constraints imposed on them by operating in this larger corporate context. The following discussion focuses on WESCO, one of the largest and best-known captive distribution systems in the world, to illustrate the issues often encountered in supporting a captive distribution operation, developing and implementing its strategy, and assessing its performance.

In 1985, with 243 branches in the United States, 37 in Canada, 4 in Saudi Arabia, and 1 in Singapore, WESCO was believed by industry observers to be the second largest of four full-line national electrical distributor chains in the United States. The other three were Graybar, with 1985 sales of approximately $1.6 billion; General Electric Supply Company (GESCO), with sales estimated at about $.94 billion; and Consolidated Electrical Distributors (CED), with 1985 sales estimated at $.69 billion. GESCO, like WESCO, was also a captive distribution system.

WESCO began operations in 1922 when George Westinghouse acquired seven bankrupt independent distributors to maintain Westinghouse's market access in these particular trading areas. WESCO continued to grow by acquiring other financially troubled distributors and by developing its own branches in some other locations. As of 1985, about one-third of WESCO's sales were of the products of its Westinghouse sister divisions; two-thirds consisted of the noncompeting lines of over 600 electrical manufacturers.

Certain role constraints in this corporate context are expressed in these comments from one WESCO manager:

> First, part of our mission is to sell Westinghouse products, and we rarely carry competing lines. Westinghouse products are high quality and well respected but not always the best, in price or per-

formance terms, in a given product area. But unlike an independent distributor, we can't shop around for another supplier.

Second, unlike an independent, we can't threaten to go elsewhere if a product division doesn't meet our requirements on payment terms or stocking levels. That decreases our negotiating leverage with the product divisions in comparison with independent distributors.

In addition, certain practices tended to impact WESCO's financial performance. One WESCO manager noted,

> I feel we incur what you might call "good samaritan" costs as a captive distributor. For example, for many product divisions we gear our purchases to complement their manufacturing cycles, not market conditions. Therefore, we carry a higher average inventory on many Westinghouse lines in comparison with the independent, who can source more opportunistically from a number of suppliers. We, however, are tied to the manufacturing cycle of the product division, which hurts our inventory levels even as it helps the product division's throughput.
>
> I also believe some product divisions tend to take WESCO for granted and give us less service support than they give to their independents. We provide more functions for the product division than the average independent distributor does.

Further, corporate policies affected WESCO. According to one WESCO manager,

> Our employees are covered by the Westinghouse benefits package, which is more generous than the benefits offered by most independent distributors. In turn, the WESCO branch pays for the benefits. Up-front salary, however, is better at many independent distributors than at WESCO. We lost some younger reps, who are generally more concerned with the cash compensation than with longer-term benefits.

Finally, a Westinghouse corporate executive suggested the following:

> Westinghouse's investment in WESCO is largely investment in inventory and receivables. As a result, our investment in WESCO is more fluid than investments made in manufacturing divisions.
>
> Overall, I believe the relevant numbers and criteria for judging WESCO's performance are different than the criteria usually employed to judge the performance of our manufacturing units. There are "hidden economies" embedded in a captive distribution operation, an intricate mingling of costs and benefits. And any specific

decision affecting the relationship between WESCO and the product divisions must be made within this context.

In effect, this last comment suggests that evaluating the performance of a captive sales operation against that of an individual profit-centered product department is like "comparing apples and oranges." While corporate managements must assess the profit performance of all business units, it is useful as well to take account, on the one hand, of the relative benefits of having a captive distribution arm and, on the other, of how the corporate context may both strengthen and constrain this operation in competing against independent resellers.

From the parent company's point of view, perhaps the single greatest benefit the captive network offers is that it is a significant distribution channel from which competing products are essentially excluded. Also to be counted as an important benefit is market feedback. In addition, the captive branches may provide valuable experience for managers destined for product department and corporate responsibilities, and for those charged with managing distribution networks. Over time, the captive system may serve as well to set performance standards for independent resellers and to be a testing ground for resale marketing programs. Finally, it may help to ensure market access if the independent reseller network in some trading area fails to secure the producer's market share and/or the independents are being acquired by competing producers to form their own captive distribution systems.

In a company that is also heavily reliant on independent resellers, however, the existence of a captive distribution arm that is perceived as competition by the independents is often a bone of contention. The independents allege that the captive reseller receives favored treatment on supply allocation, service support, market information, and prices. Managers in the captive distribution business unit believe, on the other hand, that the product departments are more often biased in favor of external channels, since the internal network can always be counted on as a market outlet.

From the captive branch's point of view, there are certain offsetting benefits. In holding its own competitively against its independent counterparts, the captive branch may count as an advantage its access to the financial and technical resources of the parent corporation. It often operates as well under the aegis of a well-known brand name. In addition, simply because of its scale, the captive unit may have opportunities for distribution efficiencies that are not available to the smaller independent, such as the computerization of its inventory management and order receipt and processing systems.

On the whole, however, sales branches in captive sales organizations face significant competitive disadvantages. First, since they may well have come into the captive network as financially troubled private businesses or as new sales locations in "uncharted" trading areas, they are frequently not located in prime market areas.

Second, the relations of the captive sales organization with its external suppliers are often strained. The latter are concerned about the captive's access to proprietary information of competitive significance, which may find its way to the captive's sister product divisions.

Another weakness of the captive is that, in building customer relationships in local trading areas, the management suffers from being considerably more transient than its counterpart in the independent business. Often the career paths of captive distributor personnel are such that their tours of duty in field branch locations are relatively short. Meanwhile the managers of the independent reseller firms and their families are closely identified with the community and benefit from any propensity on the part of the customer environment to favor local enterprise.

Managers in captive systems often feel constrained, as well, by perceived limits on their ability to set prices and even to secure sales through extensive customer entertainment. The parent corporation may impose restrictions on both counts out of concern for avoiding legal charges of price discrimination and also to maintain the integrity of its pricing practices throughout the system. According to one WESCO manager, "The independent also has more pricing flexibility than WESCO branch managers do. WESCO management sets margin guidelines. The independent can go after lower-price business on a loss-leader basis, where the WESCO branch can't."

In the captive distribution systems that we have observed, the captive distributor's response to the conditions imposed by the corporate context seems to be to orient its local marketing strategies toward segments in which its constraints are less disadvantageous and its strengths have particular value. These are often segments in which customers tend to be larger in size, require technical assistance, and buy through competitive-bidding procedures. In these circumstances, the priority of price/value considerations in the purchasing decision and of technical support may weigh more heavily than personal relationships with vendor personnel. Such an orientation may better serve the purposes of parent companies' product departments if they may, at the same time, count on their independent distribution to cover their other market segments.

Thus, captive distribution systems may be sources of internal contention over such issues as business objectives, responsibility for results, performance, and measurement. Yet they may contribute importantly to the corporation's overall marketing effectiveness.

In companies that have captive distribution business units, however, the conflict issues may not be well defined, much less resolved. Line managers are then left with little to guide their day-to-day actions in mediating the differing interests of product departments, captive sales organizations, and external reseller channels. This suggests the need at the corporate level for being explicit about the strategic roles and objectives of captive sales operations and about the measures against which their respective performances will be assessed.

9 The Gray Market Dilemma

The gray market phenomenon—the selling of goods at discounted prices through resellers not franchised by the producer—is an exacerbated form of channel rivalry. As in other forms of channels conflict, the root causes often lie in (1) the differences in cost structures between full-line, full-service distributors and minimal-service resellers and (2) the arbitrage opportunities that arise out of producers' pricing practices.

Gray markets bring into conflict the direct salesforce, producer-franchised resellers, and unauthorized resellers. The latter are not constrained by franchise terms and conditions. Nor would withholding producer support be effective in dealing with gray marketers inasmuch as such support is neither given in the first place nor sought by them.

The discussion that follows begins with a case history: the dilemmas faced by managers at Control Data in dealing with gray markets for disk drives sold to OEMs, VARs, and computer dealers. This situation will set the stage for a consideration in this chapter of the reasons gray markets develop and their costs and benefits—the latter often not openly recognized. The chapter concludes by discussing the actions producers may find useful in maintaining the delicate balance between preserving the strength of authorized channels in the face of unauthorized discounter competition, on the one hand, and maximizing sales revenues by using all possible channels, on the other. In the long run, the answer lies with customers, what product-related channels services they want, and the relative priorities they give to price, purchase convenience, and source reliability. However, the issues may be immensely clouded in the emotion and power politics of ongoing producer-reseller relationships, as well as in the sometimes conflicting measures and motivations of operating managers.

GRAY MARKETS FOR HIGH-TECH PRODUCTS

In June 1985, James Ousley, vice president for OEM Marketing at Control Data Corporation's Peripheral Products Company (PPCo) division, was reviewing with division management the performance of PPCo's disk drive product line. Prices for disk drives had fallen substantially during the past few years. At the same time, new markets had emerged, and PPCo had added indirect distribution channels to reach these markets. However, PPCo increasingly found its various selling programs not working as originally planned: channels often competed for the same customers, and a thriving gray market had emerged.

Traditionally, PPCo had sold through a direct salesforce only, but in 1979, it signed agreements with two national distributors of electronics components, Arrow Electronics and Kierulff, Inc. One executive explained,

> Our main thrust is the 30 largest computer OEMs, where a single account can generate tens of millions of dollars in annual sales revenues. However, low-end peripherals became increasingly important and also experienced dramatic price reductions. Therefore, alternate channels became a way of making sales of these products more cost efficient to all the many smaller customers that emerged as the price of peripherals and computer systems fell.

Some of PPCo's customers purchased drives at significant volume discounts, but rather than "adding technical value to the products" as required by CDC's OEM purchase agreements,[1] they resold the "raw" drives to major computer dealers, smaller systems houses, smaller OEMs, and some end-users. Hence, the gray markets. These firms, often called "pseudo-OEMs," purchased under the terms of an OEM agreement. They typically operated on low gross margins (5–10%) and undercut the prices of distributors and other resellers of PPCo's drives.

A PPCo executive commented as follows:

> We first noticed these pseudo-OEMs cropping up a few years ago, when demand was strong, and we had to put customers on al-

---

[1] Article 1 of CDC's "Terms and Conditions of Purchase" reads as follows:
ARTICLE 1 *UTILIZATION OF PRODUCTS*
　　To determine prices applicable to this Agreement, Customer certifies that the Products will be integrated, adding technical value to the Products, into the Customer's systems or subsystems which are sold or leased to others.

location for drives. Some firms began buying in larger quantities and sold a percentage of the drives "raw" to other customers. They usually sold to mid-range accounts, buying from about $100,000 to $1 million annually, and not to the major accounts, who already received big volume discounts and naturally received high priority during allocation periods.

When demand was strong, the pseudo-OEM resold the drives at a profitable premium, because customers were concerned about supply and faster delivery of product. That motivated the gray operator to purchase and sell more. And frankly, we were not overly concerned: it was a small amount of total sales being diverted in this manner, and in those market conditions, we even felt the pseudo-OEM performed an indirect service by acting as a kind of "second source of supply" for our drives at some smaller accounts. When demand weakened suddenly in late 1984, the gray market became a bigger problem. Customers became more price sensitive and the pseudo-OEM, operating with low overhead and offering little or no credit, could offer lower prices than our distributors and still make money.

This gray market grew gradually. It's difficult to know just how much is going through pseudo-OEMs, but we estimate that they account for about $30 million of our total disk drive sales.

By 1985, concerns about the gray market were widespread throughout PPCo. A PPCo distributor sales representative labeled the gray market "my number-one problem and getting bigger":

When one of our distributor's branches loses a sale to a pseudo-OEM, two things happen. First, the branch manager lets me know about it in seven different languages. Second, that distributor is less motivated to push our product. A large part of my job is motivating the distributor's salespeople to spend time on our products. Months of hard work can be shot to hell by these pseudo-OEMs.

An OEM sales rep emphasized the impact on pricing and account relationships:

I first encountered this when a customer showed me an invoice that offered our product—with exact specifications—for nearly $1,000 less per unit than the price I was offering. My first reaction was to suggest that the invoice contained a typographical error: I thought a zero had been omitted! But I checked out the order and had to go back to the customer and offer an embarrassed apology and a lower price.

District sales managers (whose bonuses were based on total district sales volume) emphasized the complexity of the situation and the impact on morale. One district manager commented, "I get different stories from salespeople about pseudo-OEMs, partly because some salespeople now depend on these resellers for a significant portion of their quotas and some do not. The result is some tension within the office, and often between people who share the same cubicle."

A regional sales manager noted that the severity of the problem differed between regions: "Many of these pseudo-OEMs are on the West Coast because that's where most drive manufacturers are located. But they sell nationwide. So often our western offices will book sales to the pseudo-OEMs, and we in the East will feel the brunt of the price competition. I sometimes feel I'm being shafted by my own people, and I don't like it one bit."

In OEM marketing, a manager explained that identifying a pseudo-OEM was not a straightforward task:

Some of these gray marketers are brokers who buy drives, stock them in a garage, and sell the lot to whomever they can find. But others are systems houses or value-added resellers who may be selling just 10–20% of their purchased drives without adding value. And in the current conditions of depressed demand in the computer industry, we've even seen some larger OEMs dumping excess drives on the gray market.

These pseudo-OEMs are a diverse group, and some are legitimate, important customers for our products. It's a tough situation. Short-term, the gray market happens incrementally with good business and it moves product. But longer-term, we wind up competing with ourselves at a lower price.

One plant manager with profit center responsibility, however, emphasized that "the gray market is a problem, but should be put in perspective":

Much of the gray activity is in the 5.25-inch drive categories, and especially in the floppy drives. This is an area where product differentiation is hard to achieve and which is quickly being superseded by new product technologies. Frankly, pseudo-OEM sales in that product category utilize capacity for products that, as far as major OEM customers are concerned, are near commodities.

When our salespeople encounter pseudo-OEM competition, they should stress to the customer that the gray marketer is not a reliable source of supply and that his units are not covered by our warranty agreements. Salespeople can also emphasize that CDC can sell the entire package of products the customer needs, whereas the pseudo-OEM cannot.

Caught in these circumstances, PPCo managers discussed three courses of action. First, some said that PPCo should identify the pseudo-OEMs and refuse to sell to them. Serial numbers could be used to track PPCo drives bought on the gray market, and sales be shut off to the pseudo-OEM. These managers argued that even though PPCo might lose some sales in the short term, if the offenders were cut off, franchised distributors would probably sell more products. In addition, shutting off gray market supplies might mitigate the price erosion and so perhaps increase profitability as well. Other managers disagreed with this analysis. Some argued that, if cut off from PPCo's drives, pseudo-OEMs would simply sell other vendors' drives. Others felt that the time and costs of tracking down gray marketers would be burdensome.

Second, some managers proposed that as the current contracts with identifiable pseudo-OEMs expired, these customers should be reclassified as "independent selling organizations," with the relevant price schedule being the same as that for distributors. If gray marketers paid the same price for drives as industrial distributors, they contended, the pseudo-OEM's competitive price advantage would be lessened.

Third, some managers suggested that purchase quantity, rather than type of customer, should be the sole criterion for determining PPCo's selling price. These managers wanted PPCo to sell its drives at the same price (for a given quantity) to all classes of customers—OEMs, distributors, large retailers, and commercial distributors. One manager explained that this pricing tactic would benefit PPCo over the long term: "The channel that's most efficient, and that can sell our product at the lowest gross margin, should gain strength over time. And that's the channel we want to use to get our products to market."

MANUFACTURERS' CONCERNS AND OPPORTUNITIES

The Control Data case illustrates the range of concerns raised when products move out of their intended channels and are sold at prices below those of authorized channels, in this instance the direct salesforce and franchised distributors. The litany of frustrations includes those of the PPCo sales reps losing business on price to pseudo-OEMs and those of marketing managers unable to control the sales of disk drives sold to OEMs. There is a concern as well that gray market prices are effectively establishing market prices at levels well below what they might have been, and in so doing are not only lowering product margins but damaging CDC sales reps' credibility with customers.

Another major concern of producers is protecting their franchised full-service channels from discount price competition. Unauthorized channels, selling at discount prices, mean lost sales for authorized distributors. In addition, full-service distributors may incur selling and promotional costs in support of the producer's product line only to find the demand thus created accruing to the benefit of gray marketers selling only on price and offering little or no service. As one computer dealer complained, potential customers become "tire kickers: they come in and breathe my oxygen, manhandle my equipment, soil my rug, and don't buy! Is it fair?" Such dealers are often strident in their demands that their supplier "do something" about the offending "free riders."

There is the further concern that user-customers, not able to rely on gray market sources for presale information, product warranties, and after-sale service, may transfer their dissatisfaction to the brand itself and be lost to competitors for repeat purchases.

Nevertheless, gray market channels may also provide the only suitable vehicle for moving out older models and those which have reached the commodity stage, in which price becomes the primary competitive weapon. In any case, while gray market pricing erodes margins, it may also respond realistically to market price competition and thus serve to maintain brand shares. Producers that closely monitor resale prices in the interest of maintaining price stability may foster some degree of price inflexibility and lack of ability to adjust to market fluctuations.

Finally, it may be argued that gray marketers' sales might not have been available in any case to authorized resellers. It may be contended that gray marketers and full-service dealers serve sets of buyers with different needs and priorities, and for the producer not to be represented in both market segments is to forgo sales revenues in one or the other.

ROOT CAUSES

While the emergence of gray markets is widely deplored among producers, the genesis of this uninvited participant in the distribution system may in part be due to producer marketing practices.

Supplier Pricing Policies. This is probably the most commonly cited cause of gray markets and the source often identified as "the real culprit."[2] Manufacturers tend to structure their discount schedules in favor of large orders, which causes their distributors and other customers to buy more than they can sell or use and then to move the rest "sideways" to unauthorized resellers. Differential pricing by type of transaction (e.g., competitive bidding, published discounted list prices) and by class of trade (e.g., reseller, OEM, VAR) creates arbitrage opportunities, difficult for purchasers to resist, especially if burdened with excess stocks. Moreover, steeply graduated quantity discount schedules encourage overordering. In addition, if goods are in short supply, the producer may require that orders be placed with lead times of six months and longer, forcing buyers to make commitments against future needs that may be difficult to forecast. Such commitments may be locked in by formidable penalty clauses for canceling orders. Control Data, for example, required disk drive buyers to place firm orders one to two years ahead, with cancellation charges of 40% if the order was withdrawn 45 days or less before the scheduled shipment date, and 15% if cancellation occurred 46–90 days before shipments.

The producer's production planning needs—the cost advantages of stable manufacturing schedules and the learning curve advantages achieved through high throughput volumes—underlie these sales conditions. The likely result, however, may be to fuel gray markets and to put downward pressure on market prices as buyers move excess supplies into unauthorized channels.

Price arbitrage may also result from differentials in currency exchange rates as products move across national boundaries. During 1984–1985, for example, when the U.S. dollar was strong and steadily appreciating relative to many foreign currencies, products

[2] See, for example, R. Howell, R. Britney, P. Kuzdrall, and J. Wilcox, "Unauthorized Channels of Distribution: Gray Markets," *Industrial Marketing Management*, November 1986, pp. 257–263.

initially sold in one country were imported by agents to other countries. These so-called parallel imports are an instance of gray markets on an international scale. Gray marketers regularly imported into the United States Caterpillar excavators and loaders that had been manufactured in plants, and sold through distributors, in Scotland, Belgium, and Japan. Even after shipping and insurance charges, these resellers were able to sell to Caterpillar's U.S. dealers, which were paying 15% more for the same equipment manufactured in Cat's U.S. plants. At prices in excess of $200,000 for a single large excavator, these price differentials—caused by the strong dollar and the company's pricing policies in its various country subsidiaries—sustained a thriving gray market for Caterpillar products.[3] As more manufacturers move toward global product strategies, with uniform goods and even multilingual packaging for all world markets, this type of gray market has more opportunity to develop.

Reseller Cost Differentials.   As well as prices, differences in resellers' cost structures facilitate gray markets. These cost differentials can be of two kinds: differences between the operating costs of full-line, full-service resellers versus narrow-line, low-service discounters; and cost differences generated by the place of a product line in a given reseller's selling strategy.

Franchised resellers tend to provide such functions as advertising, product demonstrations, and point-of-sale as well as postsale services. These functions add to the selling costs of franchised dealers and at the same time serve to support the gray marketer's free-rider strategy. It is not uncommon, for example, for gray marketers to honor manufacturers' factory rebate coupons and coordinate their activities with manufacturers' promotions as a means of "skimming" the benefits created by the authorized channel's marketing activities.

This situation is sometimes aggravated by certain manufacturers' commitments to customer service. As part of its franchise agreements with its authorized resellers, for example, IBM required these resellers to allocate a certain amount of store space for product demonstrations, keep a certain number of store personnel trained in its equipment, and stock parts sufficient to provide an acceptable level of customer service for its line of personal computers. Mean-

[3] See "The Nice, Gray Cat," *Forbes,* May 6, 1985, p. 31.

while, retailers such as 47th Street Photo might not meet these conditions, and other unauthorized resellers ran essentially cash-and-carry operations. (For example, the person reported to be Canada's single largest PC gray marketer, with total sales volume in the millions of dollars, was a practicing dentist three days a week.)[4]

IBM sought through its franchise agreements to build a store of value in its product franchise—to build brand preference and encourage repeat purchases. But the gray marketers, through their low-service/low-price selling strategies, were in effect able to "expropriate" a significant portion of this value, lowering the worth of the product franchise for the authorized resellers. Or, as one authorized reseller succinctly put it, "It's tough to compete with someone whose major selling expense is a phone bill."

Similarly, differences in reseller selling strategies can sustain a gray market. Resellers who view a product as an incremental addition to their lines may not take account of overhead costs in pricing the product. So-called loss-leader products are an extreme instance of this situation. Loss leading, the practice of inducing customers to shop at an outlet by advertising products at cost or below-cost prices, is based on the theory that although the reseller may take a loss on the advertised goods, it will recoup it on the sale of other goods to these same customers. Loss-leader tactics are common among gray marketers of a product, particularly when the product has a well-recognized brand name.

By contrast, authorized resellers, having made regular investments in the inventory and personnel required to sell and service the product, look to the name product to perform a different economic role in their operations. In addition to traffic building, the product is expected to generate a sufficient level of profits in its own right, and the resulting markup generally reflects allocated overhead and any brand-specific investments by the reseller, as well as direct costs.

Supplier Franchise Practices. Gray markets may be fostered, as well, when suppliers practice highly selective distribution, thus franchising one or only a few resellers in each trading area. In so doing, the supplier lessens any tendency toward intrabrand price competition among authorized resellers, creating a price umbrella

---

[4] Wayne Lilley, "The Graying of the Marketplace," *Canadian Business,* August 1985, pp. 46–54.

likely to draw nonfranchised dealers into the market if demand is strong.

In some cases, too, producers practice "block franchising," in which the resale enterprise is granted a franchise with the right to extend it to every sales location in the chain. The result, particularly if a number of such franchises are given to large chains, is a loss of control over the producer's representation in each trading area. In addition, the large dealers in the chain will tend to obtain their supplies of the product through gray market channels to avoid the producer markup typically put on the product.

Marketing Practices.   Gray markets can also develop when producers take on a range of marketing functions that might otherwise be assumed by full-service dealers, such as warranting the product, providing after-sale repair and maintenance services, extending credit, and fulfilling user-customer orders from their own stocks. The first two functions may be assumed by producers that seek to ensure customer satisfaction. To the extent, however, that the producer performs sales-related functions, it reduces buyer risks in dealing with minimal-service resellers. In addition, heavy advertising and promotion by the supplier is likely to benefit gray marketers in two respects: it may enhance total demand, and it is likely to build the product's brand image as something on which buyers may rely for ensurance of quality and dependability, thus decreasing their dependency on the reseller's reputation in making the purchase.

Performance Measures.   Differences in the standards applied to various producer operating units contribute to the ambivalence producers project in their gray market policies. On the one hand, sales personnel, measured in terms of sales quotas, will campaign strongly against gray marketers if the latter siphon off revenues that might normally accrue to their credit. On the other hand, those producer personnel who benefit from sales through gray markets, in Control Data's case sales reps and district managers selling to pseudo-OEMs, may sit quietly while others complain. Marketing managers charged with developing and maintaining the franchised distribution network will weigh in against the "unfair competition" of the unauthorized channels. However, plant managers with pricing authority and profit responsibility, again as at Control Data, may say, "Frankly, pseudo-OEM sales . . . utilize capacity for products that, as far as the major customers are concerned, are near-commodities."

Internally inconsistent and conflicting goals thus pose yet another obstacle in establishing a clear and consistent channels policy.

WHAT ARE THE OPTIONS?

Faced with a gray market, manufacturers have tended to adopt one or more of the following kinds of responses. Each has its rationale and limitations.

Disenfranchisement of Offenders. A common (and often emotionally satisfying) response is to disenfranchise those distributors found to be trading on the gray market. For many products, serial numbers can be used to identify the sources of units sold to gray marketers, and consistent with the terms of many purchase contracts, disenfranchisement is a legally valid alternative for the manufacturer. In pursuing this course of action, the manufacturer's management often reasons that while the firm may lose some sales in the short term, it may also sell more units through its remaining authorized distributors and also mitigate the price erosion caused by the gray marketers' sales.

Over the past few years, for example, IBM has cut off several dozen dealers found to have sold the company's personal computers to unauthorized dealers.[5] Similarly, Lotus Development Corporation established an automated tracking system in an effort to cut down on gray market sales of its popular Lotus 1-2-3 software product. To track supplies flowing to unauthorized dealers, Lotus recorded the bar code numbers of software packages as they were shipped to distributors. Meanwhile, certain Lotus managers checked print advertisements to identify unauthorized dealers and, from time to time, sent "mystery shoppers" to buy the company's products from mail-order houses and other unauthorized outlets. Eventually Lotus eliminated many dealers from its authorized list and, for a time, put a freeze on the authorization of new dealers.[6]

Such moves send a strong signal of commitment to those authorized distributors which do abide by the terms of the franchise agreement. Such full-service dealers can be especially important to a manufacturer that, in a market characterized by frequent new product introductions, will require dealer support of its market de-

[5] See Anthony Ramirez, "Blue vs. Gray: IBM Tries to Stop the Discounters," *Fortune,* May 27, 1985, p. 79.
[6] "Lotus Looks to Stop Gray Market Sales," *Computer Systems News,* March 25, 1985, p. 4; "Lotus Attacks Gray Market with Computer Sleuth," *Sales and Marketing Management,* April 1, 1985, pp. 135–136.

velopment efforts at a later date. Thus, disenfranchisement of offending distributors is often a response to the very vocal concerns of the remaining, "law-abiding" distributors.

Nonetheless, such moves also entail real costs to the manufacturer—financially, administratively, and in terms of potentially significant losses of market share. First, in many markets, the time and administrative burdens of identifying offending dealers can be onerous. In addition, "mystery shoppers" who track serial numbers can still run into the "human element" in their investigations. A manager at one manufacturer noted that many suspected gray market dealers often provided explanations such as "Oh yes, Mr. X sold those units against my orders and that's why he was fired last month." In other cases, records of sale may be falsified. Second, while contract terms may prohibit authorized dealers from selling to gray marketers, if the manufacturer disenfranchises offenders selectively, then it runs the risk of costly and protracted countersuits, based on charges of discrimination, from the terminated dealers. Third, tracking gray market sales is often costly. In the case of Lotus, for instance, the tracking system designed to label and monitor sales of its products cost more than $100,000 to develop.

Finally, disenfranchising offending dealers, while it may help to solidify relations with the rest of the distribution network, can also mean a loss of market share in a business where rapid market penetration has implications for longer-term competitive advantage.

"One-Price-for-All" Policy. Another possible response to the gray market is for the manufacturer to abandon quantity discounts. Here, the reasoning is that since the manufacturer's pricing schedule is a cause of gray markets, and since disenfranchisement of offenders is often expensive and/or ineffective, a "solution" is to sell to all customers at the same unit price.[7] This pricing strategy can eliminate an important source of the arbitrage at the heart of gray markets and allow the manufacturer to reassert a measure of channel control. Further, this one-price strategy is often viewed as a means of "rationalizing" the distributor network. As one manager in the disk drive industry explained, "Since it is different prices for different order quantities that fuel the gray market, let's eliminate

[7] For a more detailed discussion of the "one-price-for-all" option, see R. Howell, R. Britney, P. Kuzdrall, and J. Wilcox, "Unauthorized Channels of Distribution," pp. 257–263.

the differences. The distribution channel that's most efficient and that can sell our product at the lowest realizable gross margin should gain strength over time. And that's the channel we want to use to get our products to market."

In practice, however, a one-price-for-all policy often means the manufacturer sells most of its output at lower prices to all customers, big or small, regardless of the transaction costs. Further, this pricing strategy can often foreclose valid price-differentiation opportunities among different classes of customers that are purchasing very different benefits with the physical product. For some customers, for example, continuity of supply and assistance in applications development may be very important, while for others, price for the basic product may be the dominant purchase criterion. Thus, a meaningful one-price-for-all strategy must also include means for rewarding the larger, full-service dealers in the distribution network. Such dealers are usually very important to a manufacturer, and often have other supply options available.

Finally, as the comments of the disk drive manager indicate, a one-price-for-all strategy often implies a Darwinian "survival of the fittest," approach to one's distribution network. While this approach may indeed prune, or rationalize, the manufacturer's distribution network in the short term, it can limit the manufacturer's access to new segments and minimize distributors' incentives to support a product in the long term. Thus, a one-price-for-all strategy, while limiting arbitrage opportunities, may not eliminate them (since, as noted earlier, cost differentials and different selling strategies among resellers are also causes of gray markets) and may also make it more difficult for the manufacturer to react to changing market conditions.

More Intensive Distribution.   "If you can't beat 'em, join 'em!" is a time-honored response to a dilemma, and in many respects adding distributors (perhaps former gray marketers) to the dealer network is an illustration of this adage in action. By limiting distribution of its products to a few authorized dealers, a supplier with a well-recognized brand may inadvertently create demand that can only be supplied to certain segments through transshipments to unauthorized dealers. Similarly, a supplier's dealer-service criteria may be unnecessarily (or unrealistically) high, discouraging dealers who serve more price-sensitive market segments from becoming part of the supplier's network. In both of these situations, franchising more distributors might allow the supplier information and control over the flow of its product to market.

On the other hand, to the extent that the new authorized dealers would serve the same customer base as the original dealers, the increased competition would lower the value of the supplier's franchise and reduce dealer incentives to provide added value and sales support.

TAKING ACTION

Each of the remedies discussed above has its disadvantages and comes at some cost—in policing, in forgone profits, or in both. However, the producer can take actions to make manageable the problems gray markets pose in the distribution system. These actions include gathering relevant and timely information on gray market activity, coordinating to-the-market pricing, establishing franchise policies to give support to authorized resellers, and rationalizing internal performance measures.

Sources and Uses of Information. For many industrial goods producers, the magnitude of gray market selling is difficult to assess, and offending sources of gray market supplies hard to identify. It is important to have what current information can be gathered on both counts even though it is unlikely to be complete and accurate.

Many types of industrial products can be identified by serial numbers that are traceable as goods flow through channels, for purposes of identifying gray market feeder channels. In addition, warranty card returns and factory rebates may help to assess the volumes going through unauthorized resale channels and also to trace individual shipments (by coding serial numbers according to product lot numbers).

The costs and reliability of such data will vary. For IBM personal computers, for example, high transaction volumes, as well as the number of hands through which the product passes, complicate the process of tracking product flows. Tracing Caterpillar tractors would be less difficult.

Although there may be a myriad of lesser offenders transshipping sporadically to clean out excess inventories, we can speculate that the major sources of goods to unauthorized channels are few in number. If so, suppliers may be able to cut off these supplies quickly and efficiently and diminish gray market product flows—with timely information.

Coordinated Pricing

In the preceding chapter, we pointed to differential pricing by type of customer and quantities purchased as the primary source of inter- and intrachannel price competition. This assertion certainly covers price rivalry between authorized and unauthorized channels as well. The incentive to gray marketers is especially high when, within a pricing schedule, the pricing brackets are wide. In the disk drive situation, for instance, the manufacturer's pricing schedule provided for the same unit price (about $500 per unit) on orders ranging from 1,000 to 2,500 units or contracts ranging from $525,000 to $1.3 million. The wide price breaks encouraged some customers to purchase a large number of drives at a given unit price, use many for "legitimate" purposes, and still sell a significant number on the gray market at prices higher than their purchase cost but lower than authorized distributors' prices. "Tighter" pricing brackets, in which unit prices are tied more closely to specific order volumes, could mitigate the gray market problem in such situations.

Competitive-bid pricing, customer-class price distinctions, and quantity-bracket pricing inevitably create incongruities and the opportunities for price arbitrage among different channels. Producers may seek to isolate market segments by such devices as requiring buyers to "add value" to the product before selling it, proscribing sales from one reseller to another, and monitoring quantities shipped in fulfillment of contracts negotiated at low prices. These actions may be helpful in managing product flows through channels to end-users but are seldom totally effective. Field sales personnel may have neither the inclination nor the motivation to police their customers' purchases and dispositions of quantities of products.

Again, it may be useful to have in the hands of a pricing coordinator information on the producer's stream of price quotations on bids as well as prevailing market prices to users through resale and direct channels. It may then be possible to coordinate otherwise diverging price patterns across markets and to let these data inform both competitive-bid setting and revisions in discounted list prices to resellers.

Franchise Support

If authorized and unauthorized channels compete on value-added services and low price, respectively, and if the weakening of authorized channels under the ravages of price discounting are of concern, then producer support programs for full-service franchisee resellers may help to strengthen this component of the distribution system. We have cited Lincoln Electric's Flagship Distributor program in Chapter 5 as an example of qualifying selected distributors for producer support and, in return, gaining their commitment to meet certain performance standards. This approach may be particularly useful in meeting the gray market challenge.

Such programs are difficult to implement and the channels system difficult to manage, however, if the producer loses control of franchising. While qualifying each sales location of each franchise is certainly greater in personnel time and costs than block franchising, the investment may be small relative to the benefits of more rational distribution planning by trading area and producer communication with these powerful links to user-customers.

Performance Measures

Internally inconsistent criteria for measuring performance among sales, marketing, and production personnel set up organizational conflict that may often impede taking action, much less deciding what course of action would serve the business's overall best interests. It is important, then, first to identify different management viewpoints and the performance appraisal contexts out of which they come.

A second step is to lay out the range of possibly conflicting strategic objectives the business might pursue and relate them to feasible courses of action. If, for example, increasing market share remains an important objective, then the extent to which gray market outlets increase product availability and produce incremental sales volume becomes an important factor in responding (or not responding) to the complaints of authorized distributors against gray marketers. By contrast, where major account relationships are affected by gray market activity, the complaints of individual field salespeople carry special significance. In the disk drive example, a few customers accounted for the majority of sales volume, and some were adversely affected by gray market sales of drives. Hence, the company felt compelled to respond forcefully to pseudo-OEMs, despite the impact on its manufacturing operations of losing gray market volume. Another company, intending to introduce new products through its distributor network in the future, placed special weight on maintaining good dealer relations and so adjusted its pricing to dealers, despite the current volume represented by gray market sales.

If business goals and overall measures of success can then be translated into functional department goals and measures, it may be easier for managers with stakes in the outcomes to see the implications for product distribution and to choose among optional courses of action.

SUMMARY

A wide range of interests affects what to do about gray markets. In the external environment, they include those of franchised full-service resellers, unauthorized dealers, and OEM customers. In each category, too, are those who win and those who lose. The latter may include franchisees and OEMs, which lose out to discounted price competition. The winners are those who buy from their producer sources in large quantities at low prices and then transship the excess to unauthorized channels. The gray marketers, too, are among the winners; they benefit from the demand generated by producers and their full-service resellers to tap a market segment for which low price has high priority. These are the so-called free riders.

Internally there are winners and losers, as well: field sales personnel who gain credit for sales into gray market channels and those whose sales performance and customer relations are damaged by gray market pricing; plant managers responsible for manufacturing costs and product profitability; marketing managers concerned about nurturing and gaining the support of their distribution networks.

In all of this, it may be difficult to see what courses of action are in the company's overall best interests. In the context of conflicting interests and with no clear course of action, the tendencies may be either not to take action or to respond to the more powerful voices—large OEM and reseller accounts, distributor advisory councils, major plant managers, or field sales managers.

Whatever the configuration of external and internal interests, decisions about what can—and should—be done about gray market selling are better informed by gaining some factual perspectives on volumes moving through unauthorized channels. Further, choices among possible courses of action may be more easily made if functional performance measures reflect common goals, that is, if managers in sales, marketing, and production gain when the business's revenues, market share, and profits grow.

In the long run, gray market activity may be a symptom of growing product-market maturity, increasing buyer sophistication, and changing buyer priorities. If so, then the response may usefully come, at least in part, in restructuring distribution. On the one hand, some subset of full-service resellers may be given special support to serve as a vehicle for new product introductions and to address those market segments for which reseller services have value. On the other, growing the distribution network to gain greater market access as demand increases might be an appropriate strategy. Discouraging the flow of goods into unauthorized channels may be aided by some degree of price rationalization across classes of customers and purchase quantities to minimize reseller arbitrage opportunities. It may be discouraged, too, by a monitoring program and by taking direct action against major sources of gray marketer supplies, either refusing to ship to them or reducing shipments to bring the quantities supplied into line with their needs for satisfying the demand from their "legitimate" customers.

In any case, coping with gray markets seems often to involve making tradeoffs: short-run versus long-run profits and market share growth; "force-feeding" channels and OEM customers to achieve manufacturing cost efficiency versus adjusting supply to fluctuations in market demand; maximizing revenues and profits from particular classes of customers versus optimizing revenues and profits across markets. The choices are better made, we conclude, when performance measures are consistent across functional managers, when there is relevant current information on market prices and product flows through channels, and when business goals, short-run and long-run, are clearly understood.

10 Coping with Change

Of all the elements in marketing strategy, the distribution system may be the hardest to change. Adjustments in the product line, in prices, in the promotional program, or in market focus cause far less upheaval than shifts in the ways the firm goes to market. Changes in distribution almost always disrupt existing relationships among suppliers, their salesforces, independent distributors, and agents. Further, they often seem to go against patterns of past marketing success. Even when marketing managers acknowledge the need for change, many questions attend a move to new modes of distribution. What new strategies are needed? How will customers, competitors, distributors, and company sales personnel react? How should changes be implemented?

In forming distribution systems, manufacturers make certain commitments. These include understandings between manufacturers and distributors about territorial bounds, the intensity of distribution in a trading area, discount schedules, and the role of the manufacturer's salesforce. Investment in a highly trained salesforce is itself a major commitment, as are distribution schemes specifying where, when, and how the manufacturer's line may be purchased by and serviced for users. Such commitments, while essential for success, tend to limit the supplier's ability to restructure distribution systems as product technology evolves and markets change.

Our purposes in this chapter are to review (1) the factors that create the need for change, (2) the reasons that change is resisted, and (3) the kinds of changes that manufacturers make in distribution (in the context of two case studies).

The two case situations we consider involve significant changes in the market environment, which created great pressure to restructure the distribution system. At General Electric's Component Motor Operation, a market characterized by intensifying competition and downward price pressure raised questions about whether this division could afford to—or could afford not to—continue to sell through its powerful master distributor channel. At Honeywell Information Systems, managers seeking entry into the burgeoning minicomputer market recognized the need for a transition from direct selling to multichannel distribution. Their concern was how best to minimize interchannel conflict during the reorganization. In each instance, we present the facts of the case and then comment, using the conceptual framework developed earlier in the chapter.

THE IMPERATIVES OF CHANGE

The factors that create market ferment and lead sooner or later to the restructure of distribution systems include the following:

- growth and maturation of existing product-markets
- the emergence of new product-markets
- significant market decline
- escalation of competition and intensification of cost/price pressures
- changes in customer demographics and modes of buyer decision making
- changes in the channels infrastructure—some reseller institutions growing in market power, others declining
- corporate mergers and the resulting opportunities for integrating the channels systems of two or more businesses
- "one-shot" opportunities to increase market share by taking over competitors' distributors

Factors such as those listed above, which create either the need or the opportunity for constructive change, are often related to product-market changes and, in particular, to the product life cycle phenomenon. In the early stages of market development for many industrial products, it is essential to stress the development of new product applications and the education of new users. Often this requires direct selling or selling through agents. In a second stage, the imperatives of making the product widely available at competitive prices tend to reduce the value of highly technical selling and increase the need for building a distribution network. The direct sales program may continue with some narrowing of focus, generally a concentration on those customers which have emerged as large and important sources of sales revenues.

In the later stages of the product life cycle, declining market growth rates, the increasing priority of price considerations in customer purchasing behavior, and the escalation of price competition—both interbrand and intrabrand—combine to put a premium on sales cost efficiencies. This, in turn, may force a change in distribution strategy, such as adding or dropping certain levels of distribution, redefining the respective roles of the direct salesforce and resellers, reallocating the responsibility for distribution functions across channels, and restructuring discount schedules.

Throughout the product life cycle, changes in market structure can trigger strategic response, an important element of which may be a change in the distribution system. Thus, for example, new government restrictions on Medicare reimbursements to hospitals fostered the rapid development of multihospital buying groups to negotiate for supplies at the most favorable prices. This in turn led manufacturers of hospital supplies to intensify their direct sales programs, leaving to distributors the order fulfillment role for these accounts.

Finally, marketing managers are often confronted with the need to rationalize two or more networks, either when companies have merged or when the possibility exists of annexing part of the distribution network of a weakened competitor. The potential exists here for gaining additional share of local markets. However, the transition entails a two-part challenge for the supplier: assimilating the new distributors is one part; the other is dealing with the resentments of existing channels members as the supplier's distribution system becomes more intensive.

IMPEDIMENTS TO CHANGE

While the pressure for change in response to market and competitive factors is often compelling, the deterrents to taking effective, and especially timely, action can be equally powerful. Resistance forces include the following:

- the uncertainty of the outcome
- the fear that change will invite retaliation from those who perceive themselves to be the losers
- the potential costs of lost sales revenues or increased sales expenses

- the fear of losing distribution in certain market sectors and leaving opportunities for competitors
- the loss of customers that may prefer sourcing arrangements available to them through the existing system
- a lack of urgency—in the absence of crisis, the tendency may be to accept the slow erosion in competitive position rather than to "bite the bullet"
- unwillingness to write off investments in the existing system—a technically trained direct salesforce, a strong reseller network—and in the personal relationships that have developed around these institutions
- a psychological commitment to past success and an inability to assess objectively the need for change

Many of these conditions are evident in the two case situations that we consider in the following discussion.

GENERAL ELECTRIC— COMPONENT MOTOR OPERATION*

General Electric's Component Motor Operation (CMO) was the largest domestic manufacturer of electric motors in both integral and fractional horsepower sizes. Manufactured in thousands of different models, these motors were used in commercial/industrial applications as well as in consumer appliances such as furnaces, air conditioners, washing machines, dishwashers, ranges, and refrigerators.

In consumer sales, the focus of this case history, CMO sold its line of fractional horsepower (FHP) motors both directly to appliance manufacturers as components in original equipment and through distribution for replacement use. FHP aftermarket motors went through distributors, wholesalers, and electric motor shops to more than 50,000 repair shops, dealers, and contractors providing appliance repair service.

Aftermarket Distribution

CMO used four channels to serve the FHP motor consumer aftermarket: (1) industrial distributors; (2) OEM service centers

---

* Quantitative data used in this case are disguised but serve, nevertheless, to illustrate the issues.

selling via OEM-franchised distributors; (3) three master distributors selling to wholesalers, which in turn sold to dealers and service shops; and (4) a master wholesaler selling primarily to service shops and dealers but also to other wholesalers (see Exhibit 10.1).

The three master distributors accounted for half of CMO's sales to the FHP aftermarket. They were the primary source of supply for 1,250[1] heating, air conditioning, and appliance wholesalers, which accounted for 45% of total FHP consumer aftermarket motor sales. Most wholesalers carried GE motors.

The newest and fastest-growing channel to the FHP aftermarket was Addison Electrical Supply (AES).[2] This channel, classified as a master wholesaler, accounted for 20% of all FHP aftermarket sales to service shops and dealers and 15% of GE's aftermarket consumer FHP sales. AES combined the functions of the master distributor and wholesaler, buying directly from manufacturers and selling directly to wholesalers, dealers, contractors, and service shops.

Evolution of the Channels System

During the post–World War II era, original equipment manufacturers served the aftermarket for consumer motors through their dealer and service shop networks. Prices in the consumer aftermarket were high, and demand was growing for appliance parts and repair services. GE's three master distributors emerged to supply the growing network of independent dealers and repair shops. GE in effect put the master distributors in business and supplied the great bulk of their consumer FHP motors for the aftermarket. The three master distributors resold GE motors to wholesalers, which provided a broad line of motors and other items for the independent appliance service dealers and contractors. In utilizing master distributors, CMO was able to eliminate the need for a direct salesforce calling on wholesalers and thus economize on selling costs. We can speculate, too, that CMO managers did not want to compete directly with their OEM customers serving the appliance motor aftermarket through franchised distributor networks.

[1] The 1,250 wholesalers together operated approximately 4,000 branch locations.
[2] Disguised name.

Exhibit 10.1   General Electric—Component Motor Operation
GE Consumer Aftermarket Motor Distribution, 1985

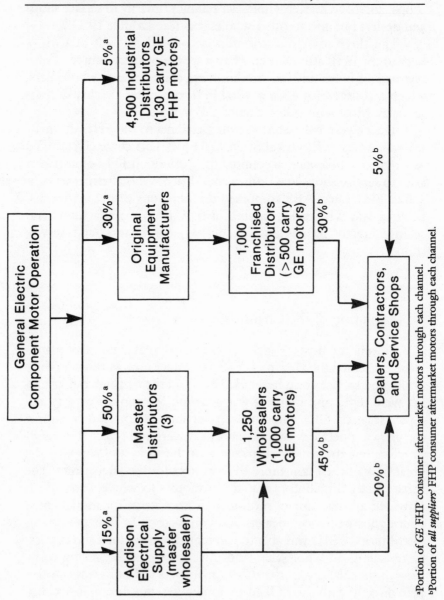

[a]Portion of *GE* FHP consumer aftermarket motors through each channel.
[b]Portion of *all suppliers'* FHP consumer aftermarket motors through each channel.

The 1970s and 1980s saw increasingly competitive pricing in the FHP aftermarket. At the same time, OEMs began to reduce the number of their suppliers. The major suppliers, GE and Emerson, became much more price competitive. Second-tier manufacturers, which lost market share in the OEM segment, turned to the aftermarket, competing for large wholesaler accounts by quoting prices comparable to GE's prices to its master distributors. Under intense pressure, master distributor prices for aftermarket motors declined markedly, dropping 13–20% in the three-year period 1983–1985.

Direct Distribution Study

In a study undertaken in 1978, GE marketing managers found that (1) GE's competitors sold motors to wholesalers at prices 5–17% below the price that GE's master distributors could sell and (2) the three master distributors were losing market share of sales to wholesalers. The study concluded by recommending a "change in sales coverage of the wholesaler aftermarket from master distributor to direct GE sales [to the wholesalers] for all General Electric [motor] products." Replacing the master distributors would mean an additional cost of sales of approximately $2 million dollars in the first year.[3] This forecast also anticipated a drop in GE's profitability for the first two years. However, in all but the worst-case scenario, profitability was expected to exceed the no-change option after two to three years.

In September of 1978, the recommendations of the study were rejected by CMO management because implementation would be risky. A second study, completed in January 1979, recommended direct sales only to the top 30 wholesalers, with the master distributors serving the remaining ones. The incremental marketing costs for this plan were estimated at $.9 million. This plan was rejected in favor of yet a third plan calling for direct sales only to the top 10 wholesalers. These 10 wholesalers operated 567 branch locations covering 40% of the FHP motor aftermarket.

CMO's management approved the "top ten" program and announced it to the master distributors in 1979. The reaction of the master distributors to the announced program was predictably adverse. "We'll make you bleed," said one. Another indicated that it

[3] This amount included the cost to GE of assuming functions previously handled by the master distributors, for example, administration, selling, order service, warehousing, inventory, credit and collections, merchandising, interest expense, and warranty.

would find a second source for FHP motors, as well as other product lines currently supplied by GE. Still another threatened to go out of the consumer motor aftermarket business.

The program met with enthusiastic support from the 10 wholesalers chosen to participate. These wholesalers felt that GE had been "living in the past" and made comments like "The master distributors won't allow you [GE] to be competitive." They felt that the move was inevitable and expected GE to offer them "substantially lower prices." Smaller wholesalers, on the other hand, were critical and resentful, concerned that they were likely to experience poor master distributor service and higher prices. It appeared that while the small distributors would probably remain loyal to GE, the supplier should expect some initial sales loss. In addition, many suggested that the small distributors would form buying co-ops to qualify for direct sales attention and lower prices.

Six weeks after announcing the program to the master distributors, CMO's management dropped it. No further changes were contemplated over the next several years, and consumer aftermarket motor sales continued to decline.

In 1985, sales to the master distributors were down, and the consumer FHP motor aftermarket continued to change. As the market became even more price competitive, it was not clear which FHP motor suppliers would survive. Many of the issues raised in the three distribution studies were still very much alive. GE was losing share to niche manufacturers selling direct to wholesalers. The product was increasingly perceived as a commodity, and the strategy of many GE competitors was to price below GE. CMO's vice president of component motor marketing, however, was sanguine: "We haven't solved the problem yet, but we're making progress."

Commentary

Because the major channels for consumer aftermarket motors were appliance OEMs during the postwar period, the costs of direct sales to independent wholesalers—supplying appliance dealers and service shops—were likely to have been prohibitive. By franchising master distributors as an intermediate level of distribution, which carried broad lines of appliance components and spare parts, GE was able to reach appliance service shops through wholesalers, both effectively and cost efficiently, and not depend on its OEM customers to get to this rapidly growing market.

By the late 1970s, however, the installed base of appliances had grown large enough to make the market for repair motors an important battleground for both appliance manufacturers and appliance motor manufacturers. The aftermarket was also highly contested turf for resellers in the multitier chain of distribution. In this intensely competitive market, distribution cost efficiency became a key determinant of market share position.

The master distributor, well suited to serve a much thinner postwar appliance service market, soon became a costly level of distribution between GE and its 1,250 wholesale customers. With the so-called second-tier motor manufacturers selling to large wholesalers at prices equal to those at which GE sold to its master distributors, the latter could remain competitive only by accepting considerably reduced margins or getting significant price reductions from GE—or both. In the meantime, the emergence of large-scale wholesalers operating nationwide so altered the economics of selling costs that the motor manufacturer could go direct to these businesses and thereby eliminate the master distributor discount.

Why, then, did GE's Component Motors business unit, even after successive studies, not move to break away from its three master distributors and sell direct to wholesalers, at least to the larger ones? These are the principal reasons:

- *Uncertainty.* With 50% of its consumer aftermarket motor business tied up with only three resellers, CMO's managers might well have been concerned about their ability to offset any loss of sales to the master distributors with sales direct to wholesalers. In addition, master distributors' threats fed apprehension as to the possible nature and extent of the punishment that these important customers might mete out.

- *Transition costs.* With predictions of lost business and additional marketing costs for selling direct to wholesalers, CMO studies forecast a 50% drop in this business unit's profitability during the first two years after the change in distribution. That was a formidable prospect for profit center managers whose performance was judged not just annually but quarterly.

- *Fear of leaving openings for competitors.* Any actions that alienate substantial sources of revenue run the risk of these accounts—the master distributors—being quickly and easily taken over by competing manufacturers.

• *Lack of urgency.* It is important to put the consumer aftermarket motor distribution issues in the context of CMO's overall mission, the manufacture and sale of motors of all sizes to a wide range of markets. Appliance motors represented only a part of the line, and of this product group, FHP aftermarket motors was a relatively small portion. As the leading supplier of motors, certainly in the domestic market, GE managers might well have been concerned about channel efficiency in this small segment of their business but not to the extent of disrupting its larger marketing program. Further, any loss of share in the aftermarket motor channels would have been gradual and not a reason for taking precipitous action.

The pattern is not uncommon. Though the signs of channels erosion may be clearly recognized, changes are not easily implemented and indeed it is not always clear what new directions the distribution strategy might usefully take.

HONEYWELL INFORMATION SYSTEMS

In 1985, Honeywell Information Systems (HIS), a division of Honeywell, Inc., made and marketed a line of computers that ranged in price from $5,000 (PC) to $8 million (DPS-88). The DPS-6 line, priced between $5,000 and $400,000, included offerings in the micro-, the mini-, and the superminicomputer categories.

In this size range, Honeywell competed with IBM, Digital Equipment Corporation (DEC), Hewlett-Packard, Wang, Data General, and Prime Computer. As of 1984, all but Prime held share positions larger than Honeywell's. Data General, DEC, and Wang had traditionally marketed through resale channels. IBM, traditionally reliant on direct selling, had recently moved toward the development of VAR, agent, distributor, and retail networks to broaden its market coverage.

Throughout most of its history, HIS, like IBM, had focused its product line, service support, and sales organization on users of large mainframe computers, to which it sold directly through its

Information Systems Division (ISD) salesforce. In 1978, recognizing that the market for minicomputers was growing rapidly, ISD managers had begun to develop "alternate" channels of distribution for Honeywell's minicomputer line to reach small- to medium-sized accounts. By year-end 1985, 15% of ISD's small systems sales were through value-added resellers (VARs) and manufacturers' reps (MRs). Both VARs and MRs developed and marketed information systems combining microcomputers and applications-specific software.

In an attempt to accommodate the change from selling direct to going through multiple channels, ISD managers had gone through three reorganizations between 1977 and 1981. As might be expected, conflicts developed at the field level between ISD's direct sales reps and the VAR accounts served by ISD's Indirect Sales Operation (ISO). According to two ISD managers,

> Our indirect business is an absolute must. It's also essential that it be integrated with the field sales operation, but it is slow in evolving because the branch managers do not yet see the benefits. . . .
>
> The issue is harmony. As prices come down and costs go up, it is important to bring the product to market through alternate channels. Traditionally, the direct sales rep has had a territory which he has regarded as his domain. Now when the product moves into the territory through other channels, the rep gets nervous.

ISD Organization

As of 1986, HIS's Information Systems Division (ISD) was organized into 3 field areas, 8 regions, and 45 branches covering the United States. Each branch manager typically supervised the work of a national account manager (NAM) and two resident marketing managers (RMMs), each of whom had three to five direct salespersons reporting to him or her. In 30 of the 45 branches, there was also an indirect sales specialist (ISS), who called on Honeywell's VARs and MRs. The ISD headquarters organization, headed by Frank Jakubik, included the National Sales Operation (NSO), which directed field sales activities; a National Accounts Program Office (NAPO), to which the national account managers in the field had a dotted-line reporting relationship; and the Indirect Sales Operation (ISO), headed by an ISD staff manager, James

Murphy, who directed the activities of the indirect sales specialists in the regions (see Exhibit 6.2 for organizational chart).

The development of the matrix organization had come in stages, with both the national account and indirect sales programs previously having had their own salesforces. Consolidating channels management at the branch level raised concerns among headquarters managers that the alternate channels program would get insufficient attention. Jakubik explained,

> As it is now, we're having a hard time getting the branch managers' attention. Normally, they are oriented toward the direct business; just about all of the branch managers came up from being very successful direct salespeople. There is another reason: When direct business is done, the branch manager gets credit both for booking and for revenue[4] but the indirect business only results in credit against a separate indirect revenue goal. Yet the whole system is supposed to be designed so that the branch manager is the territory manager and can direct a business opportunity into that channel where it can be handled best, regardless of performance and compensation crediting.

One manager thought the solution might be to have the independent sales specialist (ISS) network report directly to the ISO headquarters manager, as it had previously, rather than through the branch organization. But Jakubik thought otherwise: "I want ISO more integrated. My concept is that the branch manager is the location manager for everything that happens in that area. It is very difficult to have two or more separate salesforces located in the same place. There is no collaboration, and they tend to be competitive in a lot of little ways." His views were consistent with those of a branch manager:

> The major argument for having ISO as an independent organization is that it can be more effective when you get a dedicated organization that knows the market, and this is really a totally different market. On the other hand, when it was an independent operation, Honeywell couldn't afford the overhead. In addition, when ISO was an independent operation there were all kinds of conflicts at the field sales level.

---

[4] At HIS, as at some other computer companies, booking credit was given to the person or persons making the sale at the time the sales contract was signed. Revenue credit was accorded when the equipment was installed and payment was made. Sales goals included targets for each.

Commentary

Honeywell's experience is typical of the organizational turmoil that attends the transition from a monolithic direct sales operation to a multichannel distribution system in many industrial companies. In this case, as in others, the impetus for restructuring came with changes in the market. Like IBM and other early developers of the markets for computers, Honeywell had relied on its direct salesforce first to lease and then to sell large mainframe computers to business and government customers. In the beginning, computer applications were largely generic or "horizontal" in nature, serving such purposes as payroll accounting, credit and collections, and order processing. With the development of information systems technology, applications became increasingly customized in accordance with the specific needs of user industries, for example, airlines reservations, hospital records maintenance, retail store checkout systems, and manufacturing process controls. Honeywell Information Systems participated in this market, counting as a key competitive strength its strong technical support capabilities.

HIS managers, of course, followed the development of the minicomputer market in the 1960s but at that time elected not to participate, because "it would have required a very different distribution mode" and would have meant selling "raw iron" to systems houses that would add software programs. The comment itself implies an aversion to being postured in the market as simply a producer of standard equipment for others to customize.

However, the phenomenal growth of specialized applications markets for computerized information systems combined with the declining contributions from its large mainframe product lines ultimately led HIS to develop its DPS-6 minicomputer line, which was first introduced in 1977. Another factor influenced the decision to broaden the line: HIS's clientele, like most computer users, was moving from centralized data processing to distributed data systems in which mainframes, mini-, and microcomputers were linked to provide integrated knowledge networks. Hence, it was important to offer a full line of computers if only to meet the evolving needs of the large user market.

By the 1970s, though, the most rapidly growing segments of the computer market were not the large companies but the many small- to medium-sized users with a wide range of specialized needs that could only be satisfied by combining hardware and applications-specific software. Because of the myriad of both potential customers and innovative new uses for information systems, it would be impossible to cover the market with a direct salesforce supported by internal software development and technical service units. In addition, declining product prices and margins and rising direct sales costs combined to make direct selling alone an increasingly unattractive distribution option. As one HIS manager commented, "As the prices for the [computer] technology decreased, and as vertical applications became more important, it was essential to sell via some channel that had expertise in a given industry or with a given application."

As evidenced by Jakubik's comments, however, the addition of VARs and MRs to the distribution system created considerable culture shock. With past success attributable to direct selling and a strong support service system, the sales organization could not easily accept the need for so-called alternate channels. In businesses where technology and technical selling has acquired a high value, and where technically trained personnel enjoy high status, it may be difficult to acknowledge the need for outside technical support, much less market coverage. One HIS manager's reference to resellers as the "Vietcong" conveys not only a sense of scorn but of distrust, as well.

Gaining acceptance internally for adding VARs and MRs to the distribution system was one problem. Perhaps a greater challenge was redefining, and in effect constraining, the role of the direct salesforce in terms of the size of potential orders that salespersons might pursue. As the discussion in Chapter 8 indicated, this boundary was difficult both to define and to enforce. It left HIS field sales personnel with a distinct sense of having to compete, not only with other computer manufacturers but also with their own resale network, in making sales and meeting quotas.

Related to the boundary challenge was the task of restructuring both the headquarters and field sales organization to accommodate a multichannel strategy. HIS went through three major reorganizations in five years in attempts to integrate the reseller and direct channels, to take a coordinated approach to a wide range of customers, and to utilize each channel in a way that was both cost efficient and competitively effective. Not only structure but management processes, as well, were at issue. The lead referral system, quotas and performance measurement routines, and sales compensation all came under scrutiny.

By year-end 1985, 15% of HIS's revenues from sales of small systems, the DPS-6 line, came from its reseller channels, which had now become well assimilated into sales operations. As James Murphy, HIS vice president and head of the indirect sales program, commented,

> When the market changed so radically on us, we had to respond strategically and organizationally. What made that so difficult was that to adjust we had to significantly change the culture. Not easy! Now we've developed a strategy and a structure based on product specialization. We've altered the compensation system to make product specialization work. We have put a lot of teeth in our lead referral program, and put heavy emphasis on personal selling, telemarketing, and advertising to build our reseller program. We've developed a highly capable and specialized ISO organization to build and manage the reseller network. We've added the Porting Center[5] to make ourselves more attractive to our customers. The result: overachievement! By the end of the third quarter, we will have made our targets for the year!

At least two factors were crucial to coping with a changing market environment at Honeywell. One was top management's conviction that it was essential to compete in the market for small applications-specific information systems, along with their willingness to do this through whatever means necessary and to accommodate the transition with a new strategy. The other was a willingness to try different modes of organizing and managing a multichannel system until the desired results were achieved. In transitions of this order, it is often difficult to see clearly what new modes of management and what new structures will serve to implement new strategies, and a tolerance for trial and revision may be essential to adjust constructively to change.

SUMMARY

In both cases, the impetus for change was a growing and evolving market environment. The market for FHP appliance replacement motors had become intensely price competitive, forcing major suppliers to vacate this sector of the motor industry and forcing General Electric to reexamine its use of master distributors as an intermediate level of distribution. At Honeywell, the explosion of markets for information systems using mini-

---

[5] A technical service laboratory established to provide customer support.

computers, combined with applications-specific software prompted its move to a multichannel system using VARs and MRs to reach the myriad of potential customers.

For both GE and Honeywell, change generated strong resistance from those channels constituencies which perceived themselves as being threatened and, in all likelihood, disadvantaged. From GE's master distributors came the reaction, "We'll make you bleed!" At Honeywell, the direct sales reps' resentment of the VARs invading their turf is captured in "Vietcong" symbolism.

It is often difficult to assess the extent and power of resistance and to determine whether the retaliation strength of disadvantaged channels members is such that its cost would exceed any benefits to be gained by restructuring the distribution system. This uncertainty, coupled with what may be termed *the paralysis of past success,* often results in resistance to change. Change runs the risk, as well, of violating the strong personal ties that build up around supplier-salesforce-reseller relationships almost as implicit contracts based on each party's expectations of the other. Such relationships are the cement that holds a system together—and change becomes difficult as the cement hardens.

It seems, then, that the restructuring of distribution systems comes only if the costs of preserving the status quo in terms of sales revenues, market share, and profitability are high and the need for change is compelling. The costs of making no change were judged by the Honeywell management to be sufficiently great as to lead to reformulating distribution strategy. At GE that was not the case. Because the potential costs of realigning an obsolete channels system seemed too high in terms of transition investments and possible loss of revenues, managers in the Component Motor Operation backed off.

When significant change is implemented, the chances of success seem to depend on three factors. First, it is important to see the move in the context of a total marketing strategy intended to address specific market segments with defined product lines. In this context, what roles different elements in the channels system will play, what types of accounts each will serve, and what parts of the product line each will carry need to be planned so that any change in *how* the firm goes to market is consistent with the overall strategy.

Second, success is often contingent on the supplier's leading from strength, strength vis-à-vis both its competitors and its channels network. Honeywell did not have the market strength of such competitors as IBM or DEC, and while Honeywell did in fact undertake a major restructuring of its distribution system, it was not easy to recruit VARs and MRs to carry out a market segment entry strategy on the scale that its managers would have liked.

Third, effectively implemented change almost always has an organizational component, that is, new systems may not usually be managed effectively through unchanged organizational structures. Honeywell's experience is a case in point.

In closing, it is useful to observe that new schemes are seldom perfected in detail a priori; they almost always need revision in form and in detail as they become operative.

• PART III • Distribution in the United States—Past, Present, and Future

In Part III, we move from a focus on company strategy to a consideration of the distribution context within which managers shape distribution strategy and manage channels relationships. Chapter 11 develops the broad outlines of the distribution infrastructure in the United States. The description relies largely on U.S. Department of Commerce data. It shows, for example, the balance among direct selling, resale distribution, and the use of agents for our major industrial sectors. Comparisons among distribution patterns in these several sectors provide a further opportunity to understand the relationship between product-market characteristics and producers' modes of going to market. Chapter 11 also describes the sizes and types of enterprises that make up our channels networks.

In Chapter 12 we discuss the evolution of industrial distribution institutions in the United States going back to 1812. Chapter 12 also reviews current trends and offers certain speculations about future developments in industrial distribution. An understanding of the forces at work shaping our distribution infrastructure in the United States may help managers in two ways. First, it provides a perspective useful for understanding external factors that need to be taken into account in shaping the strategy of the individual firm. Second, it enables managers to consider their own and competitors' tactical moves in a longer-term strategic context.

Chapter 13 reviews the legal framework that establishes certain constraints in how producers and their resellers relate. This overview is too cursory to be legally authoritative; it will, however, provide managers with an understanding of the legal issues involved and a sense of the thrust of relevant law and court opinion that applies to producer-reseller transactions.

204

11   The Distribution Infrastructure

The flow of industrial goods in the United States from many thousands of plants to millions of customers is handled by three types of intermediaries: manufacturers' sales branches, independent distributors, and agents and brokers. The responsibilities of these institutions are varied and include such tasks as generating product demand, managing the physical distribution system, and handling the paperwork for order receipt, order processing, and credit and collections. This network may also take responsibility for the product in the field, often customizing it to meet user requirements and providing repair and maintenance services. In addition, the distribution infrastructure creates markets for used industrial products—taking in secondhand machinery and equipment, often in trade, and reconditioning and selling it.

In this chapter, we will first profile each of the three main channel types and then discuss some of the elements that shape the overall distribution infrastructure, such as our transportation and communication systems, economic growth, and evolving market structures. The chapter's description of the distribution infrastructure is a snapshot, at a point in time, of this constantly changing economic institution. Accordingly, we are concerned not only with its present-day structure but also with the underlying trends that have shaped it and continue to do so.

CHANNELS INSTITUTIONS OF THE UNITED STATES—A PROFILE

The U.S. Department of Commerce's *Census of Wholesale Trade* provides data with which we can profile the industrial distribution system in the United States in terms of (1) the size of the network, that is, the number of distributive organizations; (2) the relative balance of transactions accounted for respectively by distributors,

205

brokers and agents, and manufacturers' sales branches; and (3) distribution sector demographics, that is, size of firm and degree of concentration. The relevant data base is that reported for the seven largest categories of industrial goods identified by SIC (Standard Industrial Classification) code as follows:

SIC 5050   Metals and minerals, except petroleum

SIC 5063   Electrical goods

SIC 5065   Electronic parts

SIC 5084   Industrial machinery

SIC 5085   Industrial supplies

SIC 5110   Paper and paper products

SIC 5161   Chemicals and allied products

Sales at wholesale prices of goods in these seven categories amounted to almost $435 billion in 1982, representing the great bulk of industrial product sales in this country and 35% of all wholesale trade, both consumer and industrial. Of that amount, manufacturers' sales offices accounted for $196 billion, distributors for $194 billion, and agents for $45 billion (see Exhibit 11.1).

Independent Distributors

Typically, independent distributors are small firms in terms of both sales and number of employees. The great majority are privately held, being organized as sole proprietorships or as partnerships. In the seven SIC categories, about 40% of the approximately 57,000 independent distributors had fewer than 5 employees; almost 85% operated with fewer than 20 employees (see Exhibit 11.2). The average firm had annual sales of $3.3 million and operated from 1.3 sales locations.

However, firms with more than 100 employees, while representing less than 2% of the total number of firms, had average sales of almost $73 million, operated an average of 8.5 sales locations, and accounted for 40% of total sales by distributors in 1982 (see Exhibit 11.3). Many of these large distributors are publicly held.

Exhibit 11.4, based on a sample survey of publicly held firms in 1986, provides salient financial data for publicly held distributors. As it indicates, a high percentage (75–85%) of distributors' assets are current, that is, they are held as inventories and as accounts receivable. Fixed assets—buildings and equipment—repre-

Exhibit 11.1 Sales at Wholesale Prices by Channel Type and Product Group for Selected Industrial Products, 1982

Product Group	Distributors		Agents		Manufacturers		Total Sales (000,000)
	Sales (000,000)	Percentage of Total	Sales (000,000)	Percentage of Total	Sales (000,000)	Percentage of Total	
Metals and minerals, except petroleum (SIC 5050)	$ 49,009	48%	$ 8,155	8%	$ 45,526	44%	$102,690
Electrical goods (SIC 5063)	$ 24,502	45%	$ 7,415	14%	$ 22,038	41%	$ 53,955
Electronic parts (SIC 5065)	$ 16,093	40%	$11,337	28%	$ 12,935	32%	$ 40,365
Industrial machinery (SIC 5084)	$ 39,293	57%	$ 9,156	13%	$ 20,180	29%	$ 68,629
Industrial supplies (SIC 5085)	$ 19,693	49%	$ 3,167	8%	$ 16,940	43%	$ 39,800
Paper and paper products (SIC 5110)	$ 25,937	48%	$ 3,285	6%	$ 24,271	45%	$ 53,493
Chemicals and allied products (SIC 5161)	$ 19,462	26%	$ 2,768	4%	$ 53,873	71%	$ 76,103
All product groups	$193,984	45%	$45,283	10%	$195,763	45%	$435,035

Source: U.S. Census of Wholesale Trade, Geographic Area Statistics, 1982; Table 1: "Summary Statistics for the United States: 1982."

Exhibit 11.2 Independent Distributors: Selected Data on Number and Size of Firms for Seven Industrial Product Groups, 1982

Product Group	Number of Firms	Average Sales per Firm (000,000)	Average Sales per Employee (000)	Average Number of Sales Locations per Firm	Size of Firm in Number of Employees	
					Percentage of Firms with Fewer than 5 Employees	Percentage of Firms with Fewer than 20 Employees
Metals and minerals, except petroleum (SIC 5050)	5,978	$8.2	$407	1.2	41%	79%
Electrical goods (SIC 5063)	7,467	$3.2	$190	1.4	36%	81%
Electronic parts (SIC 5065)	5,579	$2.8	$193	1.2	47%	86%
Industrial machinery (SIC 5084)	14,782	$2.6	$199	1.2	42%	85%
Industrial supplies (SIC 5085)	7,979	$2.4	$155	1.4	35%	80%
Paper and paper products (SIC 5110)	9,448	$2.7	$178	1.2	40%	82%
Chemicals and allied products (SIC 5161)	6,002	$3.2	$252	1.2	48%	86%
All product groups	57,235	$3.3	$215	1.3	41%	83%

Source: U.S. Census of Wholesale Trade, Geographic Area Statistics, 1982; Industry Series WC4-22, Table 4: "Employment by Principal Activity for the United States: 1982"; Industry Series WC82-1-1, Establishment and Firm Size, Table 7, "Employment Size of Firms: 1982"; Table 1: "Summary Statistics for the United States, 1982."

Exhibit 11.3 Independent Distributors: Data for Firms with More Than 100 Employees for Seven Industrial Product Groups, 1982

Product Group	Number of Firms With More than 100 Employees		Total Sales (000,000)	Average Sales per Firm (000,000)	Average Number of Sales Locations	Percentage of Total Distributor Sales
	Number	Percentage of Total Firms				
Metals and minerals, except petroleum (SIC 5050)	200	4%	$24,208	$121.0	5.4	50%
Electrical goods (SIC 5063)	148	2%	$ 9,897	$ 66.9	12.7	41%
Electronic parts (SIC 5065)	108	2%	$ 6,899	$ 63.9	7.4	44%
Industrial machinery (SIC 5084)	196	1%	$11,413	$ 58.2	7.0	30%
Industrial supplies (SIC 5085)	132	2%	$ 4,708	$ 35.7	12.4	24%
Paper and paper products (SIC 5110)	146	2%	$10,961	$ 75.1	7.6	44%
Chemicals and allied products (SIC 5161)	65	1%	$ 5,330	$ 82.0	9.6	28%
All product groups	995	2%	$73,416	$ 73.8	8.5	40%

Source: U.S. Census of Wholesale Trade, Industry Series WC 82-1-1, Establishment and Firm Size, Table 7, "Employment Size of Firms: 1982."

Exhibit 11.4 Distributor Financial Profiles in the Seven SIC Codes, 1985–1986

	SIC 5050[a]	SIC 5063[b]	SIC 5065[c]	SIC 5084[d]	SIC 5085[e]	SIC 5110[f]	SIC 5161[g]
Assets							
Total current as percentage of total assets	78.5%	83.4%	82.3%	77.6%	81.0%	79.0%	75.7%
Total fixed	15.0%	11.5%	12.7%	15.6%	13.1%	14.3%	17.9%
Liabilities							
Total current as percentage of total liabilities	54.1%	52.6%	49.8%	52.6%	50.2%	52.8%	53.8%
Long-term debt	10.0%	10.6%	11.5%	11.8%	11.3%	11.2%	10.5%
Net worth	33.9%	34.5%	36.4%	33.6%	36.6%	33.3%	33.7%
Gross profit as percentage of net sales	22.6%	26.7%	30.2%	28.5%	29.9%	25.6%	28.1%
Profit before taxes	2.3%	2.5%	2.4%	2.2%	2.4%	1.7%	2.6%
Days' receivables	46	48	44	45	41	38	43
Days' payables	42	42	41	40	39	31	44
Inventory turns[h]	8.0	7.6	8.3	8.2	8.8	9.7	8.4
Before-tax return on net worth	16.2%	18.6%	19.9%	14.8%	15.5%	13.1%	18.8%

Source: RMA 1986 Annual Statement Studies, Robert Morris Associates, Philadelphia, 1986, pp. 193, 195, 218, 225, 227, and 233.
[a]Based on a sample of 389 publicly held firms.
[b]Based on a sample of 494 publicly held firms.
[c]Based on a sample of 256 publicly held firms.
[d]Based on a sample of 1,071 publicly held firms.
[e]Based on a sample of 512 publicly held firms.
[f]Based on a sample of 389 publicly held firms.
[g]Based on a sample of 261 publicly held firms.
[h]A ratio of sales to receivables that measures the number of times trade receivables turn over during the year.

Exhibit 11.5   Percentage of Distributors in Selected SIC
Categories Reporting Positive Net Income, 1986

	Number of Distributor Firms	Percentage Showing Positive Net Income
Machinery, equipment, and supplies (SIC 5008)	52,653	58%
Metals and minerals, except petroleum (SIC 5050)	6,382	49%
Electrical and electronic goods (SIC 5060)	17,106	59%
Paper and paper products (SIC 5110)	6,678	59%
Chemicals and allied products (SIC 5160)	7,552	68%
Total	90,371	58%

Source: Leo Troy, *Almanac of Business and Industrial Financial Ratios,* midyear 1986, Englewood Cliffs, N.J.: Prentice-Hall, pp., 204–205, 214–215, 216–217, 222–223, and 230–231.

sented a relatively small portion of total assets. Similarly, liabilities are predominantly current in nature—mostly accounts payable, notes, and short-term bank loans. Long-term debt averaged between 10% and 12%, and on average, owners' net worth amounted to about a third of liabilities. Thus, while distributor before-tax profits, 2–2½%, are low by business standards, the return on equity tends to be in the 15–20% range before taxes.

As Exhibit 11.5 shows, more than a third of distributor firms, including both privately owned and publicly held firms, reported no positive net income. It is not possible to know from these data, however, what percentage of those not reporting positive net income were in serious financial difficulty, and what number were being operated financially so as to maximize owners' personal incomes and other benefits and to minimize the income of the business for tax purposes.

Agents

The 1982 census for the seven SIC categories reported approximately 11,300 agents—about one-fifth the number of independent distributors. As measured in number of employees, agents tend to be smaller than distributors: two-thirds of the firms operated with fewer than five employees (see Exhibit 11.6). However, their average sales were $4 million, somewhat larger than the comparable number for distributors. It follows that the average sales-per-employee figure is greater for agents than for distributors: $789,000 for agents, as compared with $215,000 for distributors. Two factors tend to account for the difference. First, agents carry only a few lines, usually 10–20, and they typically sell in large quantities to a limited number of buyers. Distributors carry many thousands of items and usually sell in smaller quantities. Second, because distributors are involved in stocking and transporting goods, order processing, billing, and collections, distributor transactions require staffs of sales, office, and warehouse personnel that are large relative to those in agency operations. In the case of agents, most of the distribution functions, except for selling, are performed by the supplier that the agent represents. Thus, agencies tend to be less labor intensive than distributors, and this is reflected in sales-per-employee measurements.

Further, while agents accounted for only about 10% of sales at wholesale prices in the seven SIC categories in 1982, they were a strong factor in certain product classifications. In particular, in the sale of electronic parts, agents accounted for $11.3 billion out of a total of $40.4 billion, or 28% (see Exhibit 11.1). The explanation for this, in part, is that the electronics industry is populated by large numbers of new, small companies, many of which are not able to support an in-house sales organization. In addition, the electronic parts agents sales-per-employee figure, almost $900,000, was one of the highest in the seven industries (see Exhibit 11.6). This suggests that manufacturers find agents to be an effective sales channel in the diffusion of new technology.

Exhibit 11.6   Agents and Brokers: Selected Data on Number, Size, and Sales per Employee for Seven Industrial Product Groups, 1982

Product Group	Number of Firms	Total Sales (000,000)	Average Sales per Firm (000,000)	Percentage of Firms with Fewer than 5 Employees	Average Sales per Employee (000)
Metals and minerals, except petroleum (SIC 5050)	1,230	$ 8,155	$6.6	75%	$1,639
Electrical goods (SIC 5063)	1,788	$ 7,415	$4.1	55%	$ 660
Electronic parts (SIC 5065)	2,023	$11,337	$5.6	59%	$ 899
Industrial machinery (SIC 5084)	3,290	$ 9,156	$2.8	67%	$ 566
Industrial supplies (SIC 5085)	1,147	$ 3,166	$2.8	67%	$ 513
Paper and paper products (SIC 5110)	1,033	$ 3,285	$3.2	75%	$ 799
Chemicals and allied products (SIC 5161)	783	$ 2,768	$3.5	79%	$ 796
Total	11,294	$45,282	$4.0	67%	$ 887

Source: U.S. Census of Wholesale Trade, Industry Series WC82-1-1, Establishment and Firm Size, Table 7, "Employment Size of Firms: 1982"; Industry Series WC82-4-22, Table 4, "Employment by Principal Activity for the United States: 1982."

Exhibit 11.7   Selected Data for Manufacturers' Branch
(or Direct) Sales for the Seven Product Groups, 1982

Product Group	Total Sales (000,000)	Number of Sales Locations	Sales per Sales Location (000,000)	Sales per Employee (000)
Metals and minerals, except petroleum (SIC 5050)	$ 45,526	1,432	$31.8	$3,231
Electrical goods (SIC 5063)	$ 22,038	2,237	$ 9.9	$ 587
Electronic parts (SIC 5065)	$ 12,935	984	$13.1	$ 583
Industrial machinery (SIC 5084)	$ 20,180	3,675	$ 5.5	$ 427
Industrial supplies (SIC 5085)	$ 16,940	1,628	$10.4	$ 947
Paper and paper products (SIC 5110)	$ 24,271	1,848	$13.1	$1,305
Chemicals and allied products (SIC 5161)	$ 53,873	2,054	$26.2	$1,237
Total/Average	$195,763	13,858	$14.1	$1,192

Source: U.S. Census of Wholesale Trade: Geographic Area Statistics, 1982; Industry Series 4–22, Table 4, "Employment by Principal Activity for the United States: 1982."

Manufacturers' Salesforces

Deployed in geographically dispersed branch offices, manufacturers' salesforces accounted for the largest portion of industrial goods moving from plants to user-customers in 1982, about 45% of the total of $435 billion (see Exhibit 11.1). This figure is heavily weighted by manufacturers' direct sales of chemicals and allied products (SIC 5161), which amounted to about 70% of total sales in this category. Independent distributors represented only 25%, and agents less than 4%, of the flow of chemical products to industrial customers.

These data, combined with the relatively high sales-per-employee number of $1.2 million (see Exhibit 11.7) for chemicals manufacturers' sales offices, suggest that chemicals industry customers tend to be large, easily identified firms that buy in large dollar amounts. Thus, it becomes economical to sell direct to these accounts rather than through distributors and agents. It is also consistent with this pattern that chemicals distributors and agents report the greatest percentage of firms with less than five employees,

48% and 79%, respectively (see Exhibits 11.2 and 11.6). Chemicals producers utilize these channels to reach small users, often in less populated trading areas, and elect to deal directly with larger customers, from which they derive the bulk of their revenue.

The highest sales-per-employee number reported for manufacturers' sales branches, $3.3 million, is in the metals and minerals category, SIC 5050 (see Exhibit 11.7). The highest sales-per-employee number reported for independent distributors, $407,000, is also in this category (see Exhibit 11.2). Unlike distribution patterns in the chemical industry, with its heavy emphasis on direct sales, metals and minerals producers balance the use of direct and distributor channels to a much greater extent. In 1982, manufacturers' sales branches accounted for 44% of all sales in this SIC category; distributors for 48%.

These data indicate, first, that a significant proportion of large metals and minerals buyers can effectively be reached by the producers' salesforce. They also suggest that transactions in this product category, even in the less concentrated end of the market, are for relatively large amounts. Second, the data imply that despite high market concentration, metals and minerals producers have been less aggressive in building direct salesforces than have the chemicals manufacturers.

As the forgoing data describing the three major channels elements show, distribution networks in the United States are finely articulated complexes of small business units. The relative importance of each as a channel institution tends to vary by product category both because of the nature of the product and because of its intended markets.

TECHNICAL AND ECONOMIC DETERMINANTS

The distribution infrastructure of the United States, as profiled statistically through Department of Commerce data, is a reflection of the size and complexity of the U.S. economy. As they have evolved over more than 200 years, the institutions that make up our industrial distribution systems have themselves grown in scope and complexity. On the one hand, distribution mirrors economic growth; on the other, it makes economic growth possible.

Probably the single greatest factor shaping industrial distribution systems is market demography, that is, the size and geographic dispersion of markets and the extent of concentration among buying firms. Thus, the demographic characteristics of different markets shape the distribution channels through which industrial goods producers will serve them.

A second fundamental determinant is the availability of transportation and communication systems for shipping goods and for the exchange of information relative to ordering, stocking, product delivery, and sales financing.

A third factor is product technology; product technical complexity and nonstandardization, for example, tend to move producers toward direct sales relationships with customers.

A fourth factor, information technology, has profoundly influenced industrial distribution in recent decades. The use of computers has dramatically reduced distribution costs, in turn fostering the growth of concentration in the distribution sector. Finally, the development of a distribution infrastructure is dependent on the availability of capital to fund investments in inventories, physical facilities, and personnel employed in handling the flows of goods from factory to user.

Economic Determinants

Chapter 12, "An Historical Perspective," documents in detail the evolution of distribution in the United States. Here, we will give a brief overview of those developments and describe their impact on the distribution infrastructure. Among the salient factors shaping distribution in the early nineteenth century was the growth of rail and water transportation and telegraph lines linking coastal ports, centers of manufacturing, and agricultural regions to the emerging market areas in the interior. In this period, too, came two important developments in manufacturing process and product technology, namely, mass production and product design standardization. The rise of mass production made distribution networks imperative in moving farm and factory output to markets. The development of standardized products also made possible distributor sale and servicing of manufactured goods. Resellers could provide related supply items and standardized spare parts for these items to their user-customers as well.

These factors fostered tremendous growth in industrial distribution. The increased rapidity with which goods could be ordered, shipped, and sold considerably reduced the risks to resellers of carrying inventories. Further, as the country's population moved west, the development of trading centers created market opportunities for a broad range of industrial products. The result was that small, independent distributors became an increasingly important vehicle for economic expansion.

Between 1920 and 1985, gross national product increased from $91.5 billion to $3.7 trillion. During this same period, total U.S. population rose from 106 million to 239 million, with a marked concentration in metropolitan areas. With the rapid urbanization that followed the First World War came the development of large centers of commerce. By 1985, the largest 10 cities accounted for almost a third of the total market for industrial goods; an estimated 65% of total industrial product market potential was centered in 50 trading areas.

These developments tended to fuel the rise of multichannel distribution systems. Manufacturers developed direct salesforces to serve large and medium-sized accounts in geographically concentrated market areas. However, they relied on independent distributors to cover both smaller accounts and those outside the main centers of trade. Agents, selling on commission, typically represented manufacturers too small to afford the fixed overheads of maintaining their own field salesforce or those manufacturers needing the special expertise that agents could bring to selling—either in serving market segments with which the producer's salesforce had little experience or in the remote reaches of the producer's sales territory.

Although agents were a major channels institution in the late eighteenth and early nineteenth centuries, they became less important from the middle 1800s on, as manufacturing firms grew in size, as industrial markets became more concentrated, and as producers developed their own direct salesforces. For many firms, the choice between the costs of maintaining a salaried salesforce and paying agents' commissions came to favor the former option. As market share rivalries became more intense, it often became essential, as well, to maintain direct relations with large customers.

In the twentieth century, independent distributors came to play increasingly larger roles in taking industrial goods to market. They offered wide assortments of merchandise from many manufacturing sources to large and small industrial buyers in many industry segments. With the great diversity, both demographic and geographic, of American industry and with the increasing range of manufactured product offerings, independent distributors performed functions that few single producers found economically feasible to undertake in-house. In addition, the rise of the independent distributor was in part facilitated by the availability of capital—in the form of private savings, bank loans, supplier credit, or sometimes equity funding—to support these many thousands of often small and privately owned enterprises.

The Impact of Distribution Cost Factors

Exhibit 11.8 provides data on the relative market shares of distributors, factory branches, and agents for five of the seven SIC product categories for the years 1963 and 1982. As it indicates, industrial producers came to rely increasingly on independent distributors during this period. In all but the industrial supplies category (SIC 5085), the percentages of total sales accounted for by distributors increased significantly. (The percentage for industrial supplies remained high, representing almost half of all sales in this product group.) Total distributor sales for all five SIC categories rose from 34% in 1963 to 43% in 1982, while sales through manufacturers' factory branches declined from 54% to 45% over the same period. In both 1963 and 1982, the proportion of total sales accounted for by agents was 12%.

The increasing strength of the resale sector may be largely attributed to distribution cost factors. As noted earlier, the independent distributor was able to cluster the product lines of many manufacturers and thus to gain selling scale economies in reaching secondary and tertiary markets that could not be effectively reached by manufacturers. Further, a major factor contributing to the rise of the independent distributor has been the escalating costs of direct selling by producers to user-customers. According to one study made in 1988, the cost of personal selling rose by 160% between 1975 and 1987. Businesses selling direct to user-customers incurred the highest cost per sales call, averaging $291 in 1987; those selling to external intermediaries averaged $204.[1]

In addition, the concentration of certain trades, and of light and heavy manufacturing, in particular industrial centers provided

---

[1] See "Average Business-to-Business Sales Call Increases by 9.5%," *Marketing News*, September 12, 1988, p. 5.

Exhibit 11.8 Sales by Channel for Selected Industrial Product Groups, 1963 and 1982

	1963	1982
Electrical apparatus and equipment (SIC 5063)		
Total sales (000,000)	$ 8,478	$ 53,956
Percentage of total sales accounted for by		
• distributors	34%	45%
• factory branches	58%	41%
• agents	7%	14%
Electronic parts and equipment (SIC 5065)		
Total sales (000,000)	$ 4,296	$ 40,365
Percentage of total sales accounted for by		
• distributors	36%	40%
• factory branches	28%	32%
• agents	36%	28%
Industrial machinery and equipment (SIC 5084)		
Total sales (000,000)	$ 8,695	$ 68,630
Percentage of total sales accounted for by		
• distributors	45%	57%
• factory branches	37%	29%
• agents	18%	13%
Industrial supplies (SIC 5085)		
Total sales (000,000)	$ 7,424	$ 39,800
Percentage of total sales accounted for by		
• distributors	49%	49%
• factory branches	42%	43%
• agents	9%	8%
Chemicals and allied products (SIC 5161)		
Total sales (000,000)	$12,562	$ 76,103
Percentage of total sales accounted for by		
• distributors	16%	26%
• factory branches	80%	71%
• agents	4%	4%
All categories		
Total sales (000,000)	$41,455	$278,854
Percentage of total sales accounted for by		
• distributors	34%	43%
• factory branches	54%	45%
• agents	12%	12%

Source: U.S. Census of Wholesale Trade, *Geographic Area Statistics, 1982,* Table 1, "Summary Statistics for the United States: 1982," and *Summary Statistics, 1963,* Table 2: "United States by Kind of Business: 1963."

bases of business large enough to permit distributors to economize on selling costs and to compete through low prices and specialized market knowledge. Thus we find distributors in trading areas such as Pittsburgh stressing items commonly used by metal-working industries, those in Akron focusing on the rubber industry, distributors in the Southwest emphasizing supplies and equipment used by oil well and mine operators, and resellers in the Southeast catering to such sectors as textiles and furniture manufacturing.

Computerization

Within the past two decades, the development of distribution-related applications for computers has given further impetus to the rise of the industrial distributor. Computers have also contributed to a growing concentration in the distribution sector by creating new possibilities for achieving distribution economies of scale. For resellers, sourcing from many manufacturers and selling to thousands of customers, the computerization of order receipt, shipping, invoicing, inventory management, stock replenishment, and accounts receivable has resulted in significant cost reductions. In addition, distributors achieve cost savings by using computers to communicate with both customers and suppliers, thereby considerably reducing staffs of clerical and telephone personnel.

Distributors with multiple sales locations can also network inventories and significantly reduce their inventory-to-sales ratios. That is, they are able to draw on stocks held in any one location to fulfill orders coming out of other sales territories without any fall-off in on-time delivery performance. These are scale economies that are available to producers, as well, but not on the same order of magnitude as for resellers, which typically handle broader arrays of merchandise and serve more customers.

National Chains. Another effect of the increasing use of computerized information and communications systems is that the larger, multilocation distribution firms have tended to grow in size and to take increasing shares of the resale market. According to the estimates of managers in these several industries, the 10 largest electronic components distributors increased their combined share of the distributor-served market segment from 37% in 1972 to 55% in 1982. As of 1982, approximately 350 other electronics distributors shared the remaining 45%. In the hospital-supply industry, one national distributor—American Hospital Supply—and five large regional chains increased their share of the distributor-served segment from 23% to 55% between 1972 and 1982, with 484 local houses accounting for the rest. The sales of American Hospital Supply, the leading distributor, grew fivefold between 1972 and 1983.

According to estimates made at WESCO (Westinghouse Electric Supply Company), combined sales of the six largest electrical distributors more than doubled from 1976 to 1985, going from $2.16 billion to $4.93 billion in that ten-year period. As noted earlier, 1985 sales of the largest electrical distributor, Graybar Electric, were estimated at $1.6 billion. In the same period, the total market size for electrical equipment, according to U.S. Department of Commerce data, grew from $17 billion to $33 billion. Thus, the six largest distributors increased their share of the total market from 12.7% to almost 15% over the decade.

The growth in the number and size of national chains has been supported by the scale economies resulting from advances in information technology, market size, buyer demographics, and the nature of the product. Large national chains seem to have emerged for products that have the following characteristics:

- Users purchase many items in small amounts at one time.
- Purchase frequency is high because users seek to minimize inventories.
- Customers are geographically dispersed.
- Total purchases of the product line represent a significant cost item for the buyers (e.g., hospital supplies).

These circumstances create the setting for (1) realizing significant inventory savings by supporting nationwide sales out of a networked inventory base, (2) utilizing computer systems to manage inventories effectively and efficiently with improvements in on-time order fill rate, and (3) providing customers with information on amounts purchased, which is useful in the conduct of their businesses.

Limits of Computerization. The conditions under which computerization may yield significant economies of scale do not necessarily obtain, however, for industrial distribution generally. For example, while McKesson Chemical, the largest distributor in its field, has developed internal data systems for managing its purchasing and inventory control routines, it has not attempted to develop data links with individual customers. McKesson Chemical's customers buy small amounts intermittently, ordering only a few items at a time, and their purchases from McKesson typically represent a minor part of their costs of goods sold. Such ordering and usage patterns can easily be handled by telephone. Further, because of the bulk and high shipping costs of chemicals relative to their value, the opportunities for economizing regionally or nationally on inventories are not as great as they are for low weight-to-value products, such as hospital supplies or electronic parts.

As the cost benefits of computerized information systems have fostered the growth of large national and regional distributors in certain industries, the small independent local supply houses have become competitively disadvantaged. According to a manager in one large national chain,

> It takes as much for a one-warehouse wholesaler to load a price change into the system as it takes for us to put it into the computer for 56 distribution centers. We can do centralized purchasing; we can invest in computerization, in automated warehouses, and we can use mathematical models for order filling and delivery. We can carry a wider range of products. Also, over a period of time we have been able through computerization to accelerate the billing time, reduce inventories at the retail level, and go from 30-day to 15-day terms for accounts receivable. The resulting ROI increase gets passed on to our customers through lower prices.[2]

SUMMARY

In summary, the distribution infrastructure of the nation and of each producer industry within the nation may be described in terms of (1) the numbers of factory branches, distributors, agents, and brokers serving industrial markets; (2) the relative proportions of goods that flow through each of these channels institutions; (3) the degree of concentration among distributor firms by industry sector; and (4) the extent of product-market specialization among channels business units.

The national economic factors that shape distribution systems are (1) economic growth, (2) the stage of development of transportation and

[2] See E. Raymond Corey, "The Role of Information and Communications Technology," in *Marketing in an Electronic Age*, ed. Robert D. Buzzell (Boston: Harvard Business School Press, 1985), p. 43.

communication systems, (3) the extent of geographic market concentration, (4) the availability of capital to invest in distribution, and (5) the information technology useful for managing the flow of goods to markets. Within any given producer-industry context, the factors that influence distribution by product-market include (1) the stage of development of product technology, (2) producer industry demographics, (3) market size, and (4) the extent of geographic and demographic concentration among user firms.

As the U.S. economy has grown, its industrial product distribution infrastructure has grown in both its complexity and its diversity. Increasingly, manufacturers have developed multichannel systems including factory sales branches, independent distributors, and agents to reach market segments that are demographically diverse and geographically dispersed. Within these systems there has been an increasing reliance on resellers for reasons of distribution costs. In selling to fragmented markets, resellers have achieved scale economies through clustering assortments of goods much wider than those of any single producer. Inasmuch as computerized information systems offer particular benefits where large numbers of small transactions and broad lines of merchandise are involved, they have helped to increase the reseller cost advantage.

At the same time, factory sales branches have also grown in importance with the growth of markets and their concentration by geographic area and by industry sector. The costs to producers of direct selling in concentrated markets are typically lower than what they would give up in sales margins by going through resellers. Given the favorable economics of selling direct to large accounts, producers realize the added advantages of having close major account relationships: the benefits of direct technical interchange, direct price negotiations, and market feedback.

Relative to resellers and factory branches, agents have declined in importance as a distribution channel. However, agents continue to play an important role in introducing new technology, particularly the products of start-up ventures with limited resources. They have found a place, too, in many multichannel distribution systems as vehicles for reaching geographically remote markets and market segments in which they may have specialized knowledge and experience. Agents continue to be the primary means of market access for many companies with revenue bases too small to support direct sales operations and product lines that may require technical selling and possibly customization.

Two related trends in the U.S. distribution infrastructure are the growth of national resale chains and increasing distributor-industry concentration. These trends have been fueled to a significant extent by the development of computerized information systems for receiving and processing orders, controlling inventories, managing accounts receivable, and ordering from suppliers. Computerization has made possible important scale economies, enabling already large resellers to grow in size and placing small, local distributors at a relative cost disadvantage.

Finally, along with the expansion of the infrastructure to cover a range of markets has come increasing product-market specialization, with direct sales organizations, agents, and distributors focusing on specific segments, such as industrial construction, health care, and factory supplies. The growth of these sectors has made such specialization both economically feasible and competitively advantageous; niche-focused resellers typically offer customers a broad product assortment within their defined market segment and specialized knowledge helpful to buyers in making product selections.

This chapter has provided a statistical profile of distribution in the United States and a brief discussion of how and why it has evolved. The two chapters that follow offer an elaborated history of industrial distribution in the United States and a more detailed discussion of current and future trends. Some readers may find that Chapter 12 is not central to their interests. It may appeal to others as a record of the evolution of an important institution in American industry. For the authors, tracing this history was a fascinating exercise in archival research. The full historical detail is included in this book for the benefit of business scholars and for those business managers who wish better to understand the national context in which they develop their distribution strategies.

12 An Historical Perspective

Understanding the evolution of the industrial distribution infrastructure in the United States is useful for gaining greater perspectives on contemporary channels institutions and for anticipating the future. This chapter tracks the development of industrial distribution in America from 1812 to the present and offers some assessment of current trends and future directions. While business historians have focused largely on such topics as innovation, production process development, and changing forms of corporate organization, little attention has been paid to the history of industrial marketing and much less to that of industrial distribution. However, using primarily the selected histories of industrial manufacturing companies, we have been able to track the evolving pattern of distributive practices in the United States over 175 years. Had a richer body of information existed, it might have been possible to flesh out this description in greater detail. We were satisfied, nevertheless, that the broad outlines of historical development could be traced through these limited data and that greater amounts of data would not have significantly altered our perceptions of trends or our explanations of causality.

This chapter explains why and how the industrial distribution infrastructure developed and how different institutions within it grew and declined in relative prominence. What we see is that the need for channels systems grew with the geographic expansion and dispersion of U.S. markets. Customers in these trading areas wanted broad arrays of merchandise sourced from a multiplicity of manufacturers. Because of transaction cost considerations, many producers found it more economical to utilize agents and distributors in the more remote and less concentrated market areas and to rely on direct salesforces in selling to larger accounts and to those in the more dense geographic markets. However, many small and/

or new enterprises relied entirely on external intermediaries, for lack of resources to support the fixed overheads incurred in having a direct sales operation.

Distributors and agents also served to absorb the output of mass-produced goods, channeling them to markets in which demand fluctuated. And so the imperative for many manufacturers to maintain steady levels of production to utilize plant capacities fully and to economize on unit costs led to a greater reliance on outside intermediaries.

Forces for Change

The growth of channels institutions was supported by several conditions: First, distribution networks were, of course, made possible by the development of the railroads and of shipping, air freight, and trucking for transporting goods throughout the country. A second condition, the development of standardized products using interchangeable parts, was essential to channels development. While manufacturers and their customers had often found it necessary to have direct relationships in transactions involving customized products, product standardization permitted the uncoupling of producer and user. Distributors and agents could adapt products to user needs within a set of standardized options; they could service and repair units in the field using interchangeable parts and prescribed service procedures.

Distribution networks thrived as well with the growth of transcontinental communications—the telegraph, the telephone, and the postal system. More recently, industrial distribution channels have moved into a new evolutionary phase with the use of computerized information systems. These several modes of communication link producers, intermediaries, and customers to facilitate merchandise flows, meet their needs for tailored product assortments and prompt delivery, and yet economize on inventories moving from factory to customer.

Patterns of Growth

As channels infrastructure developed, different elements gained or declined in relative importance as measured in terms of the volume of goods moving through them. Manufacturers' sales branches grew in importance along with product-market concentration because industrial goods producers wanted to serve their large customers through direct salesforces. They found it both economical and competitively advantageous to do so because direct relationships with customers provided the context for direct price negotiations, the exchange of technical product information, and access to market information. A growing emphasis on market share objectives during the twentieth century lent increased importance to the benefits of maintaining close ties to major accounts.

Independent distributors became increasingly prominent as a channels institution because they served a critical need, that is, making broad assortments of merchandise locally available to a myriad of industrial purchasers of all types and sizes. Both manufacturers' sales branches and agents typically carried much narrower product lists.

A critical factor fostering the growth of independent distributors—entrepreneurs who bought from manufacturers and sold to users—was the development of communications. By maintaining close links with producers, distributors could coordinate product deliveries and sales to economize on inventory investments and thus bring working capital requirements within manageable limits. The lower the required investments relative to sales, the more funds became available from personal sources and commercial banks.

The most prominent of channels institutions in the early 1800s, agents, declined in importance with the rise of manufacturers' direct salesforces and independent distributors. Originally, agents served to provide market access for those firms which could not afford a salaried salesforce and for those firms needing representation in distant market areas. In addition, agents often provided financing for industrial goods manufacturers at a time when banks were reluctant to do so.

With market growth and with the growth of individual industrial firms, however, manufacturers often found it advantageous to replace agents with direct salesforces and to cover thinner markets through distributors. Agents have continued, however, to serve usefully in reaching markets in which producers have had little experience, such as those representing new and very different applications and different geographic cultures. They have also been the channel of choice for small new enterprises introducing technical products, which require relatively long sales cycles and prolonged involvement with customers for purposes of user education, product customization, and postsale service.

Important developments in the twentieth century have included the growth of national networks of manufacturers' sales organizations, the continued rise of the independent distributor, and the increasing use of complex multichannel systems composed often of direct sales branches, independent distributors, and agents. In addition, some large manufacturers have added another element to their distribution networks, a captive distribution operation functioning as a business unit in the manufacturer's organization. For the most part, captive distribution emerged as producers acquired their independent distributors to protect market coverage in cases where the viability of the independent might be in doubt. In some instances, captive branches were simply started in trading centers where the manufacturer could not find a suitable channels alternative.

Market area concentration also provided the economic basis for the development of industry-sector and product-specialized industrial distributors. We see the growth of distribution networks serving such emerging market sectors as petroleum, automotive manufacture, aircraft manufacture, and electronics. In the more dense trading centers, some distributors found opportunities in providing limited product assortments for which there was high demand at prices lower than those of competing full-line distributors. The key was to achieve high stock-turns, thereby fully utilizing inventory space and other productive assets. For example, in many large trading areas, resellers specialized in such items as industrial fasteners (nuts, bolts, and screws), welding supplies, cutting tools, and construction machinery.

There also emerged a new type of reseller organization after World War II, the so-called VAR (value-added reseller). The phenomenal growth of the computer industry greatly increased the numbers and importance of VARs, many of which had particular expertise in the development of applications-specific information systems. The VARs coupled the computers of such established manufacturers as IBM, Digital Equipment, and Hewlett-Packard with their own proprietary software programs to meet a myriad of specialized needs in areas such as hospital management, engineering design, and legal research.

The concept behind VARs, however, was not entirely new. Resellers in some product-markets had long performed such value-adding functions as assembling products, cutting materials to size, and customizing standard equipment by adding peripheral product options to the buyer's specifications. What was new about computer VARs was the greater level of value contribution achieved through developing applications for classes of buyers rather than for individual customers.

In the twentieth century, too, the rapidly growing use of computerized information systems had a powerful impact on all of industrial distribution. The use of computers, not only for the management of inventories, shipments, ordering, and billing but also to provide information links to customers, laid the economic base for the emergence of large reseller organizations operating nationwide, and thus for increased concentration in the distribution sector. With this growth in the scale of enterprise, some larger distributors have pursued backward integration strategies by making products previously sourced from outside suppliers, by acquiring manufacturing businesses, and/or by developing private brand lines made to their specifications on contract with manufacturing firms. Thus, computerization has had a significant impact across the board on the growth of individual distributor businesses, on their strategies, on their methods of operation, on their operating costs, and on the nature of the services they provide.

With this brief introductory overview, we turn now to trace the evolution of an infrastructure that has been essential to the development of our economy and, at the same time, was made possible by developments in that economy—both in technology and in market demography.

Like the broader history of economic development in the United States, the history of industrial distribution falls roughly into three stages: from the War of 1812 until the Civil War, from the Civil War until World War I, and from World War I until the present. On reflection, it is not surprising that these stages are demarcated by major wars, since war is often succeeded by periods of industrial growth and demographic change. In this chapter, we will examine developments in these three periods, focusing in particular on the relative growth of the major channels institutions—manufacturers' salesforces, agents, and independent distributors. At the close, we will offer some observations on future trends.

THE EARLY YEARS: 1812–1865

The first distribution systems in the new country evolved from arrangements established by the British to accommodate export trade with their American colonies. British goods were deposited with agents at ports along the Eastern Seaboard and then taken by vessel up and down the coast and along rivers. After the Revolution, American manufacturers utilized these distribution systems and American merchants took over many of the wholesaling networks that had been run by British commission merchants. Ports that had functioned as supply points for British goods became American depots for a trade that consisted, at first, mostly of raw materials and consumer goods.[1]

The building of canals and the development of steamboats shortly after 1815 were the first major breakthroughs in inland transportation; the second was the construction of the first railroads in the 1830s.[2] Canals opened parts of the country that were previously inaccessible and connected formerly isolated transportation systems. Steamboats and railroads cheapened the cost and shortened the time of carrying bulky freight to the interior.

Industrialization increased hand-in-hand with invention. During the first decades of the new nation, state governments granted patents to inventors of such equipment as pumping and power machinery, rice-cleaning machines, steam engines, mill machinery, and various motors.[3] Then as transport improved, small manufacturing enclaves appeared along the East Coast and in the interior—at Worcester and Springfield in Massachusetts, for instance, and at Lancaster in Pennsylvania.

While the availability of transportation was a critical factor in the development of a distribution infrastructure, the nature of the product, supply industry characteristics, and the relative dispersion of markets were equally important. Distribution networks for raw materials such as cotton, and processed materials such as iron and tinplate, for example, developed more quickly than those for machinery and equipment. The numerous markets for the former were widely dispersed along the seaboard and inland, and there was relatively little need for a direct relationship between users and producers. Further, the economics of production were such that the continuous operation of productive resources—farms and factories—created an imperative for moving these goods to market as

---

[1] Norman Sydney Buck, *The Development of the Organization of Anglo-American Trade 1800–1850* (New Haven: Yale University Press, 1925), p. 68.
[2] Alfred D. Chandler, Jr., and Richard S. Tedlow, *The Coming of Managerial Capitalism* (Homewood, Ill.: Richard D. Irwin, 1985), p. 191.
[3] Victor S. Clark, *History of Manufacturers in the United States 1607–1860* (Washington, D.C.: Carnegie Institution, 1929), vol. 1, pp. 48–49.

they were produced. For raw materials, an established network of agents was essential to bridge the distance between point of production and points of use.

In comparison, industrial machinery and equipment in the early nineteenth century tended to be highly customized, requiring close communication between buyer and seller—from product development through repair and servicing. Markets were local, and machinery was typically manufactured to order rather than produced for inventory. Marketing efforts tended to hinge on direct relationships among managers and engineers representing buyers and sellers. As time went on, however, the development of standardized equipment and the resulting dispersion of machinery markets led to a significantly greater reliance on distribution networks deployed in major centers of industrial activity.

In what follows, we consider in more detail the development of channels first for raw and processed materials and then for machinery and equipment.

Distribution Systems for Materials

Producers of basic raw and processed materials relied heavily on middlemen to move goods from point of production to points of use. Cotton, for instance, was sold to textile manufacturers, both in England and in the United States, through intermediaries. Up until the Civil War, southern planters could consign their crops directly to established cotton merchants in New England or Europe and have them sold by agents; they could ship cotton to the nearest port and have it sold there by commission agents, often called "factors"; or they could deliver cotton to merchants at assembly points on their plantations.[4] Factors (distributors) dealing in cotton, as those dealing in other staple products at that time, often gave financial assistance in the form of credit to planters, thus tending to create long-term relationships with these sources of supply.[5]

[4] Buck, *The Development of the Organization of Anglo-American Trade 1800–1850*, pp. 67–68.
[5] A contemporary analyst, Alexander Trotter, noted in 1839 the extent to which planters were dependent on factors as a result of the financial services the latter offered: " . . . the cotton planters usually obtain advances for the purchase of their slaves and the improvement of their plantations from the merchants or factors of New Orleans, or other ports on the Gulf of Mexico." (Alexander Trotter, *Observations of the Financial Position and Credit of Such of the States of the North American Union as Have Contracted Public Debts* [London, 1938], p. 33.)

Other services that factors performed for planters included acting as selling agents

Tinplate is another example of an industrial material handled principally through merchants at this time. The records of Nathan Trotter, a Philadelphia metals wholesaler, show how this trade grew during the first half of the nineteenth century. Trotter bought tinplate, copper, sheet iron, and steel from England. He imported copper by the pound and tinplate by the box and resold it in smaller quantities, at first mainly to smiths and handicrafts people and later to fledgling industrialists.[6]

Restrictive trade legislation in the form of tariffs, before and after the War of 1812, separated American buyers from English suppliers and led to the growth of domestic suppliers of industrial raw materials. New producers of processed metals, one after the other, obtained protective tariffs for their own benefit, such as the tariff levied in 1816 on imported braziers' copper. Trotter, in time, was trading widely in American-manufactured steel and sheet iron. He operated both as an independent distributor, taking title to and stocking goods himself, and as a commissioned agent. Often he was able to charge higher prices to his customers because he was willing to grant them long credits. Other services that Trotter rendered to his customers included having an assortment of goods on hand at all times and packing and shipping "with the greatest of care."[7]

Product technology and manufacturing process characteristics often strongly influenced the degree of reliance on external distribution networks. In the case of crude iron, for example, the large capital investments necessary for smelting furnaces and the high cost of shutting down and starting up production created strong economic pressures to operate the furnaces continuously. Thus, production imperatives made it essential to have access to markets that could absorb a steady flow of product. Smelting operations, however, were prone to work stoppages due to furnace breakdowns, ore pit freezing, and labor difficulties, and no one producer could ensure the on-schedule fulfillment of orders. What was needed were intermediaries who could cluster the output of a number of producing sources and channel it to foundries, bloomeries, and rolling mills in widely scattered locations.[8] Thus, after 1815, as

---

of cotton and as buying agents of plantation and personal supplies. (Buck, *The Development of the Organization of Anglo-American Trade 1800–1850*, p. 97.)

[6] See Elva Tooker, *Nathan Trotter: Philadelphia Merchant 1787–1853* (Cambridge, Mass.: Harvard University Press, 1955).

[7] Tooker, *Nathan Trotter*, p. 133.

[8] Glenn Porter and Harold C. Livesay, *Merchants and Manufacturers* (Baltimore, Md.: The Johns Hopkins University Press, 1971), pp. 41–42.

demand increased with the rise of manufacturing and as the domestic and imported supplies of iron grew to meet the needs of a growing population, merchants specializing in iron emerged and flourished. Commercial directories for the 1820s list specialized iron merchants in Boston, New York, Philadelphia, and Baltimore. Most specialized iron dealers functioned as commission agents.

Because iron merchants often sold the production of several manufacturers, they sometimes were parties to price collusion when demand exceeded supply. As William Kemble of the New York commission agent house Kemble & Warner noted in 1845, "When the manufacture of boiler iron in Pennsylvania fell short of demand . . . prices were regulated by an understanding among the mill owners, and they reaped full advantages from adopting this course."[9]

Commission agents could become quite prosperous and were often important sources of capital for their principals. In the 1840s, for instance, George and Joseph Whitaker financed the growth of Whitaker Iron (later to become Wheeling Steel) by securing capital from Baltimore merchants Thomas Garrett and William Chandler. The new firm issued 4,000 shares of stock, of which the Whitakers owned 2,400 and the merchants 1,600.[10] The steel firm Jones & Laughlin owed much to James Laughlin, a Pittsburgh commission agent dealing in both iron and groceries, for the cash he contributed to its growth.[11]

Distribution Networks for Machinery and Equipment

As the early-nineteenth-century records of New England textile machinery companies illustrate, the markets for machinery and equipment tended to be local and were reached through direct sales. Product characteristics were then, as they are today, important factors in determining modes of distribution. For these industries, to a large extent it was the technical complexity of machinery that made direct selling imperative.

Textile machinery businesses, for the most part, started as captive machine shops operated by textile mills. Eventually, these machinery-building units began making machines and equipment for other local mills. For example, the machine shop of the Boston Manufacturing Company began taking orders from other manufac-

[9] Ibid., p. 50.
[10] *Minutes of Stockholders Meeting*, 1842, 1843, 1844; *Statement of Stock, Principio Furnace*, 1941, in *Principio Furnace Papers*, quoted in Porter and Livesay, p. 65.
[11] Porter and Livesay, p. 67.

turers in 1817, according to its records.[12] Sales of the company, later to be called the Lowell Machine Shop, remained local for a long time. In 1846 and 1847, 88% of its sales were to Lowell mills, 95% were in Massachusetts, and 99% were in New England.[13] Limited in its ability to expand geographically, the Lowell Machine Shop developed a highly diversified product line between 1845 and 1860:[14] "Everything from a jew's-harp to a locomotive."[15]

These textile machine shops typically produced only upon order, with the purchase initiatives being taken by the prospective users. Such a system had no room for outside distributors, agents, or even company salespersons, as such. A letter from General Stevenson, treasurer of the Lowell Machine Shop, to George Richardson, shop superintendent, dated November 14, 1873, for instance, clearly stated "our rule [is] to build no machinery excepting such as may have been ordered and contracted for by purchasers."[16] The heads of machine shops such as Lowell and Saco & Whitin expected customers to come to their establishments to inspect their machines and to negotiate a purchase, and they often took prospective buyers on tours to see machinery in operation in local mills.[17]

Standardization and Interchangeability. Two major and related developments led to significant changes in machinery distribution: equipment design standardization and the development of interchangeable parts and components. Standardization of machinery specifications evolved out of the close initial involvement among producers and users in designing customized equipment. Such interchange set the stage for designs based on commonality of needs and thus, paradoxically, on a lower level of involvement between machinery makers and their user-customers.

As Nathan Rosenberg notes,

> [There was] an interchange of information and communication of needs to which the machinery producers responded in a highly creative way. They learned to deal with the requirements of

---

[12] George Sweet Gibb, *The Saco-Lowell Shops: Textile Machinery Building in New England 1813–1949* (Cambridge, Mass.: Harvard University Press, 1950), p. 39.
[13] Ibid., p. 191.
[14] Jonathan T. Lincoln, "Machine Tool Beginnings," *American Machinist,* August 3, 1932, p. 902.
[15] Gibb, *The Saco-Lowell Shops,* p. 192.
[16] Ibid., p. 201.
[17] Arthur H. Cole, "Marketing Nonconsumer Goods before 1917," *Business History Review,* Autumn 1959, p. 423.

their customers at the same time that the machinery user learned to rely heavily on the judgment and initiative of the machinery supplier. It was, in part, the relative harmony and mutual confidence of these relationships which made it possible for machinery makers to eliminate customer preferences that were technically non-essential or irrelevant and therefore to design more highly standardized machinery.[18]

Thus, through collaboration, machinery builders and users carried out the development of standardized products in America. At the same time, standardization reduced the need for close working relationships. The way was then open for agents and distributors to serve as intermediaries selling technical but standardized products, which could be maintained in the field through the use of interchangeable parts.

Parts interchangeability is generally considered to have been developed in the early nineteenth century under the aegis of the U.S. Ordnance Department—with its armory at Springfield, Massachusetts, playing a major role—making possible the mass production of rifles for the U.S. government.[19] Early firearms factories, built for security reasons at interior centers, such as Springfield, Massachusetts, produced entirely handmade muskets. When one part failed, an expert gunsmith was required to make the repair so that returning nonworking muskets to remote factories became an enormous problem, which fortunately plagued the enemy as well.[20] Within a few decades, however, standardized parts revolutionized the manufacture of firearms.[21]

The development of the anthracite fields in eastern Pennsylvania beginning in 1839 contributed to the widespread development of standardized machine parts. Coal provided low-cost energy for making metal parts for machinery such as the newly invented sew-

---

[18] Nathan Rosenberg, *Perspectives on Technology* (Cambridge, Mass.: Harvard University Press, 1976), p. 164.

[19] David A. Hounshell, *From the American System to Mass Production 1800–1932* (Baltimore, Md.: The Johns Hopkins University Press, 1984), pp. 3, 30–32. Modern historians now recognize Eli Whitney, inventor of the cotton gin, not as the pioneer of machine-made interchangeable parts manufacture, as was once thought, but as a promoter and publicist of the idea.

[20] Eli Whitney, in a letter to the U.S. War Department at the start of the War of 1812, reported that the British government had on hand over 200,000 stands of muskets partially finished or awaiting repairs. See Joseph Wickham Row, *English and American Tool Builders* (New York: McGraw-Hill, 1916), p. 129.

[21] Clark, *History of Manufacturers*, vol. 1, p. 420.

ing machine and reaper, as well as for firearms.[22] Concurrently, a domestic machine tool industry emerged to supply turret lathes, milling machines, planers, and shapers on which parts could be fabricated to high tolerances. By the 1850s, machine tool companies such as the Ames Manufacturing Company in Chicopee, Pratt & Whitney in Hartford, Browne & Sharpe in Providence, and Sellers & Bancroft in Philadelphia were all well-established enterprises.[23]

Finally, it should be noted that the adoption of the commercial standard inch and yard, as well as the development of more accurate measuring instruments by such firms as Pratt & Whitney,[24] was an essential factor in design standardization and in the development of interchangeable parts and components.

All of these developments served to release machinery and equipment manufacturers from confined, local markets. It allowed them to tap more distant centers of trade by using agent and distributor networks and geographically deployed sales representatives.

Design standardization and parts interchangeability also made it possible for machinery and equipment producers to delegate to their distribution channels the functions of maintaining and repairing equipment in use and supplying replacement parts. Distributors, agents, and factory branches could also make markets for used equipment, all of which generated increased demand for new equipment and supported the proliferation of product applications, markets, and customers.

With the growth of the country after the Civil War, industrial goods producers relied increasingly on the channels infrastructure to reach geographically dispersed markets. While all three major distribution institutions—independent distributors, manufacturers' direct sales branches, and agents—continued to play important roles in industrial growth, the first two increased in relative prominence, and role of the third declined. Why this happened and how producers' distribution systems evolved with product-market growth and proliferation are the subjects of the discussion that follows.

[22] Alfred D. Chandler, Jr., *The Visible Hand* (Cambridge, Mass.: Harvard University Press, 1977), pp. 76–77.
[23] Ibid.
[24] *Accuracy for Seventy Years: 1860–1930,* Pratt & Whitney Co., Hartford, Conn., 1930.

THE MIDDLE YEARS: 1865–1920

The nation that emerged from the Civil War was far different from that at the beginning of the century. By 1870, the country had almost doubled in size to 3 million square miles, and its population had increased more than five times to 39.8 million.[25] Gross national product (GNP), estimated at $1.6 billion in 1839, averaged $7.4 billion between 1869 and 1878.

Paralleling this growth in size and commercial activity was a profound change in the basic foundations of the country's economy—energy, transportation, and communication. Cheap coal provided a source of energy that made it possible to replace artisans, small mill owners, and the so-called putting-out system with factories as the locus of production in many industries. Transportation improvements, especially the increase in the railroad network from 30 miles to 30,000 miles between 1830 and 1860, made larger markets a reality. Large-scale production became economically feasible and increasingly attractive to enterprising business people.[26] Information technology, which would galvanize distribution in the twentieth century, began its influence with the telegraph, first used commercially in 1847. In conjunction with the extensive railroad network, the telegraph made it easier for manufacturers to sell their products over greater distances than before. The telephone, commercialized in the 1880s, was the third cog in this powerful transportation and communication machine.[27]

The distribution infrastructure supporting the country's industrial growth and expansion continued to evolve, with the significant developments being (1) the growing role of the independent industrial distributor, (2) the relative decline in commission agents as a channels institution, and (3) the rise of highly organized manufacturers' sales networks, selling both to user-customers and to resellers.

On the whole, the use of external intermediaries—independent distributors, agents, and brokers—grew as new manufacturing companies sprang up. These firms were too small, at least ini-

[25] Porter and Livesay, *Merchants and Manufacturers,* p. 18.
[26] Ibid., p. 79.
[27] See Alfred D. Chandler, Jr., *The Visible Hand,* part 2, "The Revolution in Transportation and Communication," pp. 79–205.

Exhibit 12.1   Lowell Machine Shop Textile Machinery Sales by Region

Date	Lowell	New England and Mid-Atlantic	South	Other
1847	90.0%	10.0%		
1866–1870	34.8%	39.2%	15.2%	10.5%
1880–1883	28.0%	46.0%	24.0%	2.0%
1886–1890	8.0%	38.0%	38.0%	6.0%
1891–1895	12.0%	35.0%	47.0%	6.0%
1896–1900	10.0%	23.0%	64.0%	3.0%
1901–1905	9.0%	30.0%	60.0%	1.0%

Source: George Sweet Gibb, *The Saco-Lowell Shops: Textile Machinery Building in New England, 1813–1949* (Cambridge, Mass.: Harvard University Press, 1950), p. 243.

tially, to support sales departments capable of reaching customers beyond their immediate trading areas. In addition, the geographic proliferation and the migration of their markets forced some manufacturers to use industrial distributors or commission agents, even in the face of their strong biases against the use of intermediaries. The textile machine manufacturers, for example, were loath to change their modes of distribution, even when their markets moved south.[28] As we have pointed out, textile machinery manufacturers, before the Civil War, typically sold to a small number of mills in geographically restricted trading areas. However, this pattern was later drastically altered, as the sales record of the Lowell Machine Shop, given in Exhibit 12.1, shows.

Before 1880, the Lowell Machine Shop had relied on the personal contacts of the treasurer, the superintendent, and the directors to reach its new southern market. Neither Lowell nor its major New England competitor, Whitin, sold machinery through agencies in the early 1880s. The treasurer of Lowell, in fact, pointedly refused an agency offer in 1883, stating that customers preferred to do business directly with machine shops.[29]

Eventually, though, the New England textile machine manufacturers had to confront the fact that a direct sales system suitable for dealing with customers in close proximity to their plant sites would not work in the distant and growing southern textile market. They also recognized the benefit of selling their machines through

[28] Gibb, *The Saco-Lowell Shops,* p. 269.
[29] Gibb, *The Saco-Lowell Shops,* p. 247.

agents who carried the products of other domestic suppliers, and they faced ever increasing competition in the South. At the end of the century, a number of the smaller textile machine shops began to draw together to offer southern buyers a complete line of machinery.[30] They funneled their combined output through resident southern agents. One of the first of these "group" agencies was the Charlotte Machine Company. By 1896, it was handling such well-known machinery as Atherton pickers and Pettee cards, and advertised widely that it carried a complete line of cotton textile machinery.

Soon, the larger New England textile machine shops began to use agents, as well. In the 1905 edition of *Dockham's Directory,* both Whitin and Woonsocket listed Stuart W. Cramer in their advertisements as their southern agent.[31] The Lowell Machine Shop opened an office in Atlanta in 1899, with a manager working there on a commission basis.

Nearer to home, the textile machinery manufacturers were also selling their goods to industrial distributors, known as mill supply houses. Mill supply houses did well in circumstances where customers preferred "one-stop shopping." An added attraction was that unlike commission agents, they sometimes dealt in secondhand machinery. One mill supply house, for example, noted in boldface type in the 1905 *Dockham's Directory,* " 'Dealer In' Means I am a Seller and a Buyer—do not lose sight of either fact, please."[32] While the 1876–1877 edition of *Dockham's Directory* had carried only a few ads for mill supply houses, the 1905 *Dockham's* carried 32 such advertisements for distributors in New England, the Eastern Sea-

---

[30] As late as the second half of the 1870s, a quarter of the textile machinery installed in American mills was British, and a decade later, mill owners in the American South were still showing a clear preference for British products. *Dockham's Directories* in this period carried many pages of advertisements of English textile machinery plants. While many of these suggested direct contact with the manufacturers in the English Midlands, many also listed U.S. agents along the Eastern Seaboard and in the South. Company records show that all the important English textile machine builders sold in the United States through commission agents. Eventually, however, competition from British manufacturers in the American southern textile machinery market dwindled, largely as the result of a protectionist tariff, at one point as high as 45%, on imported machinery. (Derek H. Aldcroft, ed., *The Development of British Industry and Foreign Competition 1875–1914* [London: George Allen & Unwin, 1968], p. 193; Stanley D. Chapman, "British Marketing Enterprise: The Changing Role of Merchants, Manufacturers, and Financiers, 1700–1860," *Business History Review,* vol. 53 [summer 1979], no. 2, p. 230; Melvin Thomas Copeland, *The Cotton Manufacturing Industry of the United States* [Cambridge, Mass.: Harvard University Press, 1912], p. 317.)

[31] *Dockham's Directory 1905,* pp. 2, 12.

[32] Ibid., p. 601.

board, and the South.[33] At the same time, manufacturers indicated in *their* advertisements that mill supply houses handled their goods. William Coupe & Company, maker of industrial belting, advised its customers in 1905, "Ask your dealer for it—take no other that may be represented to be as good; and if he does not furnish it, send it to us, and we will furnish it at the regular dealer's price."[34] (It may be noted, parenthetically, that having elected to use resellers, this company was then protecting its distribution system by pricing to users at the same prices they would pay a dealer.)

Thus, both commission agents and independent distributors continued to flourish throughout the nineteenth century. Commission agents, the older of the two institutions, had been active in a number of businesses since colonial times. They did well particularly in foreign trade, which was a very active segment of the economy except during wars.

However, a major problem with agents was calling them to account and recovering sums due. Deere & Company, manufacturers of agricultural machinery, for instance, relied heavily on agents but noted, at one point, that "there was but one agent in our whole clientage that would pay us the money for what . . . he sold."[35]

While the use of independent distributors and agents grew rapidly with the increasing industrialization of the country, the former gained in prominence over the latter. After 1865, company salesforces, covering large sections of the United States, also became increasingly important channels for selling industrial goods. The following discussion considers the growth of distributors, agents, and company salesforces as industrial marketing institutions, and the factors that tended to produce shifts in their relative prominence.

Independent Distributors

In the first decades after the Civil War, independent distributors came into their own. For the most part, prior to the war, distribution was dominated by agents, who did not take title to goods but sold on commission. But dramatic improvements in transportation and communication led to increased efficiencies in physical distribution, reduced inventory-to-sales requirements, and lowered

---

[33] *Dockham's Directory 1905.*
[34] Ibid., p. 9.
[35] Wayne G. Broehl, Jr., *John Deere's Company* (Garden City, N.Y.: Doubleday, 1984), p. 87.

inventory investment risks. Many intermediaries then elected to purchase goods and became full-line independent distributors, rather than sell on commission.

A critical measure of performance for independent distributors has always been "stock-turn," the product volume channeled through a single set of storage facilities within a specified period of time. The greater the stock-turn, the more intensive the use of existing personnel and other resources and therefore the lower the distribution cost per unit.[36] Independent distributors did well after the Civil War, as now, in situations where market demand is for a range of products going well beyond the product lines of individual manufacturers.[37]

Commission Agents

The relative decline of the commission agent after the Civil War may be attributed to the declining utility to producers of the functions it performed. One of these was financing. As Professor Alfred D. Chandler, Jr., observed in *The Visible Hand,* commission agents steadily became less important to manufacturers as sources of capital for growth and the financing of receivables. Increasingly, new plant and equipment could be financed from retained earnings. In addition, after 1850, commercial banks began providing working capital to industrialists; before that time, they had concentrated primarily on financing the movement of agricultural products.

Another reason for replacing agents with direct salesforces may have been that manufacturers wished to have more direct relationships with their larger customers, to exercise greater control over negotiations, to provide services, and to have greater flexibility in responding to customer needs. Because excesses of supply over demand developed after the Civil War, and prices fell sharply from the mid-1870s to the mid-1890s, close relationships with important accounts became all the more imperative.

Agents remained, however, a preferred channel for reaching distant markets, for example, export markets, where ability to deal in the local environment gave the agent an advantage. In addition, agents often served as transition channels for companies moving into new product and/or geographic markets. Examples of the use

---

[36] Alfred D. Chandler, Jr., *Scale and Scope* (Cambridge, Mass.: Harvard University Press, 1989), chap. 1, p. 20.
[37] Scott J. Moss, *An Economic Theory of Business Strategy,* (New York: Halsted, 1981) p. 137.

of transition distribution strategies are clearly seen in the sulphur industry. Before World War I, the Union Sulphur Company dominated the market in the United States, selling mostly direct to a few large regular buyers in the chemical industry. In 1912, another company, Freeport, began to produce sulphur. To break into the market, Freeport appointed Parsons & Petit of New York as its commissioned agent. Parsons & Petit had served as importing agents for the Sicilian sulphur cartel and had valuable customer contacts. Eventually, Freeport served this market through its own direct sales organization. Similarly, when a third company, Texas Gulf, entered the sulphur market during World War I, it appointed a long-established distributor selling to fertilizer manufacturers, H. J. Baker & Bros., as its agent to build sales in this segment of the market. Once established as a sulphur supplier to fertilizer companies, Texas Gulf switched to direct sales.[38]

Direct Sales Organizations

The growing prominence, at the turn of the century, of company-employed sales representatives deployed over wide geographic areas is related directly to industrial expansion and to the emergence of large trading centers, such as New York, Boston, Chicago, Buffalo, Cleveland, Detroit, Pittsburgh, Philadelphia, St. Louis, and San Francisco. As the nineteenth century progressed, manufacturers in a number of industries had moved to replace outside intermediaries with their own direct sales organizations.

Production economies of scale encouraged direct sales in dense markets. Manufacturers committed to large-scale production frequently integrated forward to create their own marketing and sales organizations, as well as integrating backward to control their sources of supply.[39] By these means, they were better able to control the flow of products from their origin as raw or processed materials, through manufacturing, to end-use markets. Both backward and forward integration often eliminated the need for external intermediaries.

The iron- and steel-making companies, for example, shifted reliance from independent resellers and agents to direct sales, in

---

[38] Williams Haynes, *Brimstone: The Stone That Burns* (Princeton: D. Van Nostrand, 1959), pp. 101–103.
[39] Alfred D. Chandler, Jr., *Strategy and Structure* (Cambridge, Mass.: MIT Press, 1962), p. 25.

part as a result of the backward integration of rolling mills. These mills began to produce their own crude iron to support the high-volume steel production made possible by the introduction of the Bessemer process.[40] The steel companies were thus able to meet, in particular, the railroads' need for large quantities of steel rails of high strength.[41] (The production of steel rails in the United States increased sevenfold to 4.5 million tons in 1906.)[42]

Steel manufacturers could not use the ungraded pig iron handled by distributors because of the rigid specifications for pig iron imposed by the Bessemer process. It became necessary, then, for the steel companies to control quality and to ensure supply availability by owning and operating their own blast furnaces.

Iron and steel producers also integrated forward to create their own direct sales branches. By the 1890s, a network of iron and steel manufacturers' sales offices had opened in major cities across the nation.[43] These steel producers, including Carnegie Steel Company, the largest American iron and steel manufacturer before the formation of U.S. Steel, pioneered the development of direct sales branches. In its early days, Carnegie had made many sales through iron commission merchants. In the late 1880s, it began using authorized, exclusive regional agents. Then, by the turn of the century, Carnegie had company sales offices in dozens of major North American cities.

In these capital-intensive industries, as in others, where a few large plants could meet existing demand, holding market share to achieve scale economies became a prime strategic objective. Loss of market share for one manufacturer resulted not only in increased production costs but also in decreased costs for those competitors that gained share. Manufacturers' direct links to major customers provided a measure of security in the competitive give-and-take of the marketplace.

Manufacturers of products that were technically complex, and those which required direct customer contact for service, also tended toward direct sales but, by this time, on a national scale.

[40] Peter Temin, *Iron and Steel in Nineteenth Century America* (Cambridge, Mass.: MIT Press, 1964), p. 154.
[41] Ibid., p. 131.
[42] U.S. Department of Commerce, U.S. Bureau of the Census, "Physical Output of Selected Manufactured Commodities: 1860–1970," *Historical Statistics of the United States: Colonial Times to 1970* (Washington, D.C.: GPO, 1975), pp. 693–694.
[43] Ibid., p. 140.

General Electric and Westinghouse, for example, developed sales departments to handle heavy equipment and apparatus that were made to customer specifications, such as large steam turbines, generators, motors, and switchgear.[44] The scale company, E & T Fairbanks, also followed this pattern, as did the large typewriter firms later.[45]

Increasingly, too, manufacturers began to utilize mixed systems. Bucyrus-Erie, for instance, "allowed sales agencies to promote the distribution of older types of equipment, such as steam shovels, but the sale of new and specialized apparatus was restricted to company offices."[46] The textile machinery manufacturers, although they now sold through intermediaries in the South, also used direct selling supported by offering such services as drafting mill plans for prospective mill owners.

All of these developments in distribution took place in the context of a rapidly growing economy following the Civil War. Between 1869 and 1873 and 1912 and 1916, the population of the United States increased almost two and one-half times, rising from 41 million to 100 million. From 1869 to 1903, national income more than tripled, to $21.7 billion.[47] In about the same period, wholesale trade increased sixfold to $1,300 million,[48] the production of industrial machinery and equipment quadrupled to $446.9 million, and the production of industrial and commercial electrical equipment increased from $1.9 million to $111.2 million.[49] In addition, between 1869 and 1899, value-added by manufacturers, in 1879 prices, increased over six times to about $6.3 billion.

In Summary

By the mid-1800s, the development of distribution channels for industrial products had been greatly facilitated by the release of constraints related to market access and product and manufacturing technology. The growth of water and rail transportation in the first half of the nineteenth century not only made possible the expansion

[44] Chandler, *Strategy and Structure*, pp. 28–29.
[45] Broehl, *John Deere's Company*, p. 175.
[46] Cole, "Marketing Nonconsumer Goods before 1917," p. 423.
[47] U.S. Department of Commerce, U.S. Bureau of the Census, "National Income and Persons Engaged in Production, by Industry Divisions: 1869 to 1970," *Historical Statistics of the United States: Colonial Times to 1970* (Washington, D.C.: GPO, 1975), p. 240.
[48] Ibid., p. 839.
[49] U.S. Department of Commerce, U.S. Bureau of the Census, "Value of Output of Finished Commodities and Construction Materials Destined for Domestic Consumption at Current Producers Prices, 1869–1939" (Washington, D.C.: GPO, 1975), pp. 701–702.

of markets but fostered the growth, as well, of distribution networks from east to west across the country. The development of standardized designs and interchangeable parts and components was a second factor contributing to the greatly increased use of external intermediaries for reaching distant markets.

Once established, distribution channels performed a wide range of services in supporting the growth and industrialization of America. Distributors and agents absorbed the outputs of large and small producers of agricultural and processed materials flowing steadily from farms and continuous process manufacturing facilities. They clustered the products of many suppliers to offer purchasing convenience to user-customers. They took on responsibility for after-sale service and often made markets in used equipment. Intermediaries were also important sources of capital, for both their suppliers and their customers, extending credit and often helping to finance their suppliers.

Between 1865 and 1920, reliance on external intermediaries—agents and independent distributors—increased significantly, and independent distributors emerged as the more important channel in terms of dollar value of transactions. Agents, by comparison, declined in importance as a channels institution. They were no longer essential, in many cases, as a source of capital. More important, with the growth of concentrated trading areas, producers found it increasingly cost advantageous to supplement agents with direct salesforces. Agents, however, continued to be relied upon extensively for representation in geographically remote markets and for new market entries.

Direct sales networks emerged partly for distribution cost considerations but also to gain greater control over the movement of goods to markets and the relationships with key customers. As commitments to large-scale production lent urgency to market share goals, manufacturers found security in having direct relationships with customers on such matters as price negotiation, technical interchange, and sales-related services. These became increasingly important as elements of strategy in competing for markets, many of which were characterized by an excess of supply over demand.

DISTRIBUTION IN AN URBANIZING ECONOMY: 1920–1985

Between 1920 and 1985, the country urbanized rapidly as industry and the big cities grew, and small farms merged into large, highly mechanized units. U.S. population doubled during this period to 238.8 million, and gross national product increased over fortyfold to $3.7 trillion. As of 1985, an estimated 65% of industrial product-market potential was centered in 50 trading areas; the 10 largest centers accounted for almost one-third of the total potential.[50]

The discussion that follows considers the impact on industrial distribution in the United States of economic growth, increasing market concentration, and the development of computerized information systems following the Second World War. We examine, in particular, the trend toward direct selling, the increased use of multichannel distribution systems, and the emergence of captive distribution operations as business units in producers' organizations. We then review developments in the external channels infrastructure—backward integration in larger reseller firms, the emergence of so-called value-added resellers (VARs), and the use of computerized information systems to reduce distribution costs and gain competitive advantage.

The Development of Internal Distribution Networks

By the early 1930s, a few machine tool companies had grown quite large and had begun to turn increasingly to direct sales, using their own so-called "specialists" or "sales engineers," who worked out of branch offices and warehouses in principal market centers. While the smaller companies continued to rely on independent distributors, the larger companies opted for direct sales and often came to dominate individual product-markets.[51]

In the chemical industry, distribution patterns had evolved as well—starting with direct sales, moving to the use of external intermediaries with the growth of the railroads, and returning to an emphasis on direct sales with increasing geographic concentration in the 1920s. As the chemical industry developed, it found its markets largely in the steel, petroleum, textile, paper, fertilizer, and explosives industries in the East. By the beginning of World War I, it was centered in the Atlantic coastal states close to its primary customers.[52] This concentration and the growth of markets for chem-

[50] "Demographics of Distribution Are Changing, Finds ASMMA Study," *Industrial Distribution*, vol. 74, November 1985, p. 7.
[51] Glover and Cornell, *The Development of American Industries* (New York: Simmons-Boardman, 1959), p. 546.
[52] J. F. Cady, "Note on Legal Issues and the Pricing Process," #9-578-205, rev. Boston: Harvard Business School, 1978, p. 14.

icals gradually led to an emphasis on direct selling. Large customers preferred dealing directly with manufacturers, without the intervention of distributors—for one reason, because many of the newly developed chemicals required extensive instruction in their use.

As with both machine tools and chemicals, direct selling tended to predominate in other industries where there were easily identifiable, large, and concentrated markets. In the 1920s, a considerable proportion of lumber—perhaps one-third of the virgin yellow pine, for example—was sold by the producer directly to its big consumers—the railroads, manufacturers of railroad equipment, and firms engaged in the construction of buildings, bridges, and other structures requiring lumber in large quantities.[53] Direct sales were also the rule for industrial equipment, including installations and accessory equipment, in the mid-1920s. As Melvin Copeland observed,

> . . . factory equipment such as knitting machines, spinning frames, looms, material-handling equipment, refrigerating equipment, and steam boilers are usually marketed directly. Orders for installation in factories are commonly large enough to warrant direct marketing and in numerous instances, technical sales service is required.[54]

Thirty years later, in the 1960s, a survey by Ohio State University's Bureau of Business Research showed that this same dynamic still operated: "A substantial number of individual company product lines in the area of industrial machinery, equipment, and supplies are sold through channels which do not involve the use of industrial distributors at all."[55] Sales volume appeared to be a significant variable in the choice of distribution systems, with those producers in the survey reporting both the largest and smallest sales revenues being less reliant on independent distributors. On the one hand, large companies tended to prefer direct sales; on the other hand, small manufacturers with limited financial resources selling in wide geographic markets tended to use agents rather than independent distributors.[56]

[53] Edmund Brown, *Marketing* (New York: Harper, 1925), p. 221.
[54] Melvin T. Copeland, "Securing Distribution of Industrial Goods," *Advertising & Selling*, November 30, 1927, p. 58.
[55] Ibid., p. 81.
[56] The Ohio State survey of 485 producers of various lines of industrial machinery, equipment, and supplies found that "the sales size class of under $1,000,000 tended to make greater use of manufacturers' agents as a substitute for sales branches or a factory sales force selling to industrial distributors. . . . Conversely, no agents were used in the major channel of distribution of any firm or division with an annual sales volume greater than

The Use of Mixed Systems.   Depending on product, market, customer size, and producer resource availability, many producers chose to go to market through direct salesforces, independent distributors, or agents. In addition, however, many developed multiple channel systems, using company-employed sales personnel in certain markets and distributors in others. The Ohio State study, in the early 1960s, found that it was "a common practice to reserve the immediate territory about the plant for the company salesmen or even sales executives, while distributors are used in more distant areas."[57]

Minnesota Mining & Manufacturing Company, for example, had a distribution system at the end of World War I that included "fifteen large distributor houses augmenting the 3M salesforce of five salaried men and two commissioned men who worked large territories out of Detroit, St. Paul, Chicago, Milwaukee, and Cincinnati."[58]

The Ohio State study also found that some companies varied in their reliance on direct sales or distributors not only in terms of geographic location of customers but also "to secure adequate coverage of different customer classes."[59] For example, Norton Company, the world's largest manufacturer of abrasives (including grinding wheels and sandpaper), utilized multiple channel types to reach diverse markets. In 1964, Norton sold through full-line distributors, product-specialized distributors, and distributors serving particular industries, such as steel, glass, furniture, automotive, construction, and pulp and paper, as well as its own direct sales organization, as Exhibit 12.2 shows.

Captive Distribution Systems.   While the first captive distribution systems predated World War I, they became solidly established as important channels elements in the 1920s. As noted in Chapter 8, the Westinghouse Electric Supply Company (WESCO), for example, began an operating unit in Westinghouse in 1922, when George Westinghouse acquired seven bankrupt independent distributors in order to maintain the company's distribution in cer-

---

$10,000,000." The survey suggested that often agents used by small firms functioned "primarily to sell to distributors who in turn handled the sales of the product to users."

[57] William M. Diamond, *Distribution Channels for Industrial Goods* (Columbus: Ohio State University, 1963), p. 62.

[58] Carla Anderson Hills, ed., *Antitrust Adviser,* 3d ed. (Colorado Springs, Colo.: Shepard's/McGraw-Hill, 1985), sec. 2.04, pp. 78–79, and sec. 5.04, pp. 342–343.

[59] Ibid., p. 52.

Exhibit 12.2   Norton Company's Distribution Network, 1964

	Approximate Number	Approximate % of Sales
Specialist distributors for		
• grinding wheels only	20⎤	
• cutting tools (e.g., grinding wheels, drills, milling cutters, chisels)	50⎦	7
Narrow line distributors with 15–20 lines (e.g., grinding wheels, transmission belts, gears, portable power tools)	125	20
Broad line distributors with several hundred lines	125	35
Industry-oriented distributors with broad lines	105	7
Direct sales representatives	85	30

Source: Norton Company.

tain areas of the country. Through World War II, WESCO continued to expand primarily by acquiring bankrupt or faltering independent distributors. After World War II, WESCO expanded both by developing its own branches in many locations and by acquiring smaller, local distributors. With sales estimated by industry sources in 1985 to be approximately $1.6 billion, WESCO was second only to Graybar Electric, an independent distribution firm.

GESCO, General Electric's captive distribution business unit, was third in size, with estimated sales in 1985 of $750–900 million and approximately 150 branch sales locations. Other examples of captive distribution systems include U.S. Steel's Oilwell Supply Division, American Standard's Amstan Supply Division, and Ingersoll-Rand's air centers.

As in these cases, many suppliers built captive distribution systems through the acquisition of failing independent distributors, through mergers with other manufacturers having captive sales branches, and through opening up new selling locations in areas where they did not have effective sales representation, selecting this option over a direct sales office or the appointment of an independent.

Developments in Independent Distribution

Independent distributors became an increasingly important element in multichannel systems, largely as a result of scale advantage. Further, in multichannel systems, the demarcation of roles between producer and intermediary often became less clear as some resellers integrated backward, took on private brand lines, or engaged in extensive end-product design and assembly.

Backward Integration and Private Branding. As individual distributor firms grew in size, some found profit opportunities in manufacturing high-volume products that had previously been sourced from outside vendors. More often, they persuaded their suppliers to provide products made to the distributor's specifications and identified by the distributor's brand name. Able to buy in large quantities, distributors pursuing private brand strategies for parts of their lines typically solicited bids from competing suppliers and negotiated prices below those for comparable manufacturer brand merchandise. Sometimes, private brand revenue was a substantial portion of the distributor's sales volume. For example, W. W. Grainger, Inc., a large distributor of electrical equipment with 1985 sales of over $1 billion, reported that 57% of its sales in that year came from air compressors, fans, blowers, and sump pumps it manufactured and from power transmission belts, motors, heaters, and other products made by outside producers under Grainger brand names. The remaining 43% of Grainger's sales came from over 800 vendors, mostly manufacturers.[60]

Typically, products selected for private branding were those for which there was no strong customer preference for manufacturers' brands, those at a relatively mature stage of the product technology, and often those which were purchased by users as items on a "shopping list."

VARs: An Important Force in Computers. In the early stages of the commercialization of computer technology following World War II, the major producers—firms such as IBM, Hewlett-Packard, and NCR—reached their customers through their own direct salesforces. Each potential buyer was large and easily identifiable and needed computerized information systems tailored to the size of the organization and the nature of its work.

Computer industry growth after the 1960s was characterized by rapidly evolving product technology, the emergence of many different information-processing needs, and the development of

---

[60] W. W. Grainger Annual Report, 1985.

applications-specialized software. At the same time, the markets for computers grew quickly in type, size, and number of potential accounts and segmented rapidly into a myriad of niches defined in terms of both application and user industry. The industry came to conceive of itself as developing and selling "solutions."

Computer manufacturers began, then, to develop so-called alternate channels of value-added resellers (VARs), which could combine their machines with software packages often designed by the VARs themselves. There were perhaps two primary reasons for the emergence of this VAR channel. VARs might reach small users at a lower ratio of sales costs to sales revenue than the computer manufacturers could achieve. In addition, the development of software could proceed more rapidly and effectively by tapping the inventive ingenuity of large numbers of individuals, many of whom could apply their previous work experience in fields and functions for which they then developed software programs.

The rise of the value-added reseller underscored both the rapid responsiveness of individuals to new business opportunities and the effectiveness with which the channels infrastructure was able to adapt to the needs of new and unique market segments. There is another significance: VARs demonstrated the capacity of resellers to engage in the highly technical adaptation of producer equipment to individual user needs and to interact with customers on technical product dimensions. It will be recalled that, in the past, a strong reason for the rise of producers' direct salesforces was that the need for effective technical communication between producer and user precluded the use of external intermediaries. In the case of the computer industry, however, external intermediaries have effectively related to user-customers on technical product dimensions and, in so doing, have greatly facilitated the development of computer applications technology and market growth.

The Computerization of Distribution Functions. Perhaps the single greatest development in distribution in the twentieth century has been the use of computerized information systems (CIS) for managing such functions as ordering, inventory control, shipping, billing, and credit and collections. Of particular importance has been the use of interorganizational systems (IOS) for linking distributors and manufacturers with their customers' ordering departments. A study made at the Harvard Business School in 1983 of five large national distributor businesses indicates that computerization has (1) made possible significant reductions in distribution costs, (2) enabled channels institutions to add value in the form of increased customer services, and (3) led to a growing concentration among independent distributors serving particular industrial sectors.[61]

The large nationally organized distributor businesses have used CIS to give operating personnel immediate access to such supply information as the following:

- in-stock availability by stock location
- amounts on order from suppliers and scheduled shipment dates
- items that can be substituted for those which are out of stock
- prices by quantity levels (and in some systems the price at which the last sale was made)
- sales volume, by branch, for each month in the preceding twelve months

Computerized systems are also used to obtain the following information by customer:

- volume of purchases by item each month for the preceding twelve months
- items on order and promised shipment dates
- accounts receivable and payment records in terms of average days outstanding

Using such information, inventory-to-sales ratios have been reduced by deploying field stocks selectively in regional and local stocking facilities and networking the system to give each sales area access to inventories carried at several locations. In addition, such data systems have led to the increased use of drop shipments from producers' stocks to fill customer orders directly without moving through distributor warehouses.

With computerized links to customers, it has been possible to eliminate large cadres of telephone-sales clerks who received orders and typed up purchase forms, as well as office workers who handled billing and receipts. This has been accomplished through installing round-the-clock automated order-entry systems by which orders may be transmitted from a computer at the customer's location,

---

[61] See E. Raymond Corey, "The Role of Information and Communications Technology," in *Marketing in an Electronic Age*, ed. Robert D. Buzzell (Boston: Harvard Business School Press, 1985).

received at a distributor data center, processed automatically for shipment, and invoiced without human intervention. Similarly, computer links with suppliers have resulted in significant reductions in purchasing personnel by centralizing the procurement function and automating ordering procedures.

By generating data on customer ordering patterns, CIS has also provided the means for offering customers value-added services. One large distributor serving the construction trades, for example, has supplied its contractor customers with monthly reports showing by item the quantities ordered, prices paid, number of times orders were placed, and amounts delivered to each customer location or job site. Another compilation compared these data with year-ago purchases, for use in performance reviews and forecasting. The value of such data to the customer has tended to foster long-term account relationships and the reliance of user-customers on one or a very few distributors for the bulk of their requirements.

In Summary

The driving forces in the evolution of industrial distribution in the United States since the First World War were population growth, urbanization, and the growing concentration of industry in large trading areas. A key factor shaping channels development, as well, was computerization. These forces led to rapidly increasing specialization and concentration in the distribution infrastructure. Distributors emerged whose businesses were defined in terms of user industries, such as steel, mining, textiles, and shipbuilding. Others specialized by product categories, for example, welding supplies, cutting tools, and industrial chemicals. Some specialized, in a sense, by addressing service needs of small industry in secondary and tertiary trading areas.

In addition, there emerged a finely articulated division of labor in distribution. Manufacturers have tended to use their own salesforces to reach large customers and to cover concentrated trading areas. They have left to intermediaries the smaller customers and the less concentrated markets. More important, they have relied on distributors to cluster items having relatively small dollar value, in arrays that suit the purchasing behavior of a myriad of customers. Agents have served largely as the alternative for direct selling for those manufacturers too small to afford the fixed expense of salaried salesforces or for larger manufacturers attempting to enter unfamiliar markets.

With regard to distribution functions, however, the division of roles among producers and external intermediaries has become less rigid. Some distributors, VARs, for example, are taking over product design and assembly functions from manufacturers. Some manufacturers have integrated forward by developing captive distribution organizations. Some distributors have integrated backward by developing their own private brand lines either sourced from outside manufacturers or produced in their own manufacturing facilities.

The development of computerized information systems (CIS) for distribution management has had a significant impact on distribution costs, which, in general, gives further impetus to the trend away from direct sales and toward the use of independent distributors. The economies of scale achieved by resellers buying in large quantities from many producers and selling in small amounts to many users are greatly magnified by computerization. Resellers can develop networks of field stocks with the potential for significantly reducing their inventory-to-sales ratios and improving order fulfillment performance. They can replace cadres of telephone and clerical personnel by communicating with both customers and suppliers through interorganizational computer systems.

In some important product categories, industrial components and hospital supplies, for example, scale economy opportunities have opened the way for distributor industry concentration. The first movers, in each instance, have taken advantage of new computer technology to increase significantly their shares of sales to individual accounts and of their markets overall.

FUTURE PROSPECTS

As for future directions, it is reasonable to predict a continuation of currently observable trends. On the whole, the balance of power in supplier-distributor relationships may be swinging more toward the distributor. First, the distributor-served segment of the market is growing more rapidly than the direct-served segment. Second, with growing concentration in the distribution industry, the larger firms gain in bargaining power. Third, the benefits of computerization as a factor contributing significantly to the reduction of transaction costs are likely to be of greater potential benefit to resellers than producers, inasmuch as the former handle far greater arrays of products, deploy inventories over much larger territories, and sell to greater numbers of customers. Finally, the opportunities for distributor value-added have increased markedly in the past two decades in ways we have discussed previously.

Possibly, too, we may see the growth of distributors operating multinationally, both sourcing from and selling to a range of industrial countries. The Japanese trading companies have served as a model, but one that has been difficult to emulate in the United States. Surely, though, the technology is in place to manage multinational distribution systems, and the advantages of doing so are attractive.

It seems clear, in any case, that what has been predicted in the past—the demise of the independent distributor as producers take over more and more aspects of distribution—is not likely to happen. On the contrary, the tide is moving in the other direction.

In the producer-user relationship, the relative bargaining power of the producer may also decrease as user firms (1) become larger in size, (2) move increasingly toward centralized purchasing, and (3) experience intensified cost pressures in the competition for global markets. In addition, we may see a drive on the part of large users of industrial products to lower costs by developing their own captive distribution systems as procurement arms, forcing suppliers to utilize these in-house networks. This thrust is already in evidence in the case of very large hospital groups.

In Chapters 11 and 12, we profiled the industrial distribution infrastructure in the United States and linked developments in technology, demographics, and economics to its evolution over about 175 years of our history. In Chapter 13 we will consider the relationship between the regulatory framework and the functioning of our industrial distribution infrastructure.

13 The Legal Framework

Federal and state antitrust laws, as interpreted by the courts, govern many aspects of the relationships between manufacturers and resellers. Lawyers and economists describe these as *vertical relationships*—as opposed to *horizontal relationships,* which are any involvement competitors may have with each other. The relevant laws and the court decisions that apply to vertical relationships do not differentiate between supplier-reseller relationships in industrial distribution and those of retail trade. Thus, the same body of legal doctrine applies to both, and in the discussion that follows, it has been relevant to cite court decisions in both sectors. It should be noted, as well, that the laws that are applicable to producers' relationships with resellers do not apply to manufacturer-agent relations. An agent does not take title to the goods and acts as a representative of the principal.

In vertical relationships, legal issues arise in connection with suppliers' attempts to impose certain constraints and conditions on distributors' behavior with regard to what they stock, where and to whom they sell, and at what prices. Thus, manufacturers may seek to influence the prices at which distributors resell their products. In so doing, they may risk acting illegally although price restrictions are not illegal per se. The manufacturer may also delimit the geographic area in which a distributor sells the manufacturer's products or restrict the particular customers or classes of customers to which the distributor sells. Further, manufacturers often attempt to influence distributors' product line choices by (1) prohibiting them from stocking and selling directly competing products, (2) requiring them, through so-called tying arrangements, to buy one product or service in order to obtain another, or (3) requiring them to carry the manufacturer's full line of products as a condition of the

256

franchise. Finally, certain legal considerations apply to the manufacturer's rights to terminate distributors.

In this chapter, we will discuss the ways in which suppliers attempt to influence distributors, the federal legislation that relates to these actions, and the attitudes of the courts regarding each of the major forms of vertical restraint, based on recent rulings.

WHY DO MANUFACTURERS WANT TO INFLUENCE DISTRIBUTORS' BUSINESS OPERATIONS?

In attempting to influence distributor actions with regard to resale price, territorial and account coverage, and product lines carried, the manufacturer may have a variety of objectives. It may well be in the manufacturer's interests, for example, to forestall price competition on its product lines at the distributor level in order to preserve, to the greatest extent possible, the margins for all members of the distribution network and thereby provide incentives for them to represent the manufacturer's lines effectively. If some distributors in a trading area sell at lower prices than other distributors that may provide a range of product-related services and therefore have higher selling costs, prospective customers may utilize the services of the latter but buy from the former. The so-called free riders may thus increase their market share at the expense of full-service distributors, and the latter may then find it unprofitable to devote resources to the product line, choosing either to drop it or, at best, to give it little attention.

In addition, intrabrand price competition at the distributor level often causes intensified interbrand rivalry among competing manufacturers as each seeks to support its distributors' resale price cuts by lowering wholesale prices. Thus the keen interest manufacturers have in resale price maintenance may be attributed to their concerns for (1) preserving market coverage through resellers, (2) motivating distributors to provide full product-related services to their customers, and (3) avoiding the contagion of price wars started through intrabrand price competition at the distributor level.

Efforts to impose *territorial* restrictions on distributors are also motivated by a desire to dampen intrabrand rivalry; giving distributors bounded trading areas from which other resellers may be excluded serves to reduce competitive price pressures at the resale level and to make it attractive for distributors to represent the product line effectively. Opportunities to profit from carrying a line are enhanced to the extent that a distributor does not have to compete with other resellers for the available business in a geographically defined market.

Manufacturers may also attempt to draw lines between those accounts and classes of customers to which they wish to sell directly and those which are available for distributor solicitation, thus intervening in the distributor's choice of customer. If such distinctions are made, typically the manufacturer chooses to serve the large accounts and to leave the smaller ones to distributors. In so doing, its purposes are to control large account relationships and to ensure that these customers are well served. In particular, the manufacturer will wish to control price negotiations. And most important, it will usually be more profitable to deal directly on large sales than to give up the distributor margin.

The pressures that a manufacturer may place on distributors not to carry competing lines are intended to ensure that resellers give their full support to the manufacturer's products. At the same time, by acting to keep its distributors from carrying other brands, the manufacturer is, in effect, engaged in a facet of competitive rivalry, that is, blocking its competitors' access to end-markets.[1]

In requiring its distributors to take and sell the full product line and in using tying arrangements under which the reseller must buy one product to get supplies of another, the manufacturer is using what power it has in the vertical relationship to gain market access for its total offering. Otherwise, resellers may be inclined to

---

[1] An interesting example of channel blocking is the behavior of major U.S. semiconductor manufacturers aimed at preventing Japanese competitors from entering the U.S. market through their (the U.S. manufacturers') established distribution systems. The major American suppliers—e.g., Intel, Texas Instruments, Motorola, National, and Advanced Micro Devices—have long had a tacit policy of preventing their distributors from carrying products made by Japanese competitors. Those distributors which might do so would risk losing franchises for their American-made lines. In one instance, Arrow Electronics, the second largest U.S. distributor of semiconductors, having agreed to take on products of Samsung Semiconductor, a South Korean electronics manufacturer, then suffered a cutback in its supplies from Intel. Approximately a month later, Arrow terminated its Samsung franchise.

As of 1988, only one of the ten largest American electronics resellers carried Japanese lines. The one reseller, Marshall Industries, had several Japanese lines and only one American semiconductor brand. Other American distributors of Japanese semiconductors did not carry the lines of any American semiconductor manufacturers. See "Big Chill on Asian Chips in U.S.," New York Times, September 6, 1988.

"cherry-pick" the line, taking the most popular items and leaving the rest. Often, too, the manufacturer will use some version of this so-called full-line forcing to get its distributors to take on its new products for which the demand is not yet established.

For all of these reasons, it will be to the interests of producers to exert what influence they can over the strategies of their independent distributors. The extent to which they can do so is constrained by law. But the law is not always clear, and moreover, the legal context within which vertical relationships are conducted is ever changing. The law has tended to evolve over time as the courts interpret the existing body of legislation, and as different government administrations and their administrative agencies change their policies with regard to law enforcement.[2]

We begin by identifying and describing the relevant federal legislation and then move to consider legal limitations on manufacturers' actions in the following areas:

- vertical price restrictions
- vertical nonprice restrictions, such as
  —territorial and class-of-trade constraints
  —conditions restricting distributors from carrying competitive products
  —tying arrangements that require the distributor to buy one product or service to get another
  —full-line forcing, the requirement that the reseller carry the manufacturer's complete line of products
  —the manufacturer's rights to terminate distributors

FEDERAL LEGISLATION RELATING TO VERTICAL RESTRAINTS

The federal laws that cover vertical restraints include the Sherman Anti-Trust Act, passed in 1890, Section 3 of the Clayton Act (1914), the Federal Trade Commission Act (1914; amended 1938), and the Robinson-Patman Act (1936), which amended Section 2 of the Clayton Act.

Section 1 of the Sherman Act, the primary federal statute applicable to distribution, states that "[e]very contract, combination in the form of trust or otherwise, or conspiracy, in restraint of trade

---

[2] For a discussion of current federal administration views, see Richard A. Kleine, "New Department of Justice Vertical Restraint Guidelines: A Search for Legal Certainty," *The Business Lawyer,* vol. 40 (August 1985), p. 1369.

or commerce among the several States, or with foreign nations, is declared to be illegal."

Section 3 of the Clayton Act is relevant in determining the legality of restraints on distributors with regard to selling competing products, tying arrangements, and exclusive dealing. Section 3 of the Clayton Act states that

> [I]t shall be unlawful for any person engaged in commerce, in the course of such commerce, to lease or make a sale or contract for sale of goods, wares, merchandise, machinery, supplies or other commodities, whether patented or unpatented, for use, consumption, or resale within the United States . . . or fix a price charged therefore, or discount from, or rebate upon, such price, on the condition, agreement, or understanding that the lessee or purchaser thereof shall not use or deal in the goods, wares, merchandise, machinery, supplies, or other commodities of a competitor or competitors of the lessor or seller, where the effect of such lease, sale, or contract for sale or such condition, agreement, or understanding may be to substantially lessen competition or tend to create a monopoly in any line of commerce.

The Federal Trade Commission (FTC) Act gives the FTC jurisdiction along with the U.S. Department of Justice in the civil enforcement of the antitrust laws.[3] The act established the FTC with investigating, prosecuting, legislative, and judicial powers. The FTC is empowered to issue cease and desist orders if it finds an antitrust violation, but its actions are subject to review by circuit courts of appeal. The Federal Trade Commission Act was amended in 1938 to make unfair methods of competition, and unfair or deceptive acts or practices in commerce, unlawful. This amendment also gave the FTC broad powers to determine what methods, acts, and practices it would consider unlawful.

The Robinson-Patman Act, passed in 1936, amended Section 2 of the Clayton Act by providing that "it shall be unlawful for any person engaged in commerce . . . knowingly to induce or receive a discrimination in price which is prohibited by this section."[4]

The Antitrust Division of the Department of Justice is responsible for enforcing the Sherman and Clayton acts. The Federal Trade Commission also has antitrust jurisdiction under the Clayton

---

[3] Carla Anderson Hills, ed., *Antitrust Adviser,* 3d ed. (Colorado Springs, Colo.: Shepard's/ McGraw-Hill, 1985), sec. 2.04, pp. 78–79, and sec. 5.04, pp. 342–343.
[4] 15 USC §13(f).

Act, as well as under the Robinson-Patman Act.[5] Private parties who consider themselves injured by violations of the Sherman and Clayton acts may sue to recover treble damages, but they may not bring suit under the Federal Trade Commission Act.[6] In addition, almost all states now have state antitrust statutes, which private parties may invoke where federal jurisdiction is inapplicable, as in intrastate commerce.[7]

State statutes that regulate intrastate transactions are not uniform, but many are similar to the Sherman Act. The stated purposes of most are to secure the benefits arising from competition in trade, to prevent monopolies and combinations, and to protect the people from "the oppression of combined wealth."[8]

The courts apply two different standards in assessing whether a given practice unreasonably restricts competition. The simplest, and most severe, is the *per se rule,* which describes a practice as unlawful in its very nature without inquiry as to the harm it has caused or the business justification for its use. Once a practice is judged per se illegal, no further argument is relevant. The Supreme Court has declared per se unlawful practices "which judicial experience has shown to be a naked restraint of trade with no purpose except stifling of competition."[9] The per se rule has been used, for example, in cases involving resale price maintenance, that is, vertical price fixing.

The courts, however, decide most disputes concerning vertical restraints in terms not of the per se rule but of the *rule of reason,* under which the aggrieved party, be it the government or a private party, must show that the restraint had a significant anticompetitive effect. In applying the rule of reason to assess anticompetitive effects, the courts have weighed such factors as the following:

- the supplier's market share
- the degree of concentration of industry sales among a few firms
- the relative difficulty of market entry for new firms

---

[5] James F. Rill, Kathleen E. McDermott, and Christopher J. MacAvoy, "Price, Promotion, and Service Discrimination," in *Antitrust Adviser,* p. 306.
[6] Michael L. Denger, James R. McGibbon, and Willard K. Tom, "Product Distribution," in *Antitrust Adviser,* p. 80.
[7] Ibid., p. 81.
[8] John D. Wyatt and Madie B. Wyatt, *Business Law: Principles and Cases* (New York: McGraw-Hill, 1979), p. 662.
[9] Denger et al., p. 84.

- the extent of product differentiation
- the reasonableness of the restrictions
- the legitimacy of the manufacturer's business purpose

The federal government's attitude toward vertical restraints has varied over the decades since regulation was introduced. Changes are evident both in court rulings in antitrust cases and in the willingness of the Department of Justice to bring suit. The discussion that follows is based on current court interpretations of legislation governing vertical relationships. The likelihood that these interpretations will continue to evolve, that successive government administrations will vary in the rigor with which legal actions are taken under the antitrust laws, and indeed, that new legislation will be enacted is great. In addition, the following discussion is not intended to be authoritative in its interpretations of the law. Rather, it is offered as an overview of the directions the law has taken in mediating manufacturer-distributor relationships.

VERTICAL PRICE RESTRICTIONS

The courts have traditionally considered resale price maintenance agreements between a manufacturer and a distributor to be per se violations of Section 1 of the Sherman Act. In an early case, *Dr. Miles Medical Co.* v. *John D. Park & Sons* (1911), the Supreme Court declared that once a manufacturer had sold its products to distributors, it could not thereafter lawfully restrict the freedom of those distributors to sell such products at prices they determined in accordance with their own judgment.[10]

In *U.S.* v. *Colgate & Co.* (1919), the Supreme Court declared that a manufacturer has the right to announce a price at which it expects distributors to sell, and to refuse to do business with distributors that depart from this price. In addition, the decision in this case in favor of Colgate argued that there must be some sort of concerted action or agreement between manufacturers and distributors for a vertical restraint to be considered illegal.[11] Colgate had been charged with attempting to fix resale prices for soap and toiletries by combining with wholesalers and some retail dealers to secure adherence by other dealers to resale price schedules. It was alleged that wholesalers refused to sell to uncooperative dealers. It

[10] Denger et al., p. 105.
[11] Denger et al., p. 85.

was not found or proved, however, that there was an agreement by which dealers were bound to sell only at prices stipulated by Colgate, and, in the absence of such an agreement, the Court did not find Colgate's actions to be illegal. The Court held that,

> in the absence of any purpose to create or maintain a monopoly, the act does not restrict the long-recognized right of trader or manufacturer engaged in an entirely private business, freely to exercise his own independent discretion as to parties with whom he will deal. And, of course, he may announce in advance the circumstances under which he will refuse to sell. (250 U.S., at 307)[12]

Vertical restraints arising from concerted action or agreements between manufacturers and distributors are illegal, however, and the courts have often been called upon to determine whether in fact there were such agreements, as in the 1984 case *Monsanto Co.* v. *Spray-Rite Service Corp.* In this case, the Supreme Court reaffirmed the *Colgate* principle that a manufacturer "can announce its resale price and refuse to deal" with those which do not comply and that "a distributor is free to acquiesce in the manufacturer's demand in order to avoid termination." In this case, however, the Court determined that there had been a conspiracy among Monsanto and others of its distributors to fix prices and to cut off Spray-Rite's supplies of Monsanto herbicides and that such actions constituted a per se violation of the Sherman Act.

Monsanto, holding a 15% share of the U.S. market for herbicides and competing against Ciba-Geigy with a 70% share, had attempted to increase its sales through a concentrated effort to strengthen its distribution channels. Toward this end, Monsanto announced in 1967 that distributors would henceforth be appointed for one-year terms with renewal dependent on three conditions: (1) that the distributor's primary business be the resale of Monsanto products to dealers; (2) that it employ trained sales personnel able to educate dealers and their customers, in this case farmers, in herbicide applications; and (3) that it was satisfactorily developing the market potential in the geographic area for which it had been given prime responsibility. This program was credited with contributing to Monsanto's almost doubling its market share in herbicides over the next four years.

In 1968 Monsanto terminated Spray-Rite, one of its large

---

[12] Robert Pitofsky and Kenneth Dam, "Is the Colgate Doctrine Dead?" *Antitrust Law Journal*, vol. 27 (1962), pp. 772–788.

corn-herbicide distributors, indicating that Spray-Rite had failed to train its sales personnel and to work with dealers in implementing Monsanto's program of market education. In 1972 Spray-Rite went out of business. It then brought suit against Monsanto, contending (1) that it had been cut off not for the reasons alleged by Monsanto but for selling Monsanto herbicides at discount prices and (2) that the chemical manufacturer had conspired with other of its distributors to fix the resale prices of its herbicides. Further, Spray-Rite indicated that after it had been terminated, it had sought to secure Monsanto products from other distributors but had been unable to obtain the supplies it needed. Thus, it alleged that Monsanto had encouraged its distributors to boycott Spray-Rite in the interests of supporting Monsanto's resale price maintenance conspiracy.

Evidence was introduced at the trial to show that Monsanto considered Spray-Rite to be a price cutter and that other distributors had complained about Spray-Rite's pricing tactics and had urged that Spray-Rite be terminated. Further, it was brought out that Monsanto had previously pressed Spray-Rite to raise its prices under threat of being discontinued.

The jury ruled in favor of Spray-Rite, on the grounds that there had been a conspiracy to fix prices and that there had been other nonprice restraints, such as delineating areas in which distributors were given primary sales responsibility, which served to support the price conspiracy. Judging Monsanto's actions to be a per se violation of the Sherman Act, the court awarded treble damages amounting to $10,500,000. The decision had rested heavily on the finding that Monsanto had engaged in a conspiracy with its distributors to maintain resale prices and to terminate Spray-Rite.[13]

As the Supreme Court declared in U.S. v. Parke, Davis & Co. in 1960, "When the manufacturer's actions . . . go beyond mere announcement of his policy and the simple refusal to deal, and he employs other means which effect adherence to resale prices . . . he has put together a combination in violation of the Sherman Act."[14]

---

[13] Terry Calvani and Andrew Berg, "Resale Price Maintenance after Monsanto: A Doctrine Still at War with Itself," Duke Law Journal, 1984, 1187–1204, and Cynthia Atchinson, "Spray-Rite Service Corp. v. Monsanto Co.: The Justice Department Challenges the Per Se Rule against Resale Price Maintenance," University of Pittsburgh Law Review, vol. 46 (fall 1984), pp. 180–187.
[14] J. F. Cady, "Note on Legal Issues and the Pricing Process," #578-205, rev. Boston: Harvard Business School 1978, p. 9.

However, a series of cases since the Second World War have established that it is "perfectly lawful" for a manufacturer to advertise *suggested* retail prices directly to the consumer.[15] A manufacturer, for instance, may legally preticket a product, putting on it a price tag or sticker showing the suggested retail price. The courts have upheld this practice in several cases. It can also safely do mass-media and direct-mail advertising and distribute promotional material designed for display to the public showing its prices.[16] But it cannot restrict distributors from advertising their prices as well.[17]

In addition to conspiring with resellers to fix resale prices, there are other manufacturers' actions to influence distributors' prices that are legally questionable. For example, a 1972 district court case, *Pearl Brewing Co.* v. *Anheuser-Busch, Inc.* ruled that "a practice of conditioning a price reduction on the acquiescence or cooperation of the recipient to reduce its prices can only be viewed as imposing restrictions on the reseller's freedom of decision and, as such, is an unlawful price-fixing combination."[18]

In summary, the manufacturer's key weapon in securing adherence to resale price schedules is the refusal to continue supplying nonconforming resellers. Thus, the legal battles involving vertical price restrictions have been joined where the manufacturer's right to select with whom it will deal comes up against the distributor's right to sell goods, which it owns, at whatever prices it elects to charge. The relevant case law seems firmly to indicate that to establish the illegality of refusing to deal with offending resellers, it must be shown that a conspiracy existed between the manufacturer and its wholesalers and retailers to identify and terminate price cutters. If the manufacturer has acted unilaterally, the Court has ruled that this is a proper exercise of its right to select its customers. Although the manufacturer may have received complaints about price cutters from conforming resellers, if it does not consult with other resellers and/or solicit their support in enforcing the stipulated prices but takes actions independently, it is deemed to be acting within its rights as affirmed in *Colgate* and subsequent decisions. In the meantime, the right of the manufacturer to announce and advertise "sug-

---

[15] Jack B. Owens, Lee N. Abrams, Michael L. Denger, Joshua F. Greenberg, and Harry M. Reasoner, "Antitrust Dos and Don'ts of Distribution," *Antitrust Law Journal,* 1984, p. 374.
[16] Cady, p. 17.
[17] Denger et al., p. 115.
[18] Owens et al., p. 375.

gested" resale prices is uncontested; the question of legality rests on the tactics it uses to secure adherence.

Finally, it may be observed that as a practical matter, manufacturers often find the enforcement of resale price schedules exceedingly difficult and cumbersome. Refusing to deal with nonconforming resellers is at best a rearguard action where there is widespread demand for the product and where differences between reseller costs and manufacturer-stipulated resale prices allow room for some resellers to undercut their competitors.

VERTICAL NONPRICE RESTRICTIONS

There are several categories of vertical nonprice restrictions practiced by suppliers. They are (1) territorial and class-of-customer constraints, (2) restrictions on the distributor's freedom to carry competitors' products, (3) tying arrangements, and (4) full-line forcing. A territorial restriction seeks to prevent or at least discourage a distributor from selling outside a particular geographical area, while a customer restriction prohibits it from selling to specific customers or classes of customers. As noted earlier, a manufacturer establishes territorial restrictions in order to minimize intrabrand competition in a trading area and thereby provide incentives for the authorized distributors there to support the sale of its product.

Territorial and Class-of-Customer Constraints

Territorial restrictions may take various forms. The producer may, for example, appoint only one distributor in Territory A, thus granting the selected franchisee an "exclusive." It may also attempt to restrain distributors of its products in other territories from selling in Territory A. Finally, the producer may agree that it will not itself sell direct to customers in Territory A.

The legal status of vertical nonprice restrictions has changed greatly even in the past two decades. Currently the ruling in a 1977 case, *Continental TV, Inc.* v. *GTE Sylvania, Inc.* is recognized as establishing the legal principles by which such restrictions may be judged.

In the GTE Sylvania case, Sylvania, with a 2% share of the television market, had undertaken in 1962 to improve its participation by going from a system of intensive to selective retail distribution. Terminating its independent wholesalers, Sylvania then sold directly to a network of selected retailers, each of whom agreed to sell the manufacturer's line only from locations approved by Sylvania. In trading areas of more than 100,000 in population, there were at least two such dealers.

Continental TV, a Sylvania dealer in northern California, was terminated when it began selling Sylvania TV sets in its newly opened outlet in Sacramento, although this site had not been approved by Sylvania. Continental's move, it may be noted, followed an action taken by Sylvania to franchise another retail dealer close to one of Continental's authorized sales locations.

In ruling in favor of the defendant, Sylvania, the Supreme Court noted that important efficiencies could be achieved through inducing dealers to invest in promoting and servicing new products. Further, through exercising control over the manner in which products are sold, the Court noted, the manufacturer was able to encourage the maintenance of product safety and quality standards at the point of sale.[19] The Court reasoned that vertical nonprice restraints often increased interbrand competition even though they might restrict intrabrand competition among sellers of the same brand.

Following the *Sylvania* decision, the courts have judged the anticompetitive effects of vertical nonprice restrictions in terms of the rule of reason.

Legally, the courts may challenge exclusive distribution arrangements under the Sherman Act, the Clayton Act, the Federal Trade Commission Act, and a number of individual state laws. In most cases, exclusive distributorships have been upheld by the courts. Questions may arise, however, where

- the supplier is the dominant firm, and interbrand competition is weak, or
- the distributor has also been granted territorial exclusivity by a number of the supplier's competitors and may thus be judged to have an interbrand monopoly in its trading area.

Since the *Sylvania* case in 1977, the courts have generally agreed that a restriction that eliminates intrabrand competition does not necessarily restrain interbrand competition, and they have

---

[19] Robert Pitofsky. "The Sylvania Case: Antitrust Analysis of Non-price Vertical Restrictions," *Columbia Law Review*, vol. 78 (January 1978), p. 1, and Richard Posner, "The Rule of Reason and the Economic Approach: Reflections on the Sylvania Decision," *The University of Chicago Law Review*, vol. 45 (fall 1977), p. 1.

increasingly required the establishment of a threshold to show that the manufacturer imposing the restriction had sufficient market power to restrain interbrand competition. They have usually allowed restrictions in cases where there was proof of active interbrand competition or where the manufacturer could show that the restrictions served such legitimate purposes as minimizing health, safety, or product liability concerns.[20]

The greater the impact of exclusive distribution arrangements on interbrand competition, the more likely it is that the courts will find such restrictions at fault. On the other hand, the rule of reason requires courts to analyze overall competitive effects rather than to rely on a single index such as market share. If a manufacturer can factually demonstrate that exclusive distribution is reasonable in view of market competition, this will be weighed against share-of-market considerations.[21]

As with territorial restrictions, the legality of customer restrictions—that is, restraining distributors from selling to certain classes of customers or specific accounts—rests on whether these constraints have an adverse effect on competition. Further, the courts have also taken the position in some of these cases that while such vertical nonprice restrictions may dampen intrabrand competition, there may be beneficial effects on interbrand competition.[22]

Multiple-Channel Distribution. The question of legality may become more complicated when manufacturers sell through a combination of company-owned outlets, direct salesforces, and/or independent distributors. In such cases, territorial or customer restrictions may be questioned because the manufacturer is construed to be in competition with its distributors. Under these circumstances, there is the possibility that the restrictions the manufacturer imposes on independent distributors may be considered horizontal agreements among competitors. The courts hold that horizontal agreements among competing manufacturers or competing distributors to fix prices, to allocate markets or customers, or to otherwise limit production or distribution are per se violations of antitrust laws.

[20] Ibid., pp. 100–103.
[21] J. F. Cady, "Legal Issues in Distribution Strategy, II: Dealings with Distributors," #9-579-078. Boston: Harvard Business School, 1978, p. 3.
[22] *Donald B. Rice Tire Co.* v. *Michelin Tire Corp.* 638 F2d 15 (4th Cir), cert denied, 454 US 864 (1981), and *Red Diamond Supply, Inc.* v. *Liquid Carbonic Corp.* 637 F2d 1001 (5th Cir), cert denied, 454 US 827 (1981).

The trend in recent cases, however, is for the courts to treat nonprice restrictions imposed by a manufacturer engaged in multiple-channel distribution as they would vertical nonprice restraints generally, that is, in terms of their effect on competition.

In summary, the thrust of case law emerging from legal actions that have tested the validity of territorial exclusivity is that such arrangements have generally been upheld. They may be challenged, however, if the manufacturer and/or the distributor can be shown to dominate its respective markets and if the arrangement will tend to stifle competition.

Restrictions on Distributors' Freedom to Sell Competitive Products

Manufacturers may also impose restrictions the purpose of which is to prohibit distributors from handling competitors' products. Principal among these are exclusive dealing arrangements, also known as brand exclusivity. In an exclusive dealing or requirements contract, a distributor agrees to purchase products or services solely from one manufacturer over a period of time or, more often, to sell no products in direct competition with those of the manufacturer.

Manufacturers may find exclusive dealing arrangements useful when the development of distributor loyalty and enthusiasm is particularly important in supporting new product introductions and weak brand products. They also favor exclusive dealing when distributor push is important for sales and when it may result in lower distribution costs.[23]

The courts do not consider exclusive dealing arrangements to be per se illegal, and generally observe guidelines stated in 1961 by the Supreme Court in *Tampa Electric Co.* v. *Nashville Coal Co.*, which cautioned that it was essential

> to weigh the probable effect of the contract on the relevant area of effective competition, taking into account the relative strength of the parties, the proportionate volume of commerce involved in relation to the relevant market area, and the probable immediate and future effects which pre-emption of that share of the market might have on effective competition therein.[24]

---

[23] Cady, "Legal Issues in Distribution Strategy, II: Dealings with Distributors," pp. 2–3.
[24] Owens et al., p. 383.

Most courts require proof of a significant adverse effect on competition before striking down exclusive dealing arrangements, and they consider such factors as

- market share being foreclosed to competition (arrangements foreclosing over 33% of the market are ordinarily presumed to be unlawful absent other evidence),[25]
- whether the purchaser subject to the restriction is an end-user or a distributor,
- the availability of other distribution channels, and
- the duration of the agreement.[26]

Thus, courts generally consider partial requirements contracts or minimum purchase contracts to be legal, and requirements agreements negotiated by manufacturers possessing a very small share of the relevant market stand a good chance of escaping challenge. In fact, many challenged contracts of this nature have been found legal.[27]

Tying Arrangements

A tying arrangement exists when a manufacturer conditions the sale of one product upon the purchase by a distributor of a second, less desirable product. The courts have judged tying contracts as foreclosing competition under Section 3 of the Clayton Act, Section 1 of the Sherman Act, and Section 5 of the Federal Trade Commission Act, and have viewed them more negatively than exclusive dealing arrangements or requirements contracts. Tying arrangements that involve a manufacturer's use of economic power or leverage in the market for one product to curb competition for other products, or to deny distributors freedom of choice with respect to the products they purchase, have historically been held to be per se illegal.

The Supreme Court noted, however, in a 1984 case, *Jefferson Parish Hospital District No. 2* v. *Hyde,* that for tying arrangements to be unlawful, the manufacturer must possess sufficient market power to be able to restrain competition. This instruction has been interpreted as inconsistent with a per se unlawful ruling, and subsequent

---

[25] Denger et al., p. 147.
[26] Ibid., pp. 147–148.
[27] F. M. Scherer, *Industrial Market Structure and Economic Performance,* 2d ed. (Boston: Houghton Mifflin, 1980), pp. 585–586.

to *Jefferson Parish,* the law on tying arrangements has been described by one lawyer as "opaque, theoretically unsatisfying, and even more dangerous," as "a majority of the Court was willing to leave the law in . . . a muddle."[28]

Some lawyers believe that the courts will now allow tying arrangements unless there is appreciable monopoly power involved or unless a substantial volume of sales is foreclosed in the relevant product-market, two conditions unlikely to obtain in the case of relatively small sellers of unpatented products or small companies attempting to break into new markets.[29] Others suggest that manufacturers may take as a guideline that "tying agreements are inherently anticompetitive in their impact and, thus, while not strictly *per se* illegal, are very close to it."[30]

Thus, given the lack of clarity in recent court decisions, lawyers have treated tying arrangements as legally questionable conditions in the manufacturer-reseller contract.

Full-Line Forcing

Imposing a requirement on a distributor that it carry the producer's complete line of products in a particular product category—say, air compressors or electric motors or grinding wheels—is called full-line forcing. In court tests brought against manufacturers, this particular practice has generally been deemed legally permissible. The courts have recognized that if the reseller represents itself as a Manufacturer X distributor, the reseller's customers might normally expect it to stock a complete Brand X line.

Full-line forcing may be challenged, however, if it carries with it certain requirements regarding quantities to be purchased and distributor inventory levels to be maintained and if such conditions effectively preclude the distributor from also stocking competing product lines. In a 1973 case, *Colorado Pump & Supply Co.* v. *Febco, Inc.,*[31] the court ruled in favor of a manufacturer that had required its distributors to "maintain adequate inventories of products and parts which will enable Distributor to offer for sale a complete line

---

[28] Harry M. Reasoner, a member of the Texas, New York, and District of Columbia bars, and chairman of the Economics Committee of Antitrust Section of the American Bar Association from 1978 to 1981, cited in Owens et al., p. 385.
[29] Scherer, p. 583.
[30] Louis W. Stern and Thomas L. Eovaldi, *Legal Aspects of Marketing Strategy: Antitrust and Consumer Protection Issues* (Englewood Cliffs, N.J.: Prentice-Hall, Inc., 1984), p. 318.
[31] 472 F2d 637, 640 (10th Cir), cert denied, 411 US 987.

of said products and service them after installation." The court determined that the supplier did not possess market power and that
there was not proof that the distributor was restricted from handling competing products.

Full-line forcing may also be challenged as an illegal tying
arrangement if it involves unrelated product lines. A fictitious example might be a manufacturer of electrical apparatus and components that requires that its distributors carry its switchgear line as a
condition of being given a franchise for its air conditioner line.[32]

Enforcement Trends

While this discussion has focused on court opinion in inter-
preting the relevant legislation, it may be noted that the attitudes
of political administrations toward law enforcement may significantly affect the legal climate in which business decisions are made
and implemented.

The Reagan administration, for example, took a lenient attitude toward vertical nonprice restrictions. In its *Vertical Restraints
Guidelines,* issued in 1985, the Department of Justice took the position that anticompetitive effects are not possible unless there is a
significant concentration in either the supplier or distributor market, since otherwise interbrand competition would eliminate successful collusive or exclusionary conduct. The *Guidelines* stated that
the Department of Justice will not challenge vertical nonprice restraints imposed by a manufacturer that has less than a 10% market
share. For those with larger shares, the department proposed to
apply indices based on the degree of concentration of the manufacturer and the distributor in the market in which the restraint is
imposed, and on the extent to which similar restraints are used by
other suppliers in that market. Restraints would then be analyzed
in terms of their effect on competition, starting from the presumption that the level of adverse competitive impact is not significant.[33]

The *Guidelines,* for example, construed selective distribution—
that is, franchising a limited number of outlets in a trading area—
and arrangements under which a manufacturer assigns areas of primary responsibility to distributors as "vertical restraints that are

---

[32] For a discussion of legal aspects of full-line forcing, see Carla Anderson Hills, ed., *Antitrust
Adviser,* 3d ed., pp. 151–153.
[33] Ibid.

always legal." Likewise, location clauses that establish or restrict the outlets from which the distributor may sell the supplier's products are deemed to be "always legal."[34]

The *Guidelines* were not law, however, nor were they necessarily consistent with judicial interpretations. They simply provided guidance as to the conditions under which the Justice Department was likely to bring action against suppliers imposing vertical restrictions on resellers. It is important to note, moreover, that the bulk of the litigation in this area is initiated not by the U.S. government but by private parties aggrieved in a trade relationship.

TERMINATING DISTRIBUTORS

While a manufacturer has the right to select its distributors and to determine the circumstances under which it will refuse to sell, its right to terminate a distributor may be challenged if the grounds for dismissal involve distributor nonconformance with allegedly illegal vertical restraints. A supplier may terminate a distributor, however, for failure to perform against standards specified by the former in the geographic area for which the latter has primary responsibility. Other legitimate grounds for disenfranchisement include a determination on the part of the supplier that the distributor is a poor credit risk. In addition, the manufacturer may drop one or more distributors in an area and appoint new ones in a program to strengthen its distribution network. In the 1982 *Valley Liquors, Inc.* v. *Renfield Importers, Ltd.* case, the circuit court upheld the termination of a price cutter who was benefiting from the promotional efforts of competing distributors, on the grounds that this action had the effect of encouraging nonprice competition among resellers, thereby stimulating interbrand competition.[35]

In some cases, the termination of a reseller has been upheld where the action has been prompted by a concern on the part of the manufacturer related to issues of product safety or product liability risks. In a 1979 case, *Clairol, Inc.* v. *Boston Discount Center, Inc.*, for example, Clairol refused to sell its salon products to distributors selling at retail, on the grounds that they were not suitably

---

[34] U.S. Department of Justice, Antitrust Division, *Vertical Restraints Guidelines,* sec. 2.5, January 23, 1985, pp. 12–13.
[35] Denger et al., p. 88.

labeled for safe use by consumers. The federal district court found in favor of Clairol, deeming that such a restraint was reasonable.[36]

This brief statement of legal considerations related to terminating distributors is based on the interpretation by the courts of federal statutes. It is important to note, however, that there is a growing body of state legislation in this area that manufacturers must take into account in contemplating cancellations of distributor franchises.

SUMMARY

An understanding of the legal dimensions of producer-distributor relations must begin, on the one hand, with a recognition of producers' objectives in managing their channels networks. On the other, it must take account of the spirit of the relevant regulation in the United States and the way it is being interpreted by the courts.

As for the first, producers act to preserve channel strength. They do so by attracting the most qualified distributors, motivating them to give full support to selling their product lines, and preventing them from engaging in destructive intrabrand competition. This explains producers' interests in maintaining resale price levels as a condition of preserving resale margins for all resellers in the distribution system. It also explains territorial restrictions as a way of damping the intensity of competition at the resale level in individual trading areas. In addition, it provides a rationale for franchise terms that require the distributor to carry the producer's full line.

A second objective is to preclude, to the extent possible, the intrusion of competitors in the producer's distribution channels or, at the minimum, to encourage distributors to give primary attention to the producer's line, treating competitors' brands as secondary offerings. This explains provisions sometimes found in franchise agreements that stipulate that the distributor not carry competing products. More often, however, such strictures are imposed in the context of day-to-day relationships between producer and distributor personnel.

A third broad objective is often to reserve the largest accounts for direct selling both as a way to maximize sales margins and to realize the strategic advantages to be gained through having direct selling relationships with large users, such advantages as direct technical interchange, access to market information, and direct negotiation of prices and terms.

---

[36] For a comprehensive and well-documented discussion of the legality of terminating distributors, see Deborah Scammon and Mary Jane Sheffet, "Legal Issues in Channels Modification Decisions: The Question of Refusals to Deal," *Journal of Public Policy & Marketing*, vol. 5.

This is the purpose of placing class-of-trade restrictions on external intermediaries and/or reserving named accounts for the producer's sales personnel.

The legislation that is applicable to producer-reseller relationships includes the Sherman Act, the Federal Trade Commission Act, Section 3 of the Clayton Act, and the Robinson-Patman Act amending Section 2 of the Clayton Act. These and a substantial body of court opinion make up the regulatory framework in which the so-called vertical constraints described above may be judged.

The broad purpose of the law is clearly to preserve competition and to avoid restraints of trade. Court cases have sought, therefore, to determine the extent to which producers' channels management strategies and tactics exceed the limits of legality in restraining resellers and/or in having a significant anticompetitive effect at the producer level. These judgments are rendered largely in accordance with the *rule of reason,* rather than by *per se* dicta by which an action is deemed illegal regardless of the consequences.

As for specific categories of vertical constraints, the legality of resale price restrictions hinges largely on whether or not there has been a conspiracy among the producer and some of its distributors to fix resale prices and to discipline those who refuse to conform. If conspiracy is not a factor, a producer may legally take such unilateral actions as suggesting resale prices to its dealers, announcing them to the end-user, and refusing to deal with those resellers which do not observe the suggested price schedules.

The several types of nonprice vertical constraints include

- territorial restrictions defining each distributor's sales territory and proscribing its selling in territories for which other distributors have been given primary responsibility,
- class-of-trade constraints by which resellers are advised not to sell to certain account categories or named accounts,
- exclusive-dealing clauses that require that the distributor carry only the producer's brand and not stock competing brands, and
- tying arrangements that require the distributor to stock and sell certain parts of the producer's line as a condition of being given other, more desirable parts of the line to carry.

In keeping with the broad philosophy of the courts, territorial and class-of-trade constraints have been condoned unless it can be shown that the producer has sufficient market power to affect interbrand competition. The same condition applies as well to brand exclusivity, that is, requiring distributors to carry only the producer's line and not stock competing lines. Tying arrangements, too, are judged to be legal unless there is a question of their impact on interbrand competition. But tying arrangements may be challenged if the alleged effect is to deny distributors freedom of choice with respect to the products they carry or to foreclose a substantial volume of sales in the relevant product-market. In any case, the law seems much less clear with respect to the legality of tying arrangements as compared with other vertical nonprice constraints.

Finally, the legal framework also deals with distributor termination, that is, revoking the franchise. Resellers have effectively challenged termination if the grounds for dismissal allegedly involved nonconformance to vertical constraints illegally imposed by the producer. Valid reasons for dismissal may include the reseller's inadequate performance against standards, the producer's judgment that the reseller is a poor credit risk, and a concern that the distributor is not qualified to maintain the requisite product safety standards. Producers may also legally drop certain distributors in a trading area and take on others to strengthen their distribution networks.

Those who have followed and interpreted the evolving body of law related to manufacturer-reseller relations stress that the effective defense of any actions involving vertical restraints depends critically on such actions being seen as supporting the manufacturer's legitimate business purpose. It becomes essential, then, that whatever conditions are imposed in the manufacturer-reseller contract be fairly and consistently administered. It is also important that actions taken be thoroughly documented at the time rather than reconstructed after the fact.

Taken broadly, the law strives to preserve competition and at the same time to balance the producer's right to choose those with whom it will deal and the reseller's freedom to determine from whom it will source, to whom it will sell, and at what prices. Fundamentally, the law provides a framework within which producer and reseller have freedom to negotiate the terms of their relationship without legal doctrine weighing in heavily on one side or the other.

14 Distribution: A Systems Perspective

In this study of industrial distribution systems, we have explored the factors that influence channels design and the issues that arise for producers in channels organization and management. For background data, we have cited contemporary case histories describing modes of distribution in selected product-markets. We have also drawn upon the annals of business history to describe the evolution of industrial distribution in the United States, and we have outlined the regulatory framework that mediates producer-reseller relationships. Our essential purpose has been to develop knowledge and concepts useful in designing and managing channels of distribution. We suggest, through example, methods of coping with the continuing environmental change that threatens to undermine the competitive viability of systems designed for earlier market conditions.

In focusing on issues of operational significance, we have had the opportunity to observe at close hand the interactions of suppliers, distributors, and end-users and to see distribution strategy and tactics issues in a marketing systems context. We have observed the shaping of industrial distribution systems largely through ad hoc decisions taken by producers in response to the actions of individual competitors, customers, and intermediaries.

In the discussion that follows, we offer some broad observations about the factors that mediate buyer-reseller-user relationships in industrial distribution systems. We suggest, at the outset, that the protagonists in competitive rivalry are not individual firms but the broader industrial systems of which producers and resellers are members. We identify the capacity for adaptation as critical to long-term competitive superiority and then examine those factors which alternatively inhibit or foster a company's ability to adapt.

We develop a concept of adversarial collaboration to describe the relationship between producers and resellers, pointing out the factors that tend to weigh more on one side than on the other of these seemingly opposing attributes. In considering relationships based on both collaborative and adversarial interests, the relative balance of power among the respective members of the system is relevant. We suggest the factors on which the relative power balance depends. Finally, in keeping with the precept that the effective use of power ultimately brings into consideration matters of equity and fairness, we speak to these issues as well.

Understanding how distribution works in a broader perspective may usefully inform managerial decision making. A greater conceptual understanding may have implications, as well, for public policy: if U.S. public policy makers understand the factors that contribute to the strength of our distribution infrastructure, then through legislation, court decisions, and the actions of regulatory agencies they may better create the conditions for national economic growth and competitive viability.

TOWARD A CONCEPT OF INDUSTRIAL SYSTEMS

The players in the competitive rivalry for industrial product-markets are industrial systems—in which producers and intermediaries exercise varying degrees of influence and over which neither has total control. Further, any systems member is almost invariably a member of other, often competing systems. Producers may sell through both producer-organized systems and reseller-managed systems. Resellers typically participate in multiple systems as well, often carrying the products of competing manufacturers and a myriad of other suppliers.

In networks that are producer organized, the manufacturer seeks to exercise strategic direction over market segment selection, pricing strategy, and demand generation. In reseller-managed systems, producers serve essentially as supply sources and exercise little, if any, control over these crucial dimensions of systems strategy. Producer-organized and reseller-managed systems often compete against each other for product-market share; their relative competitive strength varies by market segment, with the latter often appealing to more price-sensitive customer sets.

Industrial systems tend to orient their product lines, pricing strategies, and modes of distribution to particular product-market niches. In many cases, such niches are defined in terms of end-market demographics. Some systems focus on highly concentrated segments; others target the more fragmented end of the product-market, with markedly different strategies appropriate for each.

However, competitive superiority in one market segment often precludes comparable success in other segments. This is particularly true if the strategy appropriate for one segment is perceived to conflict with the interests of a significant body of channels members serving another market niche. Market success in one segment, then, may constrain the channel organizer's range of market segment opportunities.

Systems Rigidities and Resistance to Change

Market success, long-term, depends on *adapting to product-market change*. As markets evolve, maturing product technology, shifting market segment demographics, and intensifying competitive pressures serve to reshape buyer behavior and to revise customers' perceptions of product superiority. The requirements for customer satisfaction change as users revise their priorities among a range of product and transaction attributes. In addition, changes in market segment demographics and buyer behavior often create opportunities for reducing distribution costs by reallocating channel responsibilities.

Despite the pressures to adapt, members of an existing industrial system often resist change in the interests of protecting their investments in future revenue streams, trained personnel, physical resources, distribution skills, and the culture of past success. The will to resist change is reinforced by the considerable uncertainty that attends any prospective restructuring of an industrial system. In addition, insofar as the existing system is held together by mutual dependencies, expectations, and commitments, both explicit and implicit, change is difficult if it is perceived, as it almost invariably is, to be straining interorganizational relationships. Thus, commitments essential to the functioning of an industrial system at a point in time may eventually inhibit effective response to the pressures for change.

The power to resist change is derived to a large extent from the industrial system's customer base and from the continuing revenue stream that flows from the replacement and servicing of products in use. In general, the greater the user's costs of switching to new product brands, the stronger its adherence to traditional modes of product acquisition, the greater the conversion costs for channels members, and the greater the power of the industrial system to put off responding to pressures for change and adaptation. Change, then, comes typically as a lagged response, with the forces of resistance eventually yielding to the pressures of product-market evolution.

At a country level, economic and regulatory factors may serve to foster or inhibit change in industrial systems. There are, perhaps, three key factors. First is the pace of new product development. New and functionally superior products are often the base from which competitors may enter and take over existing industrial distribution systems. Further, new products provide the base for creating new distribution systems, particularly if they require new applications technology.

Second is the legal framework. Relative to that in many other countries, the applicable body of laws and court opinion in the United States neither locks producers and resellers into fixed long-term relationships nor intervenes to give one legal power to impose its will on the other. Instead, within the bounds of the U.S. antitrust laws, producer and reseller are free to negotiate with each other from their respective economic power bases. Attempts, for example, to allow producers legally to mandate resale prices or to restrict flows of goods across national boundaries have failed. On the other side, distributors in this country do not have the power under the law either to require that producers sell to them or to impose penalties for disenfranchisement, as they do in some countries.

A third factor that may foster or inhibit change in the distribution infrastructure is the country's mix of large and small businesses. Small manufacturing and engineering businesses have been the fountainheads of innovation. Small wholesale and retail businesses make up the great bulk of our distribution networks. On the other hand, large businesses, as the major buyers of industrial goods, utilize their formidable purchasing power to keep constant pressure on the distribution infrastructure to remain cost efficient. Improved distribution cost efficiencies achieved by large buyers tend quickly to accrue to the benefit of their smaller counterparts, just as small-business innovation eventually benefits large corporations.

ADVERSARIAL COLLABORATION

Industrial systems are held together in collaborative relationships by a mutual interest in total system rewards generated through profit margins. They are reinforced, as well, through other mutual dependencies. That is, resellers are reliant on producers for product to sell and for sales support; producers depend on resellers for market access. The adversarial nature of systems relationships lies in the contention over the distribution of value-added benefits generated by the system, with each member of the channel seeking to maximize its share of the total rewards. The adversarial quality of industrial systems relationships is enhanced because each member measures success not against other industrial systems but in terms of its own relevant markets. Resellers seek to optimize trading area market shares and profits; producers are concerned about product-market share positions. While these are not mutually exclusive optimization goals, they do not necessarily foster integrated producer and reseller strategies.

The Balance of Power

The benefits produced by the confederacy of which producers and intermediaries are members tend to flow toward that party with the greater power. Power, in this context, is defined as the relative ability of one party to determine systems price structures, to co-opt the resources of the other in support of its strategic objectives, to avoid the imposition by the other of restrictive terms and conditions, to retain control over its own operations and the allocation of its resources, and to secure preferential treatment from the other to the disadvantage of direct competitors.

Ultimately, the balance of power hinges on which party contributes the greater transaction value-added as perceived by end-users. That will vary depending on the ways in which users rank product and transaction-related values—such as intrinsic product quality, low price, purchasing convenience, technical service, and product customization—at a point in time and place of use.

Producer power is typically accrued through product differentiation along such dimensions as product functional superiority, breadth of product line, and technical service—to the extent that these attributes are of value to user-customers. The industrial producer's power tends to be reinforced by the extent of the resources it has available for marketing investments, by an extensive installed base, and by user brand preference, resulting in high reseller and user switching costs.

Intermediary power comes largely through exploiting value-added opportunities, such as contributing to product customization for specific applications, providing ordering convenience, offering data services to customers, and most important, passing distribution cost efficiencies on in the form of product price reductions. Intermediary power is enhanced to the extent that customers are brand indifferent and that their costs of switching from one brand to another are not of consequence.

MATTERS OF EQUITY

As in any relationship involving the exercise of power, questions of equity and fairness arise in the relationships between producers and resellers. Equity and fairness issues emerge, in particular, because with the evolution of industrial systems, the allocation of rewards or benefits among systems members is constantly changing.

Producers' actions that diminish the value of the reseller's franchise or producers' failures to protect franchise value raise issues of equity. Such producer-initiated actions as increasing distribution intensity in a trading area, establishing captive branches in areas served by independent distributors, reclassifying reseller accounts as direct "house" accounts as they grow in size, and disenfranchising distributors and/or agents serve to depreciate or totally wipe out the intermediary's stake in the industrial system and the worth of its investments.

Reseller actions may also significantly alter share values. One example might be developing markets for producer brand merchandise and then substituting private (reseller) brand product sourced from other suppliers at lower prices. Another is "free riding," in which some resellers benefit from the brand-specific investments of producers and competing resellers to skim off a significant portion of demand at low discount prices. Reseller transshipment of goods into unauthorized channels also serves to preempt the interests of other systems shareholders.

Thus, individual actions, taken unilaterally, pose issues of fairness as they have the effect of increasing the present rewards and future interests of some members at the expense of others. Industrial systems are ultimately built and held together not so much through contractual arrangements as by trust developed and maintained through a history of fair exchange. If industrial systems are to survive as collaborative arrangements, day-to-day actions and longer-term strategic changes, as well, must be implemented with a sense of equity and a concern for the respective interests of systems members. The alternative is a body of law that attempts to codify the interests of intermediaries and producers by establishing them as legal rights. If, as in the United States, producer-intermediary relationships are mediated more through negotiation processes than through legal restrictions, then fairness and the exercise of restraint become all the more important in ensuring the long-run viability of industrial systems.

Acting in fairness includes a concern for the stability of the terms, both implicit and explicit, that inform these relationships and for evolutionary, rather than abrupt, adaptation to changing product-market conditions. Fairness includes as well an adherence to the commitments one party makes to another and to the expectations of each with regard to their respective contributions, the sharing of rewards, and the ways in which each is obligated to protect the other's interests.

At a most basic level, systems viability is dependent on honest dealing. The absence or presence of equity, fairness, and honesty in producer-intermediary relationships ultimately establishes the balance between conflict and cooperation in the system—and thus becomes a key factor in its competitive success.

The forces of history that have shaped industrial distribution in the United States are at work today: economic growth, urbanization, market concentration, new product development, and the expansion of transportation and communications. Add to these the computer-spawned information revolution and the globalization of competition, and we have the fundamental trends that influence our distribution infrastructure now and into the foreseeable future. Demographic data underscore the increasingly important roles of distributors and company salesforces in our market economy. They evidence as well the enduring contributions of other intermediaries—agents, jobbers, and brokers. Together these distributive institutions form increasingly complex distribution networks. If they are to serve successfully our growing economy, it would seem essential that those who manage them act out of a concern for the long-term viability of the system as a whole and the respective interests of each member. Such a perspective, we believe, will serve ultimately to the benefit of all.

• APPENDIX A •

Product-Market Study I

SQUARE D: MARKET STRENGTH THROUGH DISTRIBUTION

Commonly known as fuse boxes, load centers and circuit breakers (LC&CBs) are safety devices that act to interrupt the flow of electricity in overloaded or faulty electrical circuits. Several hundred different specifications of breakers and load centers are listed in the catalogs of each major electrical supplier. They are to be found in virtually every electrified building in the United States. Approximately 90–95% of them are sold through electrical distributors to a myriad of firms that are engaged either in building construction or in the manufacture of electrical equipment that contains load centers. The undisputed leader in a field of nine LC&CB producers—a field that includes such giants as General Electric, Westinghouse, and Siemens—is the Square D Company, headquartered in Palatine, Illinois.

The purpose of this study is to describe the elements of strategy involved in marketing a mature product of low unit value such as the load center. We first describe the product briefly and then discuss its market segments and the channels through which the product flows from manufacturer to user. Next, we describe the market positioning of the three major suppliers. Then we focus on Square D's distribution strategy and on how this company was able to take a position, early in the development of the technology, as the leading supplier of load centers and to hold that position for more than half a century. Finally, we discuss Square D's strategic constraints and give a brief analysis of this market leader's strategy.

PRODUCTS AND MARKETS

Circuit breaker technology today remains much the same as it was when it was first introduced just after World War II. In the late

285

1940s, circuit breakers that could be reset replaced fusible circuit-interrupting devices in which the fuse was destroyed by a line fault. Only modest changes have been made in the product design since that time.

Load centers and breakers are subject to codes established by an independent rating agency, Underwriters Laboratory (UL). These products are also uniformly checked by municipal building inspectors. Thus, at a minimum, product acceptability is dependent on meeting UL standards. When code requirements change, manufacturers must redesign products to meet the new standards. The importance of conforming to UL specifications was demonstrated when one LC&CB manufacturer lost its UL rating in the early 1980s and suffered an immediate and substantial loss in market share.

As of 1985, the LC&CB market in the United States was approximately $700 million, and the generally recognized market segments and approximate market share each represented were as follows:

Single-family housing	45%
Multifamily housing	20%
Manufactured housing	5%
Light commercial construction	25%
Consumer	5%[1]

Single-family housing, the largest of these, was also the most diverse. It consisted primarily of building contractors, typically small businesses, that purchase load centers as part of a wide range of electrical products used on the construction site. Buyer behavior differed, depending on the size of the job. Most projects in this market were relatively small, involving one or two houses. According to a marketing manager for one of the major suppliers,

> [On small projects,] the contractor goes to one distributor not more than 20 miles from the job site, takes the brand the distributor gives him, seldom checks prices of individual items but judges the distributor's relative price levels with reference to a few key items such as single-pole 15- and 20-amp circuit breakers and popular models of junction boxes and switches.

---

[1] These data were derived from the responses of six leading suppliers to a questionnaire survey. Together the six companies represented over 90% of the total LC&CB market in 1985.

On larger jobs, the contractor was likely to shop individually for large volume items, getting several price quotations from different distributors. Most contractors, though, generally stayed with a familiar brand.

Multifamily housing was a more concentrated market. Major suppliers estimated that approximately 40% of this segment was represented by the largest 10% of contracting firms.[2] With firms in this segment, buying modes ran the gamut from multi-item shopping—with convenience a major criterion for source and brand selection—to single-item buying, where price sensitivity often plays a more important role than brand or distributor relationships. On large projects in this market, brand switching based on price awareness often occurred as the purchase was put out for competitive bids. Consulting engineers and architects also influenced brand choice on the large projects. Among other services that they provided, these professionals supplied the contractor with lists of qualified brands of major components and supplies.

The third market segment, manufactured housing (mobile homes and prefabs), was the most concentrated. An estimated 70% of this market was accounted for by the largest 10% of its customers. Although this was a small segment, 5% of the total for load centers, the average dollar value negotiated in a load center transaction for manufactured housing exceeded $100,000 and often reached several million dollars. On a single invoice, the average amount purchased might be in the range of $2,000–3,000.[3] Therefore, like large project purchasers in the multifamily housing segment, buyers in the manufactured housing segment were very price sensitive. The same was true of the light commercial sector, although this segment was less concentrated than either multifamily or manufactured housing.

Finally, the consumer do-it-yourself (DIY) market was reached largely through such retail outlets as hardware stores, home centers, and local building supply stores. Purchases were typically for small amounts in regularly patronized outlets, with load centers as one item on a shopping list of supplies for a repair, remodeling, or building extension project. Having a familiar brand name, such as

---

[2] Based on questionnaire survey responses from eight LC&CB manufacturers.
[3] Ibid. A *transaction* in this context means a buyer-seller negotiated sales contract. A single invoice may represent a transaction or may be an order against a previously negotiated contract.

Exhibit I.1   Electrical Distributors—Product Mix

Residential lighting fixtures	Circuit breakers and load centers
Lighting fixtures	Fuses
Lamps	Transformers
Building wire	Busway
Flexible cord, cord sets	Motor controls, starters, relays
Power cable (includes service entrance)	Programmable controllers
Lugs, connectors, and terminators	Motors and drives
Fasteners	Instruments and test equipment
Pole line hardware/utility supplies	Tools (power and manual)
Electrical tape, insulating materials	Electric heating equipment
Metal conduit	Sound, signal/alarm, and communica-
Nonmetallic conduit	tion equipment
Conduit fittings and bodies	Electric generator sets
Wiring devices	Electrical appliances, housewares, table
Boxes and enclosures	lamps
Panelboards/switchboards	Electronics products
Safety switches	Other
Switchgear	

Source: Electrical Wholesaling, November 1985.

General Electric or Westinghouse, was a marketing advantage for producers competing for the DIY trade. Low competitive prices were perceived as being of lesser importance than in other LC&CB market segments.

Although brand preference in all market segments was alleged to be moderate, builders and electrical contractors showed a strong penchant for continuing to buy familiar brands. Their primary concerns were ease of installation and avoidance of callbacks due to product defects. In addition, electricians and contractors tended to select among the brands their distributor sources carried and indeed may have recommended. This behavior was especially true of small residential contractors building and/or renovating a few houses at a time.

THE DISTRIBUTION INFRASTRUCTURE

Load center manufacturers use resale channels extensively to reach their markets. The product's low unit value and standard design, the fact that most purchase orders for load centers usually include a broad array of other items, and the demographics of the market (i.e., small buyers in widely dispersed geographic areas) favor distribution through resellers. For a distributor carrying the

Exhibit I.2   Electrical Distributors—Customer Mix

Electrical contractors	40.8%
Nonelectrical contractors	6.5%
Commercial/institutional accounts	7.9%
Industrials (maintenance and repair)	18.4%
Industrials (OEM)	6.9%
Utilities	3.9%
General public-retail	4.4%
Retailers for resale	4.5%
Government (federal, state, local)	3.5%
Other wholesalers	1.8%
Export	0.8%
Others	0.6%

Source: Electrical Wholesaling, November 1985.

lines of many suppliers, the costs per transaction of taking orders and delivering from stock are significantly lower than for a manufacturer with a more limited product line. The distributor's cost per sales call is also typically lower than the manufacturer's because the distributor can amortize its costs of selling and order fulfillment over a larger number of line items and greater total sales volume per order than can any of its individual suppliers.

The range of product offerings available through electrical distributors is shown in Exhibit I.1; the classes of customers they serve are noted in Exhibit I.2.

Our 1987 survey of 76 electrical distributors, all of which carried load centers and circuit breakers, exhibited the following average data:

Annual sales revenue	$ 5.7 million
Number of employees	18
Number of customer accounts	2,200
Number of suppliers	150
Number of items (specifications carried in stock)	7,500
Annual LC&CB sales revenue	$ 0.9 million
Number of LC&CB brands carried	1.9
Market area	4 cities in a 50-mile radius[4]

[4] It should be noted that the firms in this sample were, on average, larger than the electrical distributor population as a whole. Department of Commerce data cited in Chapter 11 indicate that as of 1982, the approximately 7,500 electrical wholesaling firms had sales on average of $2.3 million.

Asked to list the factors relevant to effective distribution of LC&CBs in order of importance to customers, the 76 respondents provided the following ranking:

1. Prompt and adequate supply
2. Competitive prices
3. Relationship with distributor
4. Manufacturer's brand name
5. Technical information and assistance
6. Extended credit terms
7. Incentives and promotions
8. Warranty coverage

The survey respondents placed particularly heavy emphasis on the first three factors—supply, price, and distributor relationship. From the distributors' perspective, other factors seemed to pale by comparison in their perceived importance.

THE MAJOR SUPPLIERS

In 1985, three producers accounted for an estimated 70% of load center sales in the United States: Square D's Distribution Equipment Division, General Electric's Construction Equipment Business Operation, and Siemens-ITE's Circuit Protection Division. All were divisions of larger electrical manufacturing companies. In each case, the LC&CB product line was part of a broader product portfolio that was channeled through a single distribution system. Other divisions of the three parent companies made and marketed heavy electrical equipment and sold these lines directly to utilities, OEMs, transportation companies, manufacturing plants, and government customers.

Square D was the leading supplier of distribution and control equipment to contractors of single-family housing and light commercial construction, two segments representing an estimated 70% of the total LC&CB market. Square D was believed to be the second largest supplier in multifamily housing, the fourth in manufactured housing. Square D's management attributed the weaker position in the latter two segments, in part, to its unwillingness to cut prices in these highly price-competitive segments. Overall, the company had traditionally maintained a price premium of approximately 8–10% over its competitors.

General Electric, building on its widely recognized brand name, had taken the lead in the consumer DIY market. In single-family and light commercial, it held a number-two position. Siemens-ITE was seen by its competitors as holding a strong—possibly a leading—position in the highly price-competitive multifamily housing segment, while Westinghouse, the number-four firm in load centers, was strong in manufactured housing. GE, ITE, and Westinghouse were thus positioned around the market leader, the first by providing a strong brand alternative to Square D, and the others by offering respected product lines at competitive prices in segments where Square D had taken a less aggressive stance.

Having pioneered the development of circuit breakers for residential use, Square D had a first-mover advantage in these important segments. The company focused on the single-family and light construction segments, which valued product quality and brand preference more highly than lowest price. Square D then proceeded to build a broad and well-regarded product line and a strong distribution system in those markets where it could realize the highest margins. To have attempted to build comparable market share positions in the latter would have required a more price-aggressive strategy that would have led, inevitably, to lower prices and margins in its primary markets. Its competitors presumably found these opportunities more attractive than attacking Square D in its areas of strength.

SQUARE D—A CONSISTENT STRATEGY

Square D was founded in 1903 as the McBride Manufacturing Company, later becoming the Detroit Fuse and Manufacturing Company. It made electrical knife switches, identified with the letter D in a square. With growing brand recognition for its product line, the company changed its name to Square D in 1917. In 1985, Square D's sales were $1.4 billion and the company was organized into two business groups, Electrical Equipment and Electronics. The first group included four domestic divisions: Distribution Equipment, Power Equipment, Control Products, and Utility Products. The product lines of all four divisions were marketed by the Electrical Equipment sales organization. This product portfolio included distribution and control equipment such as load centers and breakers, transformers, safety switches, switchboards, panelboards, switchgear, motor control centers, busway, motor starters, relays, connectors, and programmable controllers. Square D's line of electrical products was greater in breadth than the lines of competitors, going to market through a single sales organization.

Product Quality

Since its founding, Square D had modified and added to its line continually to meet changes in the marketplace, as well as changing UL requirements, allowing the company to maintain the dominant position. Square D pioneered in nonfusible-breaker technology with the development of its multibreaker (MB) line in 1935 for residential use. The MB was a four-breaker unit that fit into one housing to economize on manufacturing costs. The company marketed the MB through distributors to residential electrical contractors; it also supplied these units to other manufacturers of electrical equipment, including General Electric and Westinghouse. Until the late 1940s, Square D was alone in the market for nonfusible circuit breakers for residential housing.

In 1950, Federal Pacific, then called Federal Electric Company, introduced its Stabloc breaker, designed as a single unit in one housing or receptacle. This design gave Federal Electric a significant product advantage over Square D; in the event of product failure, only the single unit needed to be replaced. By comparison, Square D's entire MB model had to be replaced even though only one of the four breakers might be faulty.

To protect its position, Square D quickly designed and introduced a single breaker unit, the XO, in 1952. The design was flawed, however, primarily because the materials used in its construction led to problems of dimensional instability. After an initial upward sales surge with the introduction of the XO, the company's market share deteriorated in 1954–1955. In 1956, Square D introduced a new QO breaker line, which has remained essentially unchanged ever since. The QO breaker, not interchangeable in Square D load centers with competing breakers, quickly set the industry standard for quality and dependability.

According to one Square D manager who lived through this experience,

> The fact that we could recover from the XO breaker was due in large part to distributor loyalty. If they had not stuck with us, it would have been difficult to get back into the market. But they liked the completeness and general quality of the Square D line, and many of them had a second line of breakers they could sell. Then when the QO came out, it was so far superior to anything else on the market, they quickly got behind it.

In the early 1980s, when Siemens-ITE introduced a new load center made of Noryl, a high-performance plastic, to replace the metal box, Square D again quickly followed. While the use of Noryl did not improve functional performance, it did result in manufacturing cost savings.

Pricing and Market Selection

Given its reputation for consistently high product quality, Square D was generally regarded among competitors and distributors as being able to command a price premium at the resale level. There was evidence as well that because of the line's large installed base and its attractiveness to resellers, Square D had not had to offer distributor margins as high as those of all or most of its competitors to gain and hold distribution.

A premium pricing strategy and extensive market coverage through resellers were consistent with Square D's primary market focus on the residential housing segment, a market noted to be less price sensitive and more brand conscious than other segments, such as manufactured and multifamily housing.

It is likely that the market selection and pricing elements in Square D's marketing strategy developed early in the company's history as mutually reinforcing factors, particularly given Square D's early orientation toward the use of independent distributors as channels for reaching the highly fragmented markets for low-end (in terms of capacity) electrical equipment.

Field Organization

As of 1986, Square D's electrical equipment sales organization included 650 field sales representatives, almost all college graduates with engineering degrees, and about 350 support personnel. This field sales organization was larger than those of competitors. It was organized into 5 regions, 21 sales areas, and 126 offices—the largest office, in Chicago, having 37 sales reps. In major market areas, sales reps specialized in either the industrial/OEM market[5] or the building construction market. They called on users directly, both

---

[5] The industrial/OEM market, while not one of the five recognized LC&CB market segments, was composed of customers for other products of Square D's Electrical Equipment business group.

OEMs and contractors; they also called on distributors that themselves typically addressed one market segment or the other.

In the construction market, sales reps worked with architects and consulting engineers to ensure that Square D was specified or at least included on lists of qualified bidders, and with contractors to calculate the requirements of major jobs and often to negotiate price. The resulting business was then channeled through a distributor, which would service the order.

In calls on industrial customers and OEMs, sales reps became involved in product design and in performance specifications to customize Square D products for particular applications. Some of this business was then channeled through distributors, which handled delivery, credit, and after-sale service. Other orders were taken and handled directly by Square D.

Square D devoted more salesforce time than did other major load center suppliers in calling on both user-customers and those who influenced purchasing decisions, the architects and consulting engineers. The company also generally had higher advertising and promotional expenditures than its competitors.

Distribution

Square D's Electrical Products Group derived 85% of its sales revenues through more than 1,300 distributor locations.[6] The remaining 15% was accounted for by direct sales to large industrial and OEM customers.

Most distributors were independent 1- or 2-branch operations, although Square D's largest distributor was Graybar Electric, selling nationally through 175 branches. Distributors tended to specialize in one of the two major customer groups: They sold either to electrical contractors for residential and commercial construction or to industrial customers and OEMs purchasing electrical equipment or components.

Unlike General Electric and Westinghouse, Square D had not developed a captive distribution organization but relied entirely on independent distributors. This gave it the advantage of not being perceived by distributors as competing with them at the resale level.

---

[6] The Square D distributor network is probably the oldest in the industry. According to our distributor survey, its members had been in business longer—45 years on average—than distributors for which the primary brand was either GE or ITE.

Square D distributors were also impressed by the company management's personal attention to them. Mitch P. Kartalia, who served as Square D's chief executive from 1968 to 1983, had a strong marketing background. Having moved into headquarters marketing, and later become sales vice president and then executive vice president, he had an intimate knowledge of Square D's distribution network. Kartalia spent considerable time calling on distributors and knew many of them personally. He was an active participant in the National Electrical Manufacturers Association, serving on its board of governors, and maintained an active involvement in the National Association of Electrical Distributors.

In franchising distributors, Square D management had established the following guidelines:

1. While distributors might carry competing lines, the Square D brand must be their primary offering.

2. Franchising was done by location. Thus, in the case of some large multibranch firms, some locations might be franchised while others were not.

3. Square D sold to electrical contractors only through distributors but would sell direct to industrial and OEM accounts if the customer preferred to buy in this way.

4. All Square D distributors were required to carry its full line of distribution and control equipment.

In working with distributors, Square D sales reps stressed promotional efforts that attracted user-customers. For example, distributors were asked to hold "counter days," which featured product demonstrations by distributor and Square D personnel for distributor customers. Square D subsidized such events by funding direct-mail announcements and by covering the costs of refreshments. However, Square D's marketing managers did not generally favor the use of "spiffs," that is, bonuses, contests, and vacation trips, to motivate individual distributor salespersons. They believed that awards to individuals, as opposed to user-oriented promotions, did not produce long-term sales increases.

Another important element in Square D's distribution strategy was the Blue Chip stocking program. Distributors designated as Blue Chip dealers were required to maintain stock at specified levels. Inventory levels were established on the basis of each distributor's sales of each item over the preceding 12 months, and distrib-

Exhibit I.3 Distributor Perceptions of Square D as Compared with Other Suppliers of Load Centers and Circuit Breakers (mean values are on a scale of 1–5, with 1 representing superiority)

	Square D[a]	Other Suppliers of LC&CBs[b]
Brand preference among users	1.3	2.0
Product quality	1.4	1.8
Product availability	1.7	1.8
Breadth of product line	1.4	1.9
Price levels	2.9	2.6
Price incentives and discounts	3.1	2.7
Salesforce visits to distributors	1.9	2.7
Technical assistance	2.0	2.6
Advertising support	2.1	2.6
Order-processing and delivery schedules	2.6	2.4
Pull-through assistance and development of specifiers and end-users	2.5	2.7
Distribution sales training	2.2	2.8
Responsiveness of supplier to distributor requests and suggestions	2.8	2.8
General overall satisfaction	2.6	2.7

[a]Based on questionnaire responses from 17 electrical distributors that listed Square D as their primary supplier of load centers and circuit breakers.
[b]Based on questionnaire responses from 69 distributors listing other than Square D as their primary supplier.

utors could return up to 2.5% of their annual purchases of stock at full credit. In the event of a significant change in local demand, the specified stock levels could be altered for any one item, and up to 75% of that item could be returned for credit. In major metropolitan areas, Square D's special distributor inventory reps worked solely on implementing and monitoring the Blue Chip program.

As of 1986, 75% of Square D's distributor locations carried the Blue Chip designation, and 95% of Square D's sales through its distributors came through these outlets. The program had the effect of stabilizing ordering patterns for Square D plants and of ensuring that the brand was always in stock on distributor shelves. Furthermore, the Blue Chip program was seen as a way of encouraging distributors to take on new products, knowing that they could return the unsold quantities for credit.

Our distributor questionnaire survey indicated Square D's areas of strength vis-à-vis competitors. As Exhibit I.3 indicates,

Square D ranked well on such counts as the following: brand preference among users, product quality, breadth of product line, product availability, and distributor sales training.

The Square D line was an important contributor to its distributors' revenues. Typically, the line accounted for 20–30% of distributor sales and was the largest source of gross margin dollars of any line they carried.[7] Further, distributors that listed Square D as their primary brand reported that sales of this line, as a percentage of total revenues, averaged more than twice the sales reported by other distributors for their comparable primary lines.

In expressing their overall satisfaction with Square D as a supplier, those distributors which listed Square D as their primary brand rated the supplier about the same as other distributors rated their primary suppliers. One reason that Square D was not rated more highly, perhaps, is that the company was perceived as charging higher prices and offering fewer price incentives and discounts than other suppliers. Also, with its large field salesforce, the Blue Chip inventory monitoring program, and a reputation for having the best-managed distribution network in the business, Square D may possibly have been seen by its distributors as being more involved in their businesses than some might have wished.

Strategic Constraints

In contrast to its leadership in the single-family residential and light commercial construction market segments, Square D had been markedly less successful in the consumer market. In 1981, because the DIY consumer market was growing at a more rapid rate than its traditional market segments, Square D's managers launched an effort to build distribution through retail hardware stores, home centers, and retail lumber yards. By 1986, Square D was selling through 25 manufacturers' reps (MRs) to large chains such as True Value, Ace, Pay n' Pak, and K mart. Typically, MRs carried 10–15 lines of related hardware items—all going into consumer markets—and provided in-store product demonstrations, checked retail stock and shelf position, and prepared advertising materials.

Nevertheless, while the program had produced substantial revenues, its future was uncertain as of 1986. While Square D's do-

---

[7] Load centers and breakers represented 15–20% of distributors' sales of Square D products.

it-yourself consumer product line had included a range of lines acquired through mergers, such as Trine door chimes, Bell weatherproof electrical boxes, and Nelco baseboard and wall heaters, in addition to load centers and breakers, the acquired lines had generated losses in 1983–1986 and were being dropped. Square D's management was concerned that the reduced consumer product line would not have sufficient breadth and depth to maintain the attention of manufacturers' reps and retailers.

Another difficulty was pricing. Square D's prices to retail chains were significantly higher than its prices to distributors for two reasons: first, extra costs associated with the consumer program for packaging, displays, advertising, and promotions; second, concerns about interchannel competition. A few distributors had been selling to retail outlets, and Square D's managers did not want to be in the position of competing with them. Nor did management think it desirable to have retail outlets competing with distributors in the residential markets for the business of the small electrical contractor. Square D, for example, elected not to franchise one large retail chain because it was aggressively pursuing electrical contractor business. "To sell through that chain would be like waving a red flag at a bull," said Square D's marketing vice president. There was also some concern that low prices to consumers would inevitably undercut the remodeling business of the small electrical contractor.

Further, Square D's policy of franchising individual locations rather than all outlets in a distribution chain was difficult to apply in the consumer market. Large retail chains with many outlets were unwilling to accept franchising by location. Because Square D employed MRs, who typically called on chain store headquarters, management had no practical means of implementing a policy of franchising individual retail stores.

For all these reasons—a shrinking DIY product line, interchannel and user-producer price competition, and the difficulty of applying industrial distributor franchising practices in the retail market—Square D's future prospects in the DIY segment were clouded. Yet Square D's marketing vice president observed that this market was likely to continue growing more rapidly than the market segments served through electrical distributors. While the DIY market for load centers and breakers represented about 15% of the market total in 1980, that figure was forecast to be 25–30% by 1990.

An Analysis of Square D's Market Leadership

Square D's strategic profile seems consistent with a long-term goal of building strength in market segments that were more quality- and brand-sensitive than price sensitive. The strategy, too, was one of a market leader whose strong brand position at the user level gave it strength in its relations with its distribution channels.

The elements of Square D's strategy may be identified as follows:

Market selection	All segments with particular stress on the two largest and less price-sensitive segments
Price	Premium price; low-to-average distributor margins
Product line	Broad line of LC&CBs as part of a full line of electrical distribution equipment

Distribution
Independent Distributors

Policies	Individual location franchising; full-line stocking; Square D the primary brand; direct top management involvement with distributors
Intensity	More intensive than Square D's competitors' systems
Roles	Provide primary market access; promote the Square D line; carry inventory; provide service; extend credit

Salesforce

Organization	Specialize by contractor and industrial/OEM segments in major trading areas; functional specialization to monitor Blue Chip program
Roles	Provide distributor support; generate demand; negotiate large contracts at user level; pass all sales through distributors in the residential and light construction segments; take large sales direct in the industrial/OEM segment selectively; monitor distribution inventory levels closely

Advertising	Maintain a steady level of trade advertising to support distributors and to build brand image at user level
Promotion	Work with distributors on product demonstrations instead of providing personal financial incentives to distributor sales personnel

Square D Strengths.   Six factors contributed to Square D's market leadership. First, it had maintained product superiority. The first to introduce the nonfusible breaker, it quickly redesigned its four-unit breaker module when Federal Electric introduced a single-unit product in 1950. In 1982, Square D was the first to match Siemens-ITE's development of the Noryl load center, replacing the metal case. Other producers then followed.

Second, its broad line of electrical equipment allowed Square D to accomplish the following:

- support large and specialized field sales resources for maintaining close contact with its distributors and mounting a significant pull-through sales effort at the user level

- afford a relatively high level of advertising and promotional support to maintain its brand image at both the reseller and user levels

- command the attention of its distributors because of the importance to them of the revenues generated by the broad Square D product line

Third, the company, over its long history, built the largest installed base of any load center manufacturer. These installed units performed satisfactorily and required Square D breakers; Square D had thus protected its own replacement market from the intrusion of other brands. Further, its clientele of electricians and contractors developed a brand loyalty that did not have to be continually reexamined because of major innovations from competing suppliers.

Fourth, Square D's senior management was consistently oriented toward building and maintaining the company's independent channels. Its top managers were personally involved in developing and implementing marketing strategies in which the company's reseller network played a key role. Nurturing this distribution system became an integral part of the Square D culture. For instance, it is quite significant that Mitch Kartalia, who served as a member of

Square D's top management and its chief executive for many years, was actively involved with two major electrical industry trade associations and spent a significant portion of his time calling on distributors. Possibly no single factor may contribute to distributor loyalty so much as the personal attention of the supplier's top management.

Fifth, in developing a distribution strategy, Square D clearly emphasized two conditions of selling effectiveness that are perceived by distributors to be of primary importance. It provided a broad product line of *dependable* quality. It gave high priority as well to ensuring that distributors kept adequate stocks on hand.

Sixth, it was Square D's policy to avoid competing with its distributors in its major market segments. It would sell only through distributors in its two primary markets, residential housing and light construction. However, in the industrial market segment, it sold to users that wished to buy direct from Square D when any of the following conditions held:

- purchases were for large amounts

- technical product specifications and user-unique product mixes were negotiated in the context of competitive-bid transactions

- price competition was particularly intense

Square D did not compete with its independent distributors through a captive distribution network, as did General Electric, through GESCO, and Westinghouse, through WESCO. Thus, its policies were structured to avoid interchannel conflict in its major markets by selling direct or by attempting to establish an in-house distribution capability. In this policy and in its franchising terms, Square D was clear and consistent over the long term. Thus, its distributors were able to develop their businesses, taking as given a set of explicit supplier relationship conditions.

These factors contributed significantly to Square D's market leadership; all were consistent with, and highly salient in, its primary market segments, residential housing and light commercial construction. Here extensive market coverage through supportive distributor channels was critical. These same factors and the relatively lower price sensitivity, especially in the residential housing market, also provided the logical basis for a premium pricing policy.

Strategic Limitations.   Square D was not the market share leader in the multifamily and manufactured housing market segments and had been unsuccessful in the consumer do-it-yourself segment. In all three areas, Square D's pricing and distribution policies, and its commitment to the single-family residential and light construction segments, limited its potential for competing effectively for market segment leadership. To contest Siemens-ITE's and Westinghouse's positions in the multifamily and manufactured housing segments respectively, Square D would have needed to be more price aggressive. To do so would have meant risking margins in the large, profitable, and less price-sensitive segments in which it was clearly the market share leader, because it would be difficult to price discriminate among these different customer sets.

In the consumer DIY market, Square D was limited by its concern about competing against its independent electrical distributors, either for retail business or, through retail, for contractor business. Square D's management did not want to be perceived by its industrial channels as supplying retail channels in areas where the two were increasingly overlapping. Further, it lacked breadth of product line in competing effectively for distribution through retail outlets. Thus, success in one or a few market segments may create a set of limiting commitments that preclude comparable success in others.

CONCLUSIONS

Square D's experience in the market for load centers and breakers is useful for identifying the key success factors that are relevant in marketing products of relatively low unit value, based on mature technologies, and reaching widely fragmented markets. They include (1) consistent product quality and high performance reliability, (2) strong resale distribution with respect to both extensiveness of market coverage and local market share position, (3) a product line sufficiently broad to command distributor attention, (4) tightly controlled and administered franchising policies and practices, (5) programs to ensure product availability at distributor sales locations, and (6) a large and well-qualified direct salesforce to monitor distributor performance and field inventory levels.

A critical reliance on indirect distribution to maintain and build the installed base also lends considerable importance to policies and practices encouraging their support, such as avoidance of competition with distributors in selling to user accounts, or through a captive sales organization, or through supporting competing channels—in this case, the DIY retail market. Distributor support is also strongly encouraged by long-term stability in marketing strategy and distribution practices and by the close personal identification of top management with the distributor community.

Because such marketing strengths take a long time to build, a first-mover advantage is of particular benefit. At the same time, a going-in leadership position provides no guarantee of ongoing market dominance. Instead, to keep a leader's edge, it becomes essential to stay in the forefront of product technology, to maintain product quality, and continually to build the distribution network and defend it through franchise policies and practices and distributor incentives from competitive incursion. Finally, however, a strategy so explicitly focused on one market segment often constrains the leader's capacity for entering and competing as effectively in market segments requiring somewhat different strategies in terms of pricing, distribution, and product line.

Product-Market Study II

MARKETING IN A WHIRLWIND: DISTRIBUTION STRATEGIES AND THE DISK DRIVE INDUSTRY

Between 1978 and 1986, sales of "rigid" disk drives increased nearly 150-fold to $4 billion, while venture capital firms invested almost $400 million in 43 drive manufacturers (including 21 start-up investments during 1983–1984 alone). In contrast to load centers, the disk drive industry of the early 1980s was rapidly growing and technologically dynamic. Most drives were sold direct by manufacturers (rather than through intermediaries) throughout this period. Thus, the industry provides an opportunity to consider the conditions that tend to lead manufacturers toward direct channels. In addition, the industry provides a setting for analyzing the implications for channel design and management of short product life cycles, substantial price performance improvements with each product generation, the consequent emergence of new and different customer groups in various market segments, and the advent of foreign competition—conditions common in other technologically dynamic industrial markets, as well.

In what follows, we first provide an overview of the industry, important customer and market characteristics, and competitive developments. Next, we consider the industry's distribution demographics and compare the marketing and distribution strategies of three selected competitors: Control Data Corporation, Seagate, and Fujitsu.

PRODUCTS AND TECHNOLOGY

Disk drives, used in computer systems, record, store, and retrieve data from selected locations on a disk as it whirls within the drive. The disk media itself can be either rigid ("Winchester") or flexible ("floppy"). Floppy disks are removable from the drive mechanism. Rigid disks, fixed in place and sealed from the atmosphere, have greater storage capacity, reliability, and speed than do

304

floppy drives. By 1986, demand for rigid drives was growing much faster than demand for floppies.

At the time of this study, disk drives were manufactured in different diameters, or "form factors," ranging from 14-inch to 3.5-inch. The larger drives had been manufactured since the 1970s; the smaller, 5.25- and 3.5-inch drives, were first introduced in the early 1980s. As drives were produced in smaller diameters, their capacity had also increased. Thus, by 1986, the smaller rigid drives were replacing the larger drives in many applications. At the same time, 8- and 14-inch drives were growing in capacity to satisfy the increasing storage requirements of mainframe and "superminicomputers." In 1986, most firms distinguished among three product categories, based on drive capacity: a low-end (less than 30 megabytes of capacity), a medium performance (30–100 megabytes), and a high-performance category (more than 100 megabytes).[1]

- At the low end, in single-user computer systems, price was the dominant purchase criterion. In these systems, some floppy drives and lower-capacity 5.25- and 3.5-inch rigid drives were typically used.

- Medium-performance drives were used in applications such as small multi-user word processing systems and in "local area networks," which interconnected a number of microcomputers at a location. In these systems, the drive's capacity and access time[2] increased in importance as purchasing considerations. In newer systems, 5.25- and 8-inch drives were typically used, with the smaller diameter drives increasingly replacing the larger.

- High-performance drives were used in large multi-user systems and in mainframe computers. In these systems, data storage capacity and fast access time were critical factors, and the price of the drive was often secondary. These systems generally used larger drives, although by 1986, newer 5.25-inch rigid drives had reached levels of speed and capacity needed in many high-performance applications.

[1] A "megabyte" is 1 million bytes, and a byte is 8 units, or "bits," of information processed by a computer as a single unit.
[2] "Access time" referred to the speed with which a disk drive could respond to computer requests for data. A drive with slow access time was described by one industry executive as "a ball and chain tied to the computer system."

During the later 1970s and early 1980s, first minicomputers and then microcomputers had revolutionized the market. Exhibit II.1 indicates the growth in microcomputer shipments and the complementary growth in sales of small rigid drives during the early 1980s.

Throughout the early 1980s, newer, high-performance products were introduced with increasing frequency, and product life cycles of 12–18 months from introduction to maturity were common in drive categories. Further, competition and the scale economies inherent in disk drive production made it imperative that manufacturers manage material, labor, and overhead costs aggressively, especially in the 5.25-inch product categories. In particular, the importance of volume production for lowered unit manufacturing costs had implications for market selection and distribution policies among disk drive vendors.

Exhibit II.2 provides historical pricing trends for the major drive product categories. As can be seen, disk drive suppliers in the early 1980s were in an increasingly price-competitive marketplace, characterized by annual price decreases of 50% or more in many product categories. Further, the price competition was significantly greater among the smaller, lower-capacity drive products.

Mass storage technology (of which disk drives are one version) was changing rapidly in the 1980s. Product life cycles were short, and innovation, a priority. As one disk drive executive stated, "In this business, the ability to innovate and to understand rapidly evolving customer requirements is essential. A successful drive product is not so much a long-term franchise as it is a platform for the timely introduction of the next generation."

MARKET: OEM CUSTOMERS

In 1985, the 10 largest computer manufacturers accounted for about 50% of disk drive purchases, and the 50 largest original equipment manufacturers (OEMs) accounted for more than 80%. It was not uncommon for a drive supplier to sell to just a few large customers. In 10-K filings for fiscal 1985, for instance, Tandon and Seagate each reported that two customers accounted for almost 60% of net sales. The customer base for disk drive vendors also tended to be geographically concentrated in areas such as Silicon Valley in California or the Route 128 area in New England.

Exhibit II.1

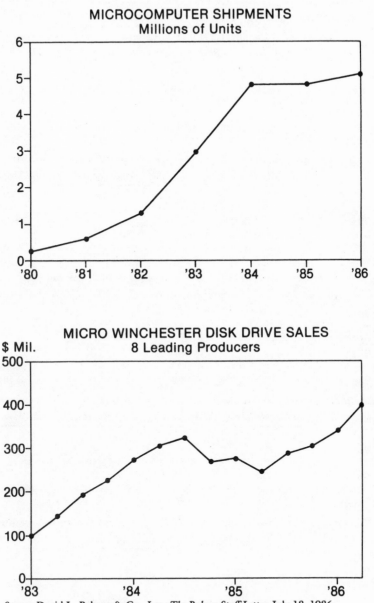

MICROCOMPUTER SHIPMENTS
Millions of Units

MICRO WINCHESTER DISK DRIVE SALES
8 Leading Producers

Source: David L. Babson & Co., Inc., *The Babson Staff Letter,* July 18, 1986.

Exhibit II.2 Industry Pricing Data for Selected Disk Drive Product Groups (prices are estimated average *unit* prices at manufacturer's selling price for the specified order quantity)[a]

	Fourteen-inch Rigid Drive (to OEM purchasing in 100-order quantities)	
	80 Megabytes	*300 Megabytes*
1981	$7,000	$11,500
1982	6,200	11,000
1983	5,800	10,500
1984	5,400	10,000
1985	5,200	9,500
1986 (E)	5,100	9,100

	Nine-inch Rigid Drives (to OEM purchasing in 100-order quantities)		
	160 Megabytes	*340 Megabytes*	*500 Megabytes*
1981	—	—	—
1982	$5,300	—	—
1983	4,800	$7,200	$8,200
1984	4,400	6,600	7,600
1985	4,100	5,300	6,300
1986 (E)	3,500	4,800	5,800

	Eight-inch Rigid Drives (to OEM purchasing in 100-order quantities)	
	160 Megabytes	*360 Megabytes*
1981	—	—
1982	—	—
1983	—	—
1984	$3,500	—
1985	3,000	$4,500
1986 (E)	2,500	4,000

	Five and one-quarter-inch Rigid Drives (to OEM purchasing in 500-order quantities)			
	10 Megabytes	*20 Megabytes*	*30 Megabytes*	*80 Megabytes*
1981	$1,250	—	—	—
1982	900	—	—	—
1983	450	$550	$1,400	$1,750
1984	300	400	950	1,450
1985	250	350	850	1,100
1986 (E)	200	250	600	950

	Three and one-half-inch Rigid Drives (to OEM purchasing in 500-order quantities)		
	10 Megabytes	*20 Megabytes*	*30 Megabytes*
1981	—	—	—
1982	—	—	—
1983	$450	—	—
1984	400	—	—
1985	325	$450	$475
1986 (E)	250	350	450

Exhibit II.2 *(Continued)*

	Five and one-quarter-inch Floppy Drives (to OEM purchasing in 1,000-order quantities)	
	.5 Megabytes (full-height)	*.5 Megabytes (half-height)*
1981	$260	—
1982	240	—
1983	210	230
1984	110	110
1985	85	50
1986 (E)	70	45

Source: Interviews with industry executives.
ᵃLarge computer OEMs often ordered in quantities much larger than those used in the exhibit's examples. Larger orders generally received volume discounts, which made the average unit price of the drive significantly lower than the prices cited in these examples.

A disk drive often represents as much as **40%** of the manufacturing cost of a microcomputer or small minicomputer and is a major determinant of the computer's performance capabilities. As a result, selecting drive suppliers is a major decision for a computer OEM and involves stringent qualification processes. In addition, many computer OEMs also manufacture their own drives.

In 1986, less than 20% of drives sold on the OEM market went through indirect channels such as manufacturers' representatives and industrial distributors, although the percentage sold through intermediaries had increased in recent years. Large OEMs generally contracted to buy large quantities of a specific drive product over some time frame, usually one to two years. If the OEM's systems were successful in the marketplace, add-on sales at other accounts, as well as replacement sales for a number of years, might follow.

Because of the importance of large customers, the senior management of a drive firm was often involved in these selling efforts. One executive of a drive firm referred to this as "head-of-state selling":

> The big sales in this business are closed at the summit conference level when I and other senior managers from this firm meet with our counterparts at the OEM. That doesn't mean the field sales representatives are not important. They must develop and maintain a strong relationship with those accounts, penetrating throughout a very decentralized and complex decision-making unit, because the product policy of a drive vendor is intimately depen-

dent upon the product plans of the larger OEMs. Also, the field sales rep must negotiate the purchase through the technical specialists and purchasing people at the OEM. And the sales rep must, of course, try to influence the OEM's make-buy decision.

But closing a major sale usually occurs at a face-to-face meeting between senior managements of the firms involved. As well as a set of technical specifications, the OEM is buying a trusted supplier. Because the product is technically complex and so important to the ultimate performance of the OEM's system, delivery, reliability, quality control, and other aspects of a good supplier are crucial.

As a result, a number of sales are concluded with the top manager from the OEM looking across the table at the top manager of the drive vendor and saying, "I understand the technical specs and the pricing arrangement, but can you *promise* me this product will in fact be delivered on time and without any 'bugs'?"

As these remarks indicate, established relationships were important aspects of the buying process in this market. Further, these relationships encompassed needs for information exchanges among a variety of different functional groups in both buyer and seller organizations—design engineers, product managers, production personnel, sales and marketing personnel, and top management—since adopting a given set of performance capabilities in a disk drive was effectively a strategic decision on both sides involved. Steady relationships with major computer OEMs provided another important benefit. As one disk drive executive commented, "It is important to get to know these major customers very well, since they effectively establish the standards for the newer computer products." Moreover, the OEM's costs of qualifying a new vendor were extensive, ranging from a battery of product tests to numerous plant visits, involving resolution of technical problems related to drive specifications and product design. Once an OEM had become familiar with a drive vendor's product, engineering team, production personnel, and top executives, the chances of the drive vendor winning contracts for new generation products were greatly improved.

DISTRIBUTION CHANNELS

While most sales of disk drives were made direct to OEM customers, drive vendors also included manufacturers' reps, industrial distributors, and value-added resellers (VARs) in their channels systems. This section explains the predominance of the direct sales channels, the role of each channel in the distribution of disk drives, and the evolution of various reseller channels for disk drives during the 1980–1986 period.

Direct Sales Force

The reasons for the heavy emphasis on direct channels in this industry include market, product, and economic factors as well as the need, in a fast-changing and technologically dynamic marketplace, to maintain good conduits of information exchange with important OEM customers.

Perhaps the most salient aspect of the industry during this period was the concentrated customer base. For most drive firms, most sales volume came from a few (often, 1–3) major accounts. For example, in 1984, one leading drive vendor made 95% of its more than $340 million in sales through 12 field sales people located in 7 offices. Clearly, the economics of selling direct to larger OEM accounts were generally very attractive for drive vendors. As a marketing executive for one firm explained, "Each of our salespeople generated an average of $10 million in sales last year (i.e., 1984) and fully burdened costs for each salesperson were less than $150,000. That's good ROI from our perspective."

In 1986, one customer bought $85 million worth of disk drives from one drive vendor. The disk drive team for this OEM customer consisted of two salespeople, one headquarters administrative person, and one administrative person part-time at the drive vendor's manufacturing plant. In addition, the vice president of sales for the drive vendor spent a substantial portion of his time in direct contact with this account. The estimate of the total direct sales-related expenses for this account (including salary, incentive compensation, benefits, and travel expenses for the relevant sales and administrative personnel) was about $450,000, or about one-half of 1% of sales revenue.

Large OEM customers also represented volume sales in an industry where manufacturing costs were especially sensitive to scale economies. A large OEM contract often made possible dramatic improvements in the drive vendor's capacity utilization, throughput, and unit manufacturing costs, as well as providing a volume base for making production plans. The result was that the manufacturing economics for a given account often affected costs and pricing to other accounts as well.

The nature of the product also influenced distribution patterns. In a market where the product development cycle was frequently longer than the product life cycle itself, direct information exchanges between buyer and seller were essential. As one marketing executive for a drive vendor explained,

> At a large OEM account, the selling cycle for a given drive product usually takes 9 to 12 months. It begins with discussions concerning market requirements and future product plans both by us and the customer. Then, when and if we find that our product development plans imply that a given form factor and performance category can be mutually targeted, we begin discussions with the account's product development engineers, system engineers, product managers, and production personnel.
>
> There are therefore three key areas of contact with OEM accounts: first, design and development engineers for the technical specifications and later the evaluation of product prototypes; second, product line management, which decides when to introduce the OEM's product and the volume of drives the OEM will want; third, purchasing managers, who negotiate prices and terms and conditions of the contract.
>
> Over time, most of these people move around a lot, and it takes continuity and frequent contact with all of these areas to manage to keep up with buying activity at these accounts.

This type of selling and account management would be much more difficult, if not impossible, to conduct through intermediaries such as distributors.

For these reasons, direct sales channels became the predominant distribution pattern within the industry. The nature of the product demanded a constant exchange of technical information, and the nature of the OEM customer base made a direct sales effort an efficient and effective manner of exchanging this information with major OEM customers. While direct channels, however, accounted for the great majority of disk drive sales throughout the 1980–1986 period, the role of intermediaries evolved in response to developments outside this customer base.

Indirect Channels of Distribution

Manufacturers' rep organizations, industrial distributors of electronics products, and VARs each performed different functions for suppliers of disk drives, and each tended to be utilized at a different stage of the industry's, and a firm's, growth.

Manufacturers' Reps. MRs were often the initial channel of distribution for drive suppliers. These suppliers were typically independent start-up companies, funded by venture capital investments that were allocated primarily to R&D and production. A direct sales channel was often initially uneconomical (except to the largest OEM customers) for these companies. One marketing executive commented, "A manufacturer's rep doesn't cost you until sales are made, and that can be very important in a business where the selling cycle generally takes 6–12 months. Initially, we couldn't afford to spend money on developing a sales organization, while the manufacturers' reps were already selling products like controllers and power supplies to OEMs."

In addition, because disk drives must be integrated in an OEM's system with complementary products, the power supply and the controller, the manufacturers' rep organization—which sold these products—was often a preferred source of supply for smaller purchasers of drives. As one drive vendor's director of marketing explained, "The manufacturers' reps were looking for disk drives to fill out their product lines to those customers, and we wanted our organization to focus its resources and money on R&D for the next generation of drive products."

For newer drive vendors in large OEM accounts, manufacturers' rep organizations also performed an important function. As the sales vice president for one drive vendor phrased it, "When we first started trying to sell to big OEM accounts, we simply didn't know which doors to knock on at a big company like IBM or DEC or Hewlett-Packard, but our manufacturers' reps did."

However, as sales to large OEM accounts increased, drive vendors usually replaced manufacturers' rep channels with direct salesforces. In addition to cost-of-sale considerations, a major reason for the switch was that once a computer OEM had committed to significant purchases of a given drive, the nature of the buyer-seller relationship often required direct, ongoing contact between the two firms.

Industrial Distributors. Whereas manufacturers' reps were often an important channel in the early stages of a drive vendor's marketing efforts, distributors tended to become part of the channel system in later stages. Distributors became most useful when the drive supplier sought to expand and diversify its customer base to include small- and medium-sized OEMs.

As drives declined in price and improved in performance during the 1980s, many new, smaller customers emerged, especially in the microcomputer segments. To reach these newer customers, many drive vendors began selling their products through industrial distributors. One drive executive noted,

> Distributors are often in a better position to identify and sell to emerging companies because they can leverage their current sales calls. They are often already calling on these small companies to sell semiconductors, controllers, power supplies, and other electronic equipment, and the purchase decision for drives at those companies is often made by the same people buying those other components.

Distributors also acted as a "buffer" against the inherent credit risks of dealing with the smaller OEMs and systems houses, which were often either spectacular successes or failures in the marketplace. During 1984, for example, Compaq Computer, a maker of portable computers that was founded in 1982, grew to nearly $350 million in sales, becoming a big purchaser of drives in the process. Meanwhile, Victor Technologies and Gavilan Computer Corporation, other makers of small computers, went bankrupt, leaving a number of suppliers with significant unpaid accounts receivable. A drive vendor explained,

> A distributor can deal with the credit risks better than a national manufacturer can. Distributors' credit departments often share account information about customers' payment patterns, and the distributor can often monitor the small company more closely than we can. A distributor, who lives or dies by working capital, often simply has the mentality that allows him to keep dunning a recalcitrant customer for accounts receivable, whereas many high-tech manufacturers are more "relaxed" in this area.

These resellers tended to be large multibranch electronics distributors. Semiconductors represented over half their sales, with the remainder typically in electromechanical devices, connectors, and various computer products (including terminals, printers, software, disks, and disk drives). A large national distributor such as Hamilton/Avnet, Inc., or Arrow Electronics (the largest and the second largest electronics distributors in the United States during the 1980s) typically stocked over 125,000 different items in 100 or more different product categories from 80–100 suppliers. Their principal suppliers were manufacturers such as Texas Instruments, Intel, Advanced Micro Devices, Fairchild, RCA, and Siemens. Through 1986, disk drives generally accounted for less than 10% of total sales of such distributors. Nevertheless, during the 1980s, some larger electronics distributors established computer products groups within their organizations to focus on computer-related equipment, including disk drives. These groups were typically given more product training and often included specialized service personnel.

Two aspects of disk drive suppliers' use of distributors are especially noteworthy, and in striking contrast to the place of industrial distributors in the channel systems for lower-value, low-tech components, such as load centers. One is the relatively subordinated role of distributors in the channel systems of most disk drive manufacturers throughout the early 1980s. As one drive executive noted, "In this business, we face an extreme version of the 80–20 rule, with the lower 50% of the customer base accounting for less than 10% of our total revenues. Our basic goal for distributor sales was to allow the direct salesforce to provide more support to large OEM customers, while distributors focused on the many small accounts." As a result, distributor channels often did not receive significant attention and support from many drive suppliers during the early 1980s. Indeed, the marketing vice president for one large drive supplier commented,

> Our company, like most drive vendors, has historically taken a "trash-can" approach to distributor channels: When demand is slow, we've tended to put our worst-selling products into our distributors; when demand is strong, and allocations are required, we've given our direct customers much higher priority than distributors; and when selling expenses come under scrutiny at budget time, distributor support resources are among the first to get cut. However, this is changing as distributors account for increasing amounts of sales volume, and as the decreasing prices for drives make direct sales calls less economical at more accounts.

A second noteworthy aspect is that, whereas load center suppliers utilized a large number of smaller, local distributors, drive suppliers typically utilized only a few large national or regional distributors. A number of factors influenced this pattern.

Product-related factors played a part. In the early 1980s, disk drives were more technically complex and had longer selling cycles than most other products carried by distributors. As a result, both supplier and distributor had to devote attention to training distributor sales representatives. For suppliers, such training was easier to do for a limited number of distributors, and only the larger distributors could typically afford to make the necessary investments in in-house product specialists.

The pace of technological change in the industry also favored a supplier's reliance on a few large distributors because inventory and product life cycle transitions could be managed more easily if the supplier worked with only a few distributors. Short product life cycles for disk drives meant that transitions from one generation of products to the next, and the consequent risks of obsolescent inventory, were frequent issues. When they began selling through distributors, many drive suppliers followed practices established by semiconductor manufacturers, which protected their authorized distributors against potential inventory write-downs due to product obsolescence or manufacturers' price reductions. Under the terms of these agreements, distributors usually had the right to return to the manufacturer for credit a defined portion of those inventory items purchased within a designated period of time.

In addition, customers generally looked to upgrade their drive purchases over time and so preferred to deal with distributors that stocked a range of disk drives in different performance categories. National distributors and the few regional distributors for which computer products were an especially important revenue source were more likely to make these necessarily large inventory investments.

Finally, organizational issues also played a role. Because only a few salespeople were typically allocated to the development and maintenance of distributor sales programs, managers responsible for distributor sales often found it both necessary and convenient to concentrate their efforts on a limited number of distributor accounts. As one distribution sales manager for a disk drive firm explained, "When I became head of distribution sales, I was the entire management team for this channel and had seven junior distributor sales reps to cover the country. It quickly became evident that working with only two national distributors would provide us with market coverage while also cutting down on the required number of contacts at distributor organizations."

Value-Added Resellers. VARs were a third intermediary utilized by drive suppliers. As the market for drives expanded, customers in different industries or occupations often required specialized computer applications or upgraded systems with increased computing capacity. VARs became an important channel to these niche markets, providing manufacturers with access to accounts that were either too small or too specialized to be reached directly or through computer retail channels.

VARs also played a role in the growing add-on market for hard disk drives, which consisted primarily of two types of customers. One type was the microcomputer owner who sought to upgrade the memory capacity of an existing machine. VARs purchased the disk drives from drive vendors, integrated the drive into a subsystem or plug-in card format, and resold the product to small businesses or to retailers such as Computerland or Businessland.

A second type was the microcomputer buyer who sought systems of either greater capabilities or lower price than those generally available from major manufacturers. Two factors were important in creating this market: One was the establishment of technical standards, and the other was the substantial markup typically taken by computer OEMs on drives. The standards were in large part the result of IBM's successful introductions of its PC and its PC-XT models in 1981 and 1983, respectively. These products established a standard operating system for microcomputer software as well as standard electrical interfaces between the drive and other hardware components. This meant that any of a number of manufacturers' disk drives could be installed in an IBM or IBM-compatible personal computer. The substantial OEM markups left opportunities for VARs to assemble computer systems from components and to sell at lower prices than those of established brand manufacturers. Many VARs bought the basic hardware from IBM or from a so-called clone manufacturer of microcomputers; then bought disk drives from a drive vendor, the controller and other circuitry from other companies; and finally assembled the computer system, either adding software for a specialized vertical market or simply selling the system to retailers at a lower price than that of IBM or of another major manufacturer. These VARs represented a particularly important channel for newer drive vendors not yet established as suppliers to large OEMs.

Thus, where manufacturers' reps were typically part of drive vendors' channel systems in the initial stages of a firm's marketing efforts, and distributors were typically added to reach smaller customers, VARs emerged at a later stage. They became important channels as microcomputers proliferated among a variety of end-use segments, as an aftermarket for disk drives developed among a dispersed installed base of customers, as technical standards were established for microcomputers, and as customers became educated in the product's uses and components.

COMPETITIVE DISTRIBUTION STRATEGIES

The competitive environment in disk drives was complex and challenging in the 1980s. In 1985, there were more than 50 U.S. drive manufacturers and nearly 20 foreign companies in the U.S. market. In addition to competition from other drive vendors, manufacturers of drives typically faced potential competition from their major customers as well. As volume for a given drive product grew, the motivation for an OEM to take more of that drive's production in-house also grew. A number of the largest OEM customers for drives—IBM, Hewlett-Packard, and Digital Equipment Corporation, for example—had traditionally manufactured removable disk-pack products for their larger computer systems; and IBM, which had pioneered disk drive technology, manufactured a substantial portion of 5.25-inch drives for its line of personal computers.

Within the disk drive industry, competition differed by product class, that is, size and capacity of disk drive. The major players in 8-, 9-, and 14-inch drives, for example, were not necessarily the key competitors in the smaller, 5.25- and 3.5-inch, product categories. Within a given form factor, moreover, competition and market shares differed according to whether a company focused on the low-performance, medium-performance, or high-performance drives within that product category. Among manufacturers of 5.25-inch rigid disk drives, the largest product category in 1985, market shares in 1985 were those shown in Exhibit II.3.

The leading players in this product category illustrate the range of competitors in the industry. Control Data, Quantum, Micropolis, and Priam had entered the medium-performance 5.25-inch rigid drive category from a base in larger form factors. As large

Exhibit II.3 5.25-Inch Rigid Disk Drives Market Share Summary, 1985 U.S. Shipment of Noncaptive Disk Drives

Less than 30 Megabytes			30–100 Megabytes		
Drive Manufacturer	Units (000)	%	Drive Manufacturer	Units (000)	%
Seagate	533.0	36.0	Control Data Corp.	115.4	23.9
Computer Memories	291.2	19.7	Quantum Corporation	87.0	18.0
Miniscribe	212.0	14.3	Micropolis	63.9	13.2
Tandon	131.9	8.9	Priam	39.0	8.1
Microscience	63.0	4.2	Seagate	27.6	5.7
Rodime	27.0	1.8	Hitachi	27.6	5.7
Other U.S.	114.5	7.7	Fujitsu	20.9	4.3
Other Non-U.S.	56.4	3.8	Other U.S.	76.6	15.8
Total	1,429.0	96.4	Other Non-U. S.	25.0	5.2
			Total	483.0	99.9

More than 100 Megabytes

Drive Manufacturer	Units (000)	%
Maxtor	22.8	88.0
Fijutsu	0.9	3.5
Hitachi	0.7	2.7
Other U.S.	1.6	5.8
Other Non-U.S.	—	—
Total	26.0	100.0

Source: Market share figures cited in James N. Porter and Robert H. Katzive, *1986 Disk/ Trend Report: Rigid Disk Drives* (Los Altos, Calif.: Disk/Trend, Inc.).

form demand decreased, these companies introduced a succession of 5.25-inch products at comparable performance levels. By contrast, Seagate entered the medium-performance segment from a solid position in the less-than-30-megabytes segment of 5.25-inch rigid drives. Hitachi and Fujitsu were Japanese companies that entered the U.S. market later from a base in high-performance drives in Japan. Hitachi and Fujitsu, like Control Data, were full-line, vertically integrated computer manufacturers that produced a range of other computer and electronics products in addition to disk drives. By contrast, Quantum, Micropolis, Priam, and Seagate manufactured only disk drives. These latter companies were all less than ten years old, and originally venture capital-funded start-ups that had made their initial public offerings of stock relatively recently. (Seagate, the oldest, went public in September of 1981.)

All of these factors—the established product base of the company, its original niche within the disk drive market, its status as a U.S. or non-U.S. supplier—helped to shape the distribution strategies of these companies. In what follows, we focus on three companies—Control Data Corporation (CDC), Seagate Technology, and Fujitsu—to illustrate the different roles of distribution within the marketing strategies of disk drive suppliers.

Control Data Corporation: Peripheral Products Company

Control Data Corporation's Peripheral Products Company (PPCo) division had been a major supplier of disk drives to OEM customers since 1970. In total, disk drives accounted for more than 80% of PPCo sales. For non-IBM mainframe systems, its 14-inch disk drive was an industry standard and helped to fuel a nearly 50% per annum sales growth rate for PPCo throughout the 1970s. The division's goal was to maintain leadership in the higher-performance/higher-capacity segments of each form factor. In 1985, high-performance 9- and 14-inch drives accounted for the majority of the division's approximately $1.5 billion in revenues. As Exhibit II.3 indicated, PPCo later introduced 5.25-inch drives. The firm was a minor participant in the low-end segment of 5.25-inch rigid drives but was the leading supplier of medium-performance drives in that product category.

PPCo's channel system was a reflection of both its product line history and the traditional importance of large OEM customers for an established drive vendor. As a manufacturer of computers and peripherals for nearly two decades, CDC had developed a broad line of disk drives and related product categories, a reputation for quality in higher-performance drive categories, and an extensive service and support network. In addition, PPCo had for years sold its drives to the larger computer OEMs.

In 1985, PPCo's 70 direct sales representatives called on OEM accounts. The sales reps were assigned geographically, and on average, each generated about $10 million in revenues. Separate from the field sales organization was a national programs unit, which focused on five major accounts that together bought nearly 50% of PPCo's drives. In 1979, PPCo signed agreements with two national electronics distributors, Arrow Electronics and Kierulff, Inc.

Nonetheless, by 1986, sales to distributors still accounted for less than 10% of PPCo's total sales of drives, while direct sales to larger OEMs accounted for 90%.

PPCo added indirect channels primarily for selling expense efficiencies as prices of drives dropped throughout the early 1980s, as the potential number of drive buyers expanded, and as the division's product line grew to include smaller, less expensive drives. In addition, the indirect channel provided a PPCo product presence in potentially successful new accounts while minimizing direct credit risks with these accounts and allowed PPCo to "conserve" its direct selling resources for the approximately 100 large accounts that represented more than 80% of its drive sales. One executive noted,

> With this customer base, the question becomes, "What's the most efficient way of serving those hundreds of customers who buy small but, in the aggregate, can add up to a significant figure?" In addition, turnover among accounts in this business is very high; new OEMs constantly emerge. We need a presence among those smaller customers who may grow big but initially buy small. . . . Our basic goal for distributor sales was to allow the direct salesforce to provide more support to large OEM customers, while distributors focused on the many small accounts.

PPCo's pricing policies reflected the respective roles of its direct and indirect channels. On order quantities of 1,000 or more, the direct price to OEM accounts decreased, while the distributor price remained constant. Thus, as the order quantities increased, distributors were progressively priced out of the large buyer market.

This pricing structure had been established for two reasons: One was that as an OEM purchased more, it generally demanded a volume discount that made selling through an intermediary (which necessarily marked up the price of the drive) economically unattractive for PPCo. Another reason was that as a customer's purchase volume of PPCo drives increased, the account became a potential buyer of the division's line of related peripheral products. Management believed that this cross-selling potential was better developed through its full-line direct sales channel than through distributors, which sold a variety of competing peripheral products. Thus, the pricing structure was also intended to "migrate" a growing account to the direct sales channel.

Seagate Technology

Seagate was founded in 1979 by Alan Shugart, who had previously worked for IBM and Memorex and had then founded Shugart Associates, which introduced the first 5.25-inch floppy disk drive in 1976. Shugart began Seagate with venture capital funding, and in May 1980, Seagate became the first firm to build and sell 5.25-inch rigid disk drives. As other producers entered the market, Seagate's initial product, the ST506, became the industry standard for 5.25-inch rigid drives. Between 1982 and 1983, Seagate introduced a steady stream of new generations of its original drive, at increasingly lower costs and for new applications. These pioneering products gave Seagate an early leadership position in the low-performance, 5.25-inch category. By 1983, according to the company's annual report, Seagate "supplied well over 40% of 5.25-inch rigid disk drives shipped worldwide." Similarly, sales revenues grew dramatically. From less than $10 million in the company's first fiscal year (1981), sales reached $343 million by 1984.

Throughout this period, Seagate sold its products predominantly to OEMs that manufactured smaller minicomputer and microcomputer systems. Three national electronics distributors accounted for a small proportion of Seagate's sales throughout this period. A direct salesforce of about 12 people served DEC, IBM, and Apple, which accounted for 27%, 25%, and 15% of Seagate's net sales, respectively, in 1983. Seagate became one of the first drive vendors to move production facilities offshore. In this way, they could lower the labor cost component of production costs and pursue an aggressive, volume-driven manufacturing strategy. One executive noted,

> Our initial philosophy was to commit manufacturing capacity significantly in excess of current orders because we realized that scale economies and experience effects would lower costs and because we knew it was important in this business to service our large and growing OEM accounts quickly and reliably. This philosophy helped us to gain important market share early on; other drive suppliers were skeptical about the forecasts for drive demand, while we were more willing to commit capacity and price aggressively. In addition, another factor drove this philosophy: Seagate was begun with a little more than $1 million in venture capital, and so we were funding R&D, working capital requirements, and marketing expenses from operating sales revenues.

However, Seagate, with only a few main customers and a limited product line, was vulnerable. During 1983, sales of IBM personal computers increased dramatically, and the computer OEM placed larger orders with Seagate for 1984 delivery of rigid drives. In early 1984, though, IBM found that forecasted demand was greater than actual demand for its PC line. It stopped placing new orders for drives and stretched out deliveries on existing purchase commitments. This cutback had a ripple effect, as it reduced orders from manufacturers of IBM-compatible microcomputers that also utilized these disk drives. As a number of computer OEMs cut back on their purchases of disk drives for microcomputers, drive vendors in the relevant product categories (primarily lower-capacity 5.25-inch rigid drives) were suddenly faced with a leveling off in shipments and with declines in sales revenues, as indicated in Exhibit II.1.

Seagate's sales decreased from over $343 million in 1984 to $214 million in 1985, while gross margin decreased from 27% in 1984 to 12% in 1985. The company decreased the prices of its major product, the ST412, by more than 45% in one year. Seagate was forced to lay off more than 700 employees and, according to its 1985 annual report, to "cut absolutely all unnecessary costs." The company reduced marketing and administrative expenses by 32%.

These events motivated a revision in the company's marketing strategy and a change in its distribution system. Seagate actively sought out other, smaller customers for its drives and a cost-effective means of reaching these smaller customers to diversify its customer base and thus decrease its reliance on major OEM customers. In 1985, Seagate canceled its agreements with its three current electronics distributors in favor of an exclusive agreement with Hamilton/Avnet, Inc., the nation's largest distributor of electronic products. "Then," said Shugart, "we put a lot of attention into making Hamilton/Avnet distribution work."[3] Seagate revised its sales compensation plan so that salespeople received the same commission whether a customer bought direct or through the distributor.

In addition, whereas advertising had been minimal when Sea-

---

[3] Quoted in "Seagate Rebounds from IBM's Big Cutbacks," *Business Marketing*, May 1986, pp. 21–22.

gate was selling primarily to a few large OEM accounts, the company increased its ad expenditures in 1985 and also established a telemarketing staff to develop leads for both its direct and its indirect channels. Seagate also initiated direct-mail campaigns to computer dealers and resellers, instructing them on how to install Seagate disk drives in IBM and IBM-compatible microcomputers. The company revised its credit policies to reflect the nature of its customers. Said Shugart, "We got creative on how to extend credit and how to collect. It's not just being careful. We have to have a lot more contact between them and our credit people."[4]

Value-added resellers also became increasingly important for Seagate. Whereas Seagate's distributors resold the drive in the form in which it was received, the company required its VARs "to add value, by combining Seagate product with other materials having a cost . . . of not less than 25% of the Seagate product." By fiscal 1986, according to Seagate's 10-K, the VAR channel represented approximately one-third of the drive producer's sales. Shugart said, "One and a half years before that, it was zero."[5]

While Seagate's primary strength was in the low-end drive segment, and CDC's in the medium-performance class, their distribution strategies may nevertheless be usefully compared. Indirect channels became a much more important part of Seagate's distribution system during the 1980–1986 period than they were for CDC. Seeking to diversify its customer base after a dramatic fall in orders from its large OEM customers, Seagate turned to indirect distribution as a means of reaching many smaller customers for a product line coming under increasing price and margin pressures.

Product line breadth also affected the companies' distribution patterns. CDC had long manufactured drives in a number of form factors. However, Seagate, first as a thinly capitalized start-up venture and later as a firm under severe margin pressures, sought to focus its R&D and production investments in the 5.25-inch category. Unlike CDC, Seagate could not afford "across-the-board" competition. During the 1980s, however, many large OEMs had broadened their product lines to include computer systems of varying capabilities, putting Seagate at a growing disadvantage in this customer segment. By contrast, VAR customers, focused on certain

[4] Ibid., p. 22.
[5] Ibid.

products, became more attractive and important for a narrow-line producer such as Seagate.

Fujitsu

As of 1986, Fujitsu Limited was Japan's largest computer company and a worldwide supplier of a variety of electronic and telecommunications equipment.[6]

Like CDC, Fujitsu was a vertically integrated supplier of drives, other peripherals, and computer systems. In contrast to Seagate, it produced drives in a range of form factors. Further, disk drives represented one among many different products manufactured and sold by Fujitsu and accounted for only 20% of total corporate sales.

In the 5.25-inch drive category, Fujitsu was an important but smaller competitor in terms of market share, as indicated in Exhibit II.3. In other drive categories, however, Fujitsu was a more powerful presence. For example, in 8-inch drives with 100–300 megabytes of capacity, Fujitsu was the leading supplier in the United States, with a 36% market share (compared with 9% for CDC in this product category). In 8- and 9-inch drives with 300–500 megabytes of capacity, Fujitsu, with a market share of approximately 33%, trailed CDC (47%) in 1985; and in the 300–500 megabyte/ 10.5- and 14-inch product category, Fujitsu was again the leading supplier, with an estimated 65% of the U.S. noncaptive market in 1985.[7]

To sell its products in the United States, Fujitsu established a wholly owned subsidiary, Fujitsu America, Inc. (FAI), in 1976. FAI was staffed with experienced marketing and sales personnel from American peripherals suppliers. As one manager for a competing firm noted, "Fujitsu was a later entrant in the U.S. market and lacked an established product base and relationship with com-

---

[6] In 1986, Fujitsu's total sales were $9.4 billion, and it was organized into four major business segments: Data Processing (mainframe, small-business, and personal computers, as well as disk drives and other computer peripheral equipment; 66% of total sales volume in 1986); Telecommunications (PBXs, network systems, telephones, and other equipment; 15% of sales); Semiconductors and Electronic Components (13% of sales); Car Electronics (car radios, compact disc players); and Mobile Communications Equipment (4% of total sales volume).

[7] Market share figures are based on data in *1986 Disk/Trend Report: Rigid Disk Drives*. The noncaptive market is composed of firms that purchase from outside suppliers; the captive market, in contrast, is measured in terms of disk drives made by computer OEMs for use in their own end-products.

puter OEMs. They bought a number of contacts, and knowledge of which doors to knock on, when they hired experienced people from other vendors."

FAI's Computer Product Group had 50 salespeople in 1986, selling to OEMs and to various intermediaries—including VARs, smaller regional distributors, and manufacturers' rep organizations known as high-tech distributors. In this respect, Fujitsu was unlike most other drive vendors, whose distribution systems were generally composed of a few national electronics distributors, rather than of a variety of smaller distribution organizations. The reasons for the difference include the nature of Fujitsu's market entry, the constraints imposed on available marketing resources, and (until recently) the relative unattractiveness of the national distributor channel for a vendor like Fujitsu.

As a later entrant into the U.S. drive market, Fujitsu did not have a significant installed base of drives among computer systems manufactured by large U.S. OEMs. Further, various factors made the establishment of such a base a protracted process for FAI. As one FAI manager noted,

> Fujitsu was conservative in its approach to the U.S. drive market. When I joined FAI in 1979, there was only one sales manager on board, and we weren't allowed to hire more salespeople until we could demonstrate that we had orders. In effect, we had to bootstrap our marketing and sales efforts for a number of years due to resource constraints. In fact, we did not have a major account sales effort at FAI until 1984.
>
> In addition, we initially had a brand recognition problem for our drives among U.S. companies. I remember more than one potential customer mistaking us for "Fuji," and wondering why a film maker was in the disk drive business.

Fujitsu also faced difficulty in selling drives to larger U.S. OEMs because of the company's geographical distance from the U.S. market. The Japanese firm did not establish a U.S. manufacturing facility for drives until 1986, and this was a disadvantage in a product category like disk drives, where the marketing process required the continuing exchange of information between buyer and seller. One FAI manager explained, "For a number of years, large computer OEMs were impressed with our drive products but reluctant to do business with a company whose engineering group was located thousands of miles away and which spoke a different language." In addition, as the sales subsidiary of a major, worldwide supplier of computer systems, competing with large U.S. computer OEMs in a number of product categories, FAI naturally ran into resistance from some OEMs. Finally, Fujitsu chose to direct its efforts away from the larger OEMs. As an FAI manager noted, "CDC, the competitor with a product line most similar to ours, was focused on the larger OEMs, and had strong, long-established relationships there. Our marketing philosophy was 'hit 'em where they ain't.' It seemed a better use of our resources to try and get established with a different group of companies."

National electronic distributors were not a part of FAI's channel system until 1986 for the following reasons: First, FAI's management believed Fujitsu's product line during this period was not suitable for those distributors. Michael Gluck, senior vice president of FAI's Computer Product Group, recalled, "Our product line in drives was focused on the high-end, higher-performance drives. These products have a selling cycle and technical service requirements unsuitable for most large national distributors who, in my opinion, are good at providing time and place utility for customers seeking less technical, frequently ordered products."

Second, without significant presence among large computer OEMs, Fujitsu's product line was in turn less attractive to national electronics distributors, whose drive sales were in part derived from servicing large OEMs on an as-needed basis. As one manager commented, "You need a certain established sales volume with large accounts before using national distributors makes sense. Without that installed base and without good name recognition, a vendor will get ignored at those distributors."

Third, FAI's limited marketing resources at this time meant that it lacked the resources to support a national distribution effort. David Krevanko, vice president of FAI's Computer Product Group, noted, "Selling our product line through national electronics distributors requires dedicated technical support for the distributor. We lacked the resources to provide that kind of support. At the same time, our own direct sales program was only gradually expanded to include offices throughout the United States."

Fourth, national electronics distributors, with more than half their sales derived from U.S.-manufactured semiconductors, were probably reluctant to add a supplier like Fujitsu to their vendor list. The major U.S. semiconductor suppliers generally franchised only a few national distributors that did not carry the competing lines of foreign suppliers. Fujitsu, a leading Japanese supplier of semiconductors, would have been a significant and visible addition to a distributor's vendor list, creating potential conflict in the distributor's relations with its major U.S. suppliers. One FAI manager recalled that, "At one point a few years ago, we were negotiating with a large electronics distributor when its president began a highly publicized 'buy American' campaign. That effectively halted our talks with that distributor for a number of months."

In response to its limitations, Fujitsu's strategy was to turn to the smaller local and regional electronics distributors, an action that was not without precedent in the electronics industry. As noted by one trade publication,

> The reluctance of U.S. chip makers to franchise local distributors encouraged them (i.e., local distributors) to seek other suppliers. Meanwhile, this alliance between the U.S. semiconductor and distribution giants encouraged the Japanese to seek alternate outlets.
>
> Virtually thrown into each other's arms, Japanese suppliers and small distributors have found marriage both convenient and profitable. Ironically, the U.S. semiconductor giants appear to have accomplished inadvertently what they have most feared: the penetration of American markets by the Japanese firms.[8]

In line with the character of its distribution, its product line emphasis, and resource constraints, FAI also tended to direct its marketing efforts to smaller OEMs, located primarily on the West Coast. Gradually as its sales increased, FAI added direct sales offices and distributors in other parts of the country. By 1985, FAI employed a number of regional and local distributors, VARs, and manufacturers' rep organizations to sell its broad line of drives and related peripheral products. The distributors focused on different geographical areas. The VARs, which tended to focus on the personal computer market, bought computer systems from lower-priced suppliers and bought FAI 5.25- and 3.5-inch drives for those systems, which they then configured for a given application.

FAI's use of the manufacturers' rep (MR) channel was unique

[8] Norman Alster, "Local Heroes: Distributors under the Giant's Shadows," *Electronic Business,* April 15, 1985, pp. 120–124. See also *Business Week,* November 10, 1986, p. 45: "Hardly any Japanese chips are handled by the major distributors, because U.S. chipmakers often threaten to take their business elsewhere if that happens."

in the industry and helps to indicate how channel design and management reflect individual competitive circumstances. MRs had been a part of FAI's distribution system since the inception of the subsidiary, although as one FAI manager noted, "Initially, we were not very successful selling through manufacturers' reps: we entered the market later and our competitors already had many of the best MR firms carrying their lines. But our corporate parent liked the variable-cost structure of the MR channel, and so we kept at it." In the late 1970s, FAI added an MR organization experienced in selling related peripheral products and, according to management, "We started having more success." In 1981, moreover, a former sales representative for a competing drive firm established an MR firm and agreed to carry FAI's line on an exclusive basis. FAI's agreement with this firm, Dallas Digital, became a model for its management of the MR channel.

As noted earlier, a fledgling disk drive firm would typically use an MR to sell its products. But as sales volume grew and as the drive vendor began selling to larger OEM customers, this channel was generally dropped. FAI, however, continued to pay commissions to its MRs even as sales volume grew beyond the point at which a direct relationship with a given account became economically feasible. Further, FAI, which did not have a U.S. service facility for its drives until 1984, worked with its MRs throughout the early 1980s to help develop and augment their technical service capabilities for FAI's line. In return, FAI required that its MRs (1) not carry competing drive lines, (2) focus on certain OEM products in certain geographical areas, (3) carry a specified inventory of spare parts, and (4) maintain service, demonstration, and certain "value-adding" capabilities (e.g., configuring a drive for certain applications). In effect, as one FAI manager noted, "We worked to trade up the capabilities of our manufacturers' reps. Initially, this was simply necessary for us. But over time, these firms have become a key part of our distribution system and we've made a number of decisions aimed at nurturing this channel." By 1985, FAI had agreements with six MR firms, referred to as its "high-tech distributors," which accounted for approximately 30% of FAI's drive sales in that year. In turn, FAI usually accounted for 50% or more of each MR's sales.

In 1986, FAI began selling 5.25-inch floppy drives and established its first agreement with a national distributor, Schweber Electronics. Commenting on this addition to its channel system, an FAI manager noted that,

> Timing is everything with this kind of agreement. Schweber did not have a drive line in that category at the time, and we did not want to dedicate sales resources to that line. In addition, Schweber, like many other national electronics distributors, had a lot of experience selling low-end products by 1986 and, conversely, we didn't want to dilute the attention of our high-tech distributors and direct salesforce with those products.

As of 1987, an issue facing FAI's channel system was the potential overlap between its MR and industrial distributor channels: The latter were seeking to become full-line FAI resellers, while the former naturally resisted the possibility of lower-service, lower-priced competition at the reseller level. One FAI manager emphasized that

> The industrial distributor channel only became attractive and viable for us when our sales volume and name recognition developed, and when our product line expanded to include the lower-priced floppy drive products. We've tried to align our channels with different products and levels of value added, but that becomes increasingly difficult as the channels themselves change in terms of their capabilities and customer targets.

CONCLUSION

This study has discussed four aspects of disk drive distribution: (1) the factors affecting the choice between direct and indirect channels, and the role and relative importance of each in the channel system of drive suppliers; (2) the factors affecting the choice of reseller type—manufacturers' reps, industrial distributors, or value-added resellers—as the market developed; (3) the factors affecting the choice among distributors—national, regional, or local; and (4) the competitive distribution strategies of selected leading companies in the industry.

In many respects, the disk drive industry of the 1980s has been a paradigmatic instance of a high-tech market environment, characterized by rapid technological improvements, short product life cycles, a product that acts as a critical component in a technically complex system produced by other manufacturers, and rapid evolution in terms of both the numbers and nature of prospective customers.[9] Thus, while simple generalizations to other industries are inadvisable, the factors affecting distribution for disk drive manufacturers are nevertheless germane to managers in a variety of high-technology market environments.

Exhibit II.4 provides a framework for considering relationships among the various determinants of distribution strategy discussed in this chapter. In this construct, three factors directly impact a manufacturer's distribution strategy: (1) the relative importance of close buyer-seller information exchange, (2) the distribution costs involved in selling and delivering the product to a set of customers, and (3) the accessibility of a channel to a given competitor. In turn, these three factors are affected by the nature of the product and buyer behavior in a market segment, the market-segment emphasis of the supplier, and each supplier's business profile.

The business profile refers to a supplier's product portfolio (e.g., product line breadth within disk drives and complementary peripheral products, the relative emphasis on high- or low-performance drive categories), the extent of the installed base, and the supplier's financial situation. In addition, status as a domestic or foreign supplier was an important business unit factor in this industry. These elements affected the market segment choices of drive suppliers, the accessibility of channels, and the distribution cost structure of each supplier in a segment.

A supplier's product line largely determined whether or not large OEMs were potential customers. With its broad product line in different sizes and capacity categories and its history of focusing on the higher-performance categories of each form factor, CDC's primary segment was the larger computer OEMs. By contrast, Seagate, with its product focus on the lower-performance (less-than-30-megabyte) segment of 5.25-inch rigid drives, found it both necessary and desirable to develop a more diversified customer base. Fujitsu, like CDC, produced a broad line of peripheral products.

---

[9] The term *high technology* is notoriously elastic, but it usually includes computer products like disk drives. One monthly investment letter on "high-tech" stocks explicitly includes disk drives, and the product category would certainly appear to fulfill the criteria for a "high-tech business" cited by William L. Shanklin and John K. Ryans, Jr., in their book, *Marketing High Technology* (Lexington, Mass.: D.C. Heath, 1984), p. 29:

> The business requires a strong scientific/technical basis; new technology can obsolete existing technology rapidly; and as new technologies come on stream their applications create or revolutionize markets and demands.

Exhibit II.4 Determinants of Distribution Strategy: Disk Drive
Industry

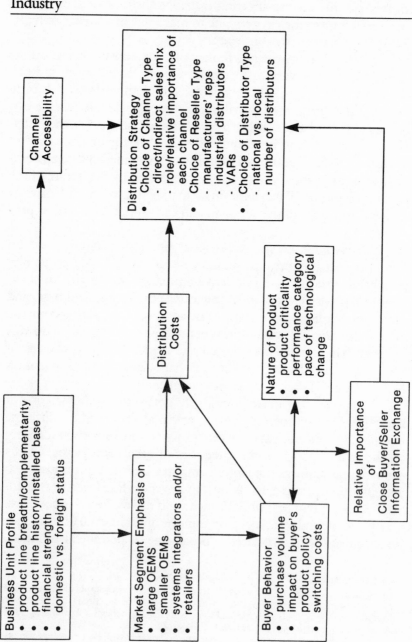

But Fujitsu, as a foreign supplier without dedicated U.S. manufacturing and service facilities and as a worldwide competitor in a variety of computer segments, naturally encountered reluctance among some OEMs to specify its drives in their computer systems. Some American OEMs feared not only that such a move would mean sourcing a critical component from a supplier with long lead times but also that it would probably be viewed as lending "comfort and support" to a direct competitor. As a result, Fujitsu initially tended to focus on segments other than large OEMs, and its channels of distribution reflected this focus.

The business profile also affected distribution costs and channels accessibility. Because they produced and sold other products besides disk drives (and, within disk drives, a broad line in a number of product categories), companies like CDC and Fujitsu could spread selling expenses over a larger product base. Similarly, because their drive lines had an installed base among a variety of customers, CDC and Seagate were attractive vendors for national electronics distributors. In contrast, because Fujitsu's drive line lacked an installed base, and because its status as a worldwide competitor in semiconductors could jeopardize distributors' relationships with their other important suppliers, this large Japanese supplier initially found national electronics distributors a relatively inaccessible channel.

In turn, depending on a supplier's market segment emphasis, marketing programs had to address different buyer needs. For large OEMs, the choice of a drive vendor established a major strategic relationship because the vendor's ability to deliver large quantities on schedule and the drive's impact on the performance of the OEM's product were critical. For the drive supplier, sales to large OEMs directly affected manufacturing costs and the direction of product line development. In contrast, smaller OEM and systems-integrator customers purchased drives in smaller volumes with IBM-compatible specifications; thus, the relationship with an individual drive vendor was less crucial. For retailers and other outlets that sold add-on drives to end-users, the emergence of standards also made a direct relationship with a particular drive vendor less important than matters such as price, delivery, and reliability of performance. Together, the interaction between the nature of the product and the buyer behavior by segment determined the relative need for close, ongoing buyer-seller information exchange between the drive manufacturer and its customers.

Exhibit II.5   Determinants of Distribution Strategy:
Major Disk Drive Suppliers

	CDC	Seagate	Fujitsu
Business profile factors	Product line breadth Product line complementarity Product line history	Product line focus Financial constraints Account vulnerabilities	Foreign supplier Later entrant: lack of installed base Blocked access to certain channels and OEMs
Market segment choice factors	Emphasis on large OEM accounts	Need to reach new segments and diversify customer base	Need to reach smaller accounts with a system
Channels configuration	Primarily direct sales channels	Emphasis on diverse reseller channels as well as direct sales	Emphasis on "high-tech" distributor MRs and VAR channels as well as on direct sales

All of these factors affected the distribution strategies of disk drive suppliers. Exhibit II.5 outlines the factors affecting the different channels configurations of the three companies highlighted in this chapter.

CDC, with its broad line of disk drives and other peripheral products and a history of close working relationships with large computer vendors, focused on the large OEM segment. Both the company's product line and order volumes in this segment made a direct sales channel the primary means of distribution in CDC's channel system. For this company, indirect channels were added primarily to lower selling expenses to smaller customers and played a relatively subordinated role.

Seagate had a narrower product line in disk drives, the financial constraints of a smaller company, and the vulnerabilities that accompanied selling to a concentrated customer base with a narrower line in a product category subject to rapid technological "leapfrogging" by competitors. Thus, Seagate sought to reach new segments and to diversify its customer base after the drop in demand from its largest customers in 1984. As a result, indirect channels, including both industrial distributors and VARs, became more prominent parts of Seagate's distribution system.

Fujitsu was a foreign supplier and later entrant into the U.S. drive market. It lacked an installed base among OEM customers and found national distributors a relatively inaccessible channel for its products. In addition, as a competitor in other product segments, selling directly to large OEMs was also more difficult for Fujitsu. As a result, indirect channels of a certain type became an important part of Fujitsu's distribution system. MRs with service capabilities and VARs that integrated the drive into a system played important roles for Fujitsu, enabling the company to sell economically to the smaller customers that *were* viable accounts for its product line.

Thus, responding from its particular circumstances, each company focused on a different market segment. Having made this choice, the nature of buyer behavior in a market segment affected the transaction costs implicit in serving that group of customers, while the nature of the supplier's product line and its status as a foreign or domestic supplier affected its distribution options and the functions required of its channels of distribution.

Product-Market Study III

ACCESS TO MARKETS AND BARRIERS TO ENTRY—THE STATIONARY AIR COMPRESSOR INDUSTRY*

Load centers are sold largely through resellers, and disk drives, mainly through direct salesforces. Distribution channels for stationary air compressors are more complex and include a variety of intermediaries. Suppliers' direct salesforces, independent distributors, subdistributors, captive distributors, and manufacturers' representatives are all involved in taking this product to market. In 1985, for the leading supplier, Ingersoll-Rand, 30% of sales were made direct to users; 35% through independent distributors, many of which sold to subdistributors as well as to user-customers; 20% through Ingersoll-Rand's captive distribution network; and 15% through agents to retail chains.

In comparison with load center distribution, our study found that air compressor channels networks included fewer distributors, each with a relatively large and often exclusive territory, and each typically carrying one manufacturer's line. One reason for these exclusive relationships was the large investment in inventories, relative to territory sales potential, required of a distributor representing any one manufacturer. Another was the importance of distributor revenues derived from machine maintenance, repair, and spare parts sales, with after-sale service demand dependent on the size of a supplier's installed base of operating equipment in the distributor's territory. Given their revenue dependency on the sale and servicing of one supplier's line, distributors tended to have close and long-term relationships with their suppliers. For example, distributor turnover in the air compressor industry, at 5% a year on average, was low in comparison with 20% for load center distribu-

* The product category, stationary air compressors, is to be distinguished from portable compressors of the type often used in road construction to power pavement-breaker tools. The term *air compressors* or *compressors* as used in this case study refers to stationary machinery.

336

tors.[1] As a consequence of these exclusive relationships, established networks of distributors became a major competitive advantage for the leading producers, and a formidable entry barrier for the challengers.

This case history considers the evolving distribution strategies of the major players in this mature industry. We focus on Ingersoll-Rand, the market leader in the United States, and on how it has retained and improved its market position; we examine the attempts of two challengers—Sullair, a U.S. firm, and Atlas-Copco, a Swedish company—to enter the American market. We also describe the response of Joy, long the number-two supplier in this country, to changes in the competitive environment. For all four firms, having access to and building the strength of distribution channels were important determinants of success.

The discussion begins with a description of product technology, product-market segmentation, and industry distribution modes before addressing the strategies of the four companies. Certain themes emerge, regarding the determinants of change in an industry like this one, and are offered in a concluding section.

PRODUCT USE AND TECHNOLOGY

Compressed air, like electricity and gas, is a utility with many applications. Unlike electricity and gas, however, it must be produced and consumed at the work location. Compressed air is used to power equipment such as drills and screwdrivers, riveting and chipping hammers, grinding and polishing tools, and paint sprayers. It is used, as well, to provide power for air hoists, winches, air brakes, jet engines, and other pneumatic machinery. Pharmaceutical and food manufacturing, high-technology plant systems such as microchip assembly, and textile weaving also require compressed air in the production process.

Stationary air compressors are generally classified by

- their horsepower (hp) rating: small (less than 25 hp), medium (25 to 300 hp), or large (above 300 hp); and

- type of compression technology: reciprocating (recip), rotary screw (rotary), or centrifugal.

---

[1] According to our 1985 surveys of producers of air compressors and load centers, respectively.

Exhibit III.1 Cost Comparisons for the Three Types of
Medium-Horsepower Compressors

	Recip	Rotary	Centrifugal
Cost of compressor per hp	$400	$175	$225
Estimated life	10–12 yrs.	5–7 yrs.	10–12 yrs.
Installation cost as a percentage of purchase price	20%	5%	12%
Spare parts and maintenance cost per year as a percentage of original cost	6%	2.5%	2%
Energy costs at 7 cents/kwh and 300 days' operation in a year	$400/hp	$450/hp	$400/hp

Source: Adapted from V. Kasturi Rangan and E. Raymond Corey, "Ingersoll-Rand," # 9-587–045. Boston: Harvard Business School, 1980, p. 2.

Of the three technologies, reciprocating is the oldest, and uses a piston moving within a cylinder to compress air. A recip lasts about 10–12 years, with some larger machines (above 300 hp) reportedly in use for over 20 years. Over the life of a recip compressor, the spare parts requirements would be about 60–70% of the original equipment's cost.

Rotary compression involves two intermeshing rotors. The estimated life of a rotary is 5–7 years, but it has fewer moving parts than does a recip. As a result, the spare parts requirements over the life of a rotary are only 15–20% of the original equipment's cost. However, the energy cost of operating a rotary is 12–17% higher than that of a two-stage recip of equivalent horsepower.

Centrifugal compression is created by an impeller rotating at very high velocity. A centrifugal machine has an estimated life of 10–12 years, with a spare parts requirement over this period of approximately 20–24% of the cost of the original equipment. However, the centrifugal technology was not well developed for applications in small- and medium-horsepower compressors.

Centrifugal compressors generate air free of oil contamination, whereas in recips and rotaries, the small particles of oil that lubricate the bearings enter the air stream through the compression

Exhibit III.2 Market Size and Available Technologies, 1985

	Small (less than 25 hp)	Medium (25–300 hp)	Large (above 300 hp)	Market Size
Recips	800,000 units (mostly 0.5– 5 hp)	5,000 units	500 units	$230 million
Rotaries	Few models	132,500 units (mostly 15– 150 hp)	1,000 units	$150 million
Centrifugal	No models	Few models	1,000 units	$ 80 million
Market Size	$175 million	$175 million	$110 million	$460 million

Source: Adapted from "The Industrial Stationary Compressor and Dryer Market in the U.S.," Frost and Sullivan report no. A1612, April 1986.

chamber, a condition that prohibits using these machines in pharmaceutical and food processing, textile weaving, and electronics manufacturing. To serve those market segments, a few compressor manufacturers had, at the time of our study, designed oil-free recips and rotaries to prevent contamination of the compression chamber.

Exhibit III.1 provides the cost comparisons for the three different types of compressors in the medium-horsepower range.

MARKET SEGMENTS AND DISTRIBUTION CHANNELS

The total market for compressors (excluding spare parts) peaked in 1981 at about $555 million, and then wavered around $450 million through 1985.

Market Segments

Recips, rotaries, and centrifugals represented about 50%, 35%, and 15% of the compressor market, respectively, in 1985. Exhibit III.2 indicates estimated market size in units by compressor type and size range for 1985, and market size in dollars for each size class and product type. Accessories such as dryers and spare parts represent an additional 10% and 35% of the market, respectively.

Buyer Behavior

To a large extent, market segments and channels were identified with different air compressor size ranges.

Small Compressors. The 25-horsepower-and-under compressors were sold to about 50,000[2] customers (not including individual purchases from retail) through about 3,500 mill supply
houses, many of which served as subdistributors for air houses, and
75 manufacturers' representatives. Compressors with less than 5-
horsepower capacity were often sold to tradespeople and consumers through hardware stores, mail-order catalog stores, and home
improvement centers. In general, convenience was an important
factor driving purchase decisions for small compressors. Product
availability, price, brand name, and localized service were also important.

Medium Compressors. An estimated 25,000[3] industrial
purchasers of medium-sized compressors bought their equipment
through independent air houses, manufacturers' captive branches,
or direct salesforces. Compressors were bought on the basis of
price, product availability, technical specifications, parts availability,
and technical service support. In addition, buying criteria differed
across the range of sizes in this category. For machines in the 25-
to 100-horsepower range, buying decisions were generally made by
the business owner or the plant engineer, and off-the-shelf availability and local after-sale service were important buying considerations. On the other hand, for machines in the 100- to 300-
horsepower range, lead times of 6–8 weeks were often acceptable
to end-users. Further, these units were priced at about $50,000 and
above, and the buying decision was most often made by a group
rather than by an individual.

Large Compressors. There were about 2,500[4] end-users for
the machines larger than 300 horsepower. The demand was primarily for centrifugal compressors, though there was also some recip
demand. These large compressors were generally used in custom-
designed applications. Contractors like Bechtel and large manufacturers like General Motors routinely issued requests with specifications for price quotations for air compressors. Transactions

[2] The numbers indicated here are average estimates obtained from four leading manufacturers in 1985.
[3] Ibid.
[4] Ibid.

typically involved negotiating product features, delivery and installation schedules, warranties, prices, and payment terms.

Channels of Distribution

In general, the greater the horsepower rating and degree of product customization, the higher the transaction amounts, the more complex the purchase process, and the longer the delivery lead times, the more manufacturers relied on direct selling. In contrast, as horsepower ratings became smaller, user markets more fragmented, products more standardized, and purchase convenience of greater importance, compressor manufacturers tended to resort to distributors and subdistributors.

According to the 1982 Census of Manufactures, manufacturers' sales to distributors accounted for 38% of the sales of compressors. Of that portion, 5% went through captive distribution, and the balance, through independents. Direct customers, including OEMs and independent construction companies, accounted for 43% of sales. Other customers, such as government agencies and large retail chains, made up the balance, 19%.

In a 1986 survey, Morton Research Company[5] listed about 900 primary distributors of stationary air compressors in the United States. These distributors, often called air houses, had $2–8 million in annual revenue, of which about 80% represented sales of medium-horsepower machines. Air houses in the $5–8 million size range reported investments in inventories and facilities at about 30% of sales. Due to that financial commitment, distributors rarely carried multiple and competing brands of air compressors. In turn, manufacturers tended to be selective in their franchising policies to help ensure distributor income streams that would be adequate to support distributor costs.

On average, distributors made about 25% gross margin on the sale of compressors and related products and services. Net margins were about 5–10%. As Exhibit III.3 indicates, almost half (or even more for medium compressors) of a distributor's profits on the compressor line were made from aftermarket sales of parts, service, and rentals.

The Morton survey also listed an additional 3,500 distribu-

[5] "The U.S. Distributor of Air and Gas Compressors: A Distributor Prospect Manual," prepared by Morton Research Company, Merrick, New York, 1986.

Exhibit III.3 Stationary Air Compressor Distributor Margins, 1985

	Complete Units	Accessories	Parts	Service	Used Equipment and Rentals	Total
25 hp and under						
Percentage of sales	61.0%	8.0%	18.0%	10.0%	2.0%	100.0%
Gross margin as a percentage of sales	20.7%	27.6%	32.4%	36.5%	34.6%	24.3%
25–300 hp						
Percentage of sales	53.0%	7.0%	22.0%	10.0%	8.0%	100.0%
Gross margin as a percentage of sales	15.0%	23.0%	28.9%	36.8%	36.5%	22.5%
Over 300 hp						
Percentage of sales	57.0%	3.0%	26.0%	11.0%	3.0%	100.0%
Gross margin as a percentage of sales	16.3%	20.0%	26.1%	32.0%	56.7%	21.9%

Source: 1986 air compressor distributor survey (79 responses).

tors, often referred to as mill supply houses, which sourced a considerable portion of their requirements from air houses and from manufacturers' reps (MRs). These subdistributors often carried a smaller-horsepower product line and generated most of their revenues from a range of other industrial products. Unlike air houses, mill supply houses often carried multiple and competing brands of small compressors. Other channels for these 5-horsepower-and-under compressors were retail hardware stores, mail-order catalog stores, and home improvement centers. Calling on these retail chains on behalf of compressor manufacturers were about 75 manufacturers' rep organizations.

An important dimension of air compressor distribution is the spare parts supply business. As of 1985, distributor spare parts revenues amounted to about $150 million. Manufacturers of the original compressors accounted for about two-thirds of the parts volume. Because manufacturers' spare parts margins were relatively high (30–40%) as compared with complete machines (15–20%), this business attracted so-called pirate parts suppliers, which accounted for a steadily increasing share of the spare parts volume. Parts entrepreneurs offered their products to any and all distributors, many of which found these sources to be attractive because

Exhibit III.4 Manufacturers Listed by Product Segments in Order of Estimated Market Share, 1986

	Small (below 25 hp)	Medium (25–300 hp)	Large (above 300 hp)

Recip	Campbell-Hausfeld
	Ingersoll-Rand
	Quincy
	Kellogg-Compair
		Ingersoll-Rand
		Joy
		Quincy
		Gardner-
		  Denver
			Ingersoll-Rand
			Joy
			Atlas-Copco
			Gardner-
			  Denver

Rotary	Few models	Ingersoll-Rand Sullair Atlas-Copco Joy	Sullair Ingersoll-Rand Atlas-Copco Joy
Centrifugal	No models	Few models	Joy Ingersoll-Rand Elliot

Source: Adapted from V. Kasturi Rangan and E. Raymond Corey, "Ingersoll-Rand," # 9-587-045. Boston: Harvard Business School, 1980, p. 4.

they were lower priced than compressor manufacturers' parts. In addition, by carrying parts for different brands of compressors, air houses had a means of gaining entry into competitors' accounts.

COMPETITION

Exhibit III.4 provides the 1985 ranking of major suppliers in each market segment. Overall, about 60 manufacturers, both domestic and foreign, competed in this mature industry, which was expected to grow only at about 2% in the next five years.

In summary, the industry context may be described in these terms: The three basic product technologies were all relatively mature, and the market, while more than doubling in 1975–1981, had peaked and declined. Market segment concentration and product applications varied with air compressor size categories as defined by horsepower ratings. The major suppliers tended to compete in selected product categories. Ingersoll-Rand alone competed in all categories and was either the number-one or the number-two supplier in each one. Joy was in the market segments for medium- and large-horsepower machines for rotaries, recips, and centrifugals. Sullair focused on rotary air compressors in the medium- and large-sized ranges, while Atlas-Copco was a major competitor in the markets for medium and large rotaries and large recips.

Salient aspects of the air compressor distribution infrastructure were the following: First, different distribution channels were used for different market segments and product size categories. Direct salesforces sold larger machines to large users. Air houses were the primary channels for 25- to 300-horsepower (medium-sized) units to medium-sized industrial users. Mill supply houses served smaller businesses and tradespeople with machines in the less-than-25-horsepower category. Manufacturers' reps reached retail chains selling to the do-it-yourself home market. Three-tier distribution was common in some segments as well; for example, air houses and MRs sold to smaller resellers classified as subdistributors.

Second, a significant share of reseller revenue came from the sales of spare parts and repair and maintenance service. Third, air houses were often given exclusive territories and often carried only one brand. This was due to the relative thinness of the air compressor market, the high level of reseller investment needed to support a supplier's product line, and the distributor's dependence for parts and service revenues on the supplier's installed base.

With stagnant demand and mature product technologies, competition among rival producers focused sharply on gaining and holding distribution. Ingersoll-Rand was the clear leader in this $650–700 million U.S. market (including accessories and spare parts) with an estimated share of 25–30%, followed by Joy, Sullair, and Atlas-Copco, a Swedish company, each with an estimated share of 8–15%.

By implementing distribution strategies to reflect its marketing goals, each company had come to occupy a unique competitive position. Faced with a mature market and a recession (1981–1983), Ingersoll-Rand appears to have made an all-out effort to hold its market leadership. This involved building and managing a multiple-channel network to maintain a diverse set of customers. Joy, on the other hand, responded to the same set of market factors by focusing on its most important and profitable product lines and by innovating in both product and process design. Both Sullair and Atlas-Copco have been formidable challengers for market share, using innovation as a wedge. Sullair innovated, both in product technology and in distribution. Atlas-Copco aggressively worked at penetrating competitors' distribution networks, using its "Z series" oil-free rotary line, superior among comparable competitive offerings, as an entry wedge. We now review the strategies of these four important producers.

Ingersoll-Rand

With sales of $2.64 billion in 1985, Ingersoll-Rand (I-R) had manufacturing operations in 16 countries, sales offices in 40, and distribution arrangements in 80 countries. The company's Stationary Air Compressor Division (SACD) began manufacturing air compressors in 1930 and marketed its broad line of recips, rotaries, and centrifugal compressors through four distinct channels. Its direct salesforce sold recips rated above 250 horsepower, rotaries rated above 450 horsepower, and centrifugal compressors of all sizes. All other compressors, except the less-than-5-horsepower recips, were sold by the company's 100 independent distributors and 22 company-owned I-R air centers. Manufacturers' representatives sold the less-than-5-horsepower compressors to OEMs (e.g., to Johnson Controls as components in pneumatic thermostat devices) and to large retail chains such as Grossman's.

In terms of channel types and the products that each channel carried, Ingersoll-Rand's multichannel system had evolved over a quarter-century. The captive distributor operation (air centers) originated during the 1971–1973 recession, primarily to maintain I-R's presence in markets where independent distributors were liquidating their businesses. By 1985, however, the I-R air centers were an integral part of the company's distribution strategy.

A senior company executive believed that I-R's current allocation of sales and market potential between its independent distributors and its air centers was ideal. Increasing the number of air centers, he argued, could demoralize the independent distributors. On the other hand, decreasing their number would make the air centers less economical to operate and thus ineffective as an alternate channel to independent distribution.

The addition of the manufacturers' rep channel in 1984, rather than being an attempt to maintain share in a recession, appeared to be a conscious attempt to gain strength in a traditionally weak market for Ingersoll-Rand—the retail market. The "rep channel" appeared to understand and to be able to service this market better than any of I-R's existing channels. In the retail market, the nature of the reselling task, often involving extensive advertising and promotion, was markedly different from selling to industrial customers.

As Exhibit III.5 indicates, I-R's distributor-class products

Exhibit III.5   Allocation of Sales Responsibility for Stationary
Air Compressors by Type and Size to Sales Channels for 1960,
1973, and 1984

	1960	1973	1984
Direct sales force	Recips 50 hp and over	Recips 150 hp and over Rotaries 150 hp and over All centrifugal	Recips 250 hp and over Rotaries 450 hp and over All centrifugal
Distributors	Recips under 50 hp	Recips under 150 hp Rotaries under 150 hp	Recips under 250 hp Rotaries under 450 hp
Air centers		Recips under 150 hp Rotaries under 150 hp	Recips under 250 hp Rotaries under 450 hp
Manufactur- ers' reps			Recips 5 hp and under

grew significantly as increasingly larger units were given to resale
channels between 1960 and 1984 and as MRs were recruited to
reach retail chains. Four factors influenced this trend: (1) the rising
costs of direct sales calls; (2) the growth of the market in terms
both of dollars and numbers of potential customers for medium
and small units; (3) the fact that competitors were using resellers
extensively; (4) the growing technical competence of I-R independent distributors; and (5) the need to support Ingersoll-Rand's distribution network, once established, with broad product lines so as
to provide a stream of sales and service revenues adequate to sustain
these networks over the economic cycle. While increasing reliance
on resellers undoubtedly increased I-R market share and enhanced
its revenues, it also meant turning over to distributors an increasing
portion of the profitable parts and services business generated by
Ingersoll-Rand's installed base.

The steadily growing reliance on independent distributors and
company-owned air centers was explained by a company manager:

We realized back in the 1960s that any machine that sold for
less than $5,000 was not economical for a direct sales approach.

Part of this rationale may also have been purely historic. Back
in the early 1970s when there were separate profit centers for rotaries and recips, the general manager in charge of rotaries was a dis-
tributor-oriented person, and the general manager in charge of recips had a direct sales bias. This might partly explain the higher

horsepower cut-off for the rotaries (450 hp as opposed to 250 hp for recips). Now, of course, all compressors are under one general manager and distribution policies are likely to be uniform.

Another manager added, "Probably the most important factor has been the differences in buying behavior among our several types of customers and our attempts to service their different needs through different channel systems."

No other competitor approached the market through as many channels. In addition, Ingersoll-Rand offered independent distributors a product line broader than its competitors and, as the market share leader, a large installed base for generating aftermarket revenue streams. It supported the distributor network and monitored its performance through a specialized salesforce responsible only for serving this channel. In terms of market coverage and sales revenues, Ingersoll-Rand's key channel has been its independent distribution system. The captive network, though important, has played a secondary role.

Finally, SACD considerably eased the natural competitive rivalry that might develop between its distributors and its direct salesforce by differentiating by product size and type between the lines each may carry. Thus, because of its broad product line and channels mix, Ingersoll-Rand has been able to cater to diverse customer needs and to steadily build share in a declining market.

Joy Manufacturing

Originally in the mining equipment business, Joy entered the air compressor industry through its acquisition of Sullivan Machinery in 1946. One of five business units, the air machinery division accounted for 1985 sales of about $120 million out of Joy's total revenues of approximately $800 million in that year. Joy held the leadership position in the large centrifugal air compressor segment and was second to Ingersoll-Rand in the medium- and large-recip category. In terms of applications segments, Joy enjoyed a strong share of the narrow but important process air market.[6]

Joy sold its centrifugals and the above-400-horsepower reciprocating compressors through a direct salesforce of 25–30 persons. Nearly all of its rotaries and half of its recip line went to market

---

[6] Compressed air used in the manufacturing processes for such products as pharmaceuticals, packaged foods, and semiconductors.

through a network of 75 independent distributors. The company salesforce was responsible for sales to both user-customers and Joy distributors. Unlike Ingersoll-Rand, Sullair, and Atlas-Copco, Joy had not attempted to build a captive distribution arm, a strategy that was perceived as paying dividends in terms of distributor loyalty. As one manager claimed, "Everybody else was sending confusing signals except us. We have not had a single distributor cancel us in the last ten years."

One industry observer commented on Joy's distribution policies in this way:

> Joy's consistency reflected the parent company's reluctance to experiment with distribution. As a corporation, Joy had a direct salesforce mentality. After all, its primary line of business was in mining equipment, which has traditionally been sold direct. Joy moved reluctantly into distribution in the early 1970s. That represented the sales philosophy of one or two individuals in the air compressor division, but was certainly not representative of its corporate philosophy.

For lack of comparable product line breadth and the worldwide manufacturing resources of Ingersoll-Rand and Atlas-Copco, Joy had to undertake major cost-cutting, redesign, and product line consolidation and focus programs to survive the 1981–1983 recession. The number of parts on many compressors was reduced by 50%, and the weight by as much as 33%, without affecting machine efficiency or performance. The company's 1986 annual report described the company's strategy:

> Both the Michigan City and Buffalo plants are experiencing a period of transition to newer technology and more sophisticated manufacturing capabilities. A commitment has been made to install a major CAD/CAM system in fiscal 1986 that will provide considerable efficiency in design, engineering, and machining. As the air machinery business matures, it is imperative that Joy provide its customers with state-of-the-art technology at a competitive price.

In addition, the company appeared to be concentrating on its strongest product line, centrifugal compressors, sold through its direct salesforce.

Another move was to strengthen, but not expand, its distribution channels. A company manager described these efforts:

> We did several things for our distributors. For the first time in the history of the company, we offered several trade promotions and

incentives. We increased our distributor training efforts. And, of course, our product rationalization program benefited our distributors. The new Twistair line of compressors, for example, were high-quality, price-competitive machines that significantly increased our distributors' competitiveness in their markets.

In 1985, Joy divested its portable compressor line, and in June 1987, Joy was bought by an investment firm, which in turn sold the compressor division to Cooper Industries, a $3.4 billion manufacturer of electronic equipment, wiring, cable, hand tools, and air compressors. Cooper merged the Joy acquisition with its Gardner-Denver division,[7] acquired in 1979, which made pumps, air and gas compressors, blowers, mining equipment, and industrial transmissions and gears.

Sullair

Unlike Ingersoll-Rand and Joy, Sullair operated solely in the air compressor business and with a narrower product line: rotaries in the 25-horsepower-and-larger size range. The company was founded in the mid-1960s when Donald Hoodes, a former Joy employee, designed the first rotary screw stationary air compressor to compete with reciprocating compressors. The initial price of the rotary was lower; it had fewer moving parts and required less maintenance than the recips. Sullair first introduced the rotary in 1965, and sales grew steadily, amounting to $179 million in 1981.

When the company had attempted to recruit independent distributors in early 1965, the effort failed because of Sullair's narrow product line, lack of an installed base, and unpromising prospects for distributor aftermarket parts and service revenues. Unable to develop an independent distributor channel, Sullair then concentrated on building a network of company-owned stores.

By 1975, the company sold its stationary and portable air compressor lines entirely through 80 such stores. These outlets, riding the crest of a construction boom, did exceedingly well. It soon became evident, however, that the stores could not be profitably operated as "one product line" businesses. Sullair then inte-

---

[7] The Gardner-Denver division made a full line of recip and rotary compressors, though its strength was in the recip line. The division had a strong customer base in the petroleum, mining, and locomotive industries. Through its 44-person salesforce, it sold directly to its sister divisions and to OEMs and large users such as construction companies, which purchased customized units. Direct sales had amounted to 30% of the Gardner-Denver division's total sales in 1983.

grated backward into assembling and selling such lines as the following:

- generating sets
- sandblast boxes
- waterblast rams
- Jacuzzi pumps
- fractional (horsepower) motors
- alternators
- light towers
- downhole steam generators

Nevertheless, the 1981–1983 recession had a devastating effect on the company's revenues and profits. Demand for its products fell sharply, just as it was expanding its product line and distribution network, leaving its captive distributors holding large inventories. Retrenching, Sullair again focused on its core business unit, the air compressor division, eliminating most of the lines that had been added to broaden the company-store revenue base.

There followed, beginning in 1984, a phased withdrawal from the captive sales outlets and another attempt to build independent distribution. In that same year, Sullair was acquired by Sundstrand, a diverse line industrial supplier. By this time Sullair had established a broad product line of rotaries and a sizable installed base and was successful in its second effort to develop external resale channels, reporting a modest profit on sales of $120 million in 1985. As of 1986, the company had fewer than ten captive stores and had recruited over a hundred independent distributors. In the product segments in which it competed, Sullair was second in market share to Ingersoll-Rand.

Recalling the effort to rebuild independent distribution, the company's marketing vice president commented,

> Distributors have long memories. They remember us as a captive distribution company. It was not easy to get their confidence and trust. We systematically went after the strong distributors of our weak competitors like Sullivan Machinery. We picked several good ones. But we have accomplished only half the job of rebuilding. We have organized several distributor meetings and conferences to assure them that we are here to stay, listen, and grow with them. We have begun to see encouraging results.

Atlas-Copco

A Swedish company with worldwide sales of $1.75 billion in 1985, Atlas-Copco was the world's largest manufacturer and marketer of stationary air compressors. A wholly owned subsidiary, Atlas-Copco of North America (ACNA), began marketing in the United States in 1950. Manufacturing operations in this country began in 1980 with the acquisition of compressor factories from Worthington and Turbonetics, long-established U.S. manufacturers of air compressors and other heavy industrial equipment. As of 1985, Atlas-Copco sold a broad line of recips, rotaries, and centrifugal machines and had an overall U.S. market share of about 10%. Its share of the oil-free rotary compressor product segment was substantially higher, at 30–40%. The company marketed its large compressors (over 300 hp) through its own salesforce, but all other products were sold through its network of 85 independent distributors.

Atlas-Copco's ability to enter the U.S. market and to establish an effective sales and service organization was unique among foreign manufacturers of compressors. This entry was accomplished in three distinct phases.

Phase One. In the early 1970s, while Atlas-Copco had substantial market shares in Sweden, Norway, France, Germany, Italy, Spain, Brazil, the Philippines, South Korea, and Japan, it held a very minor share of the U.S. market, about 1%. At that time, Atlas-Copco made a wide range of reciprocating compressors, but most were designed to meet European engineering standards. By 1975, the company had also developed several compressor models to conform to U.S. standards. The first attempt to establish distribution was a complete failure. ACNA management cited the lack of an installed base of machines as the single most important reason.

Phase Two. Beginning in 1976, the company altered its strategy; instead of offering prospective distributors a broad line of products, the company focused on its foremost product, the Z-series oil-free rotary screw compressor. The Z-series rotor, a key component, was made in Sweden to high-quality standards resulting in more reliable machine performance and lower energy consumption than in comparable American-made compressors. In addition, as a worldwide supplier of oil-free rotaries, Atlas-Copco was the low-cost producer.

ACNA offered prospective distributors a 15% margin on list price, which compared favorably with the 8–10% margins they realized on oil-lubricated recips and rotaries. The company also promoted the idea of end-user maintenance contracts to enhance aftermarket revenues for distributors. ACNA's focused efforts to penetrate air compressor distribution networks were further aided by its willingness to share distributor shelves with other brands. Since other suppliers did not provide oil-free rotary compressors required to serve all processed air applications, distributors could carry Atlas-Copco's Z series without confronting their principal suppliers' lines with direct competition. Finally, some distributors viewed Atlas-Copco as a future challenger to Ingersoll-Rand and Joy and recognized an opportunity to gain local market share. As of 1978, ACNA had built a network of about 40 distributor firms.

Phase Three. In 1979, Atlas-Copco purchased a part of Worthington's compressor business and, along with it, gained access to 40–50 Worthington distributors. Atlas-Copco also negotiated the right to manufacture certain models of Worthington machines and spare parts and to market them under the Worthington name for a period of two years. This arrangement enabled ACNA to ensure its newly inherited network of Worthington distributors of parts availability to service the installed base. With distribution in place, ACNA steadily expanded its U.S. product line to include a broad range of recips, rotaries, and centrifugals.

Thus, the key to ACNA's successful strategy for building U.S. distribution lay in using its Z series as an entry wedge. This line did not compete directly with distributors' primary brands and offered incremental revenue opportunities further enhanced by generous margins. At the same time, ACNA systematically went after the distribution systems of its weaker competitors and did not directly take on the industry leaders. The 1979 Worthington acquisition converted a beachhead into a solidly entrenched market position strengthened further by expanding the Atlas-Copco line of air compressors going through distribution.

CONCLUSIONS

In comparison with the distribution of load centers and disk drives, stationary air compressor distribution systems are built around a high degree of producer-reseller interdependency. Each distributor typically represents only one producer, with other manufacturers possibly supplying noncompeting, fill-in items. This close link between producer and reseller is a function of product technology, the size of distribution investments required to support one supplier's line, the importance of brand-specific parts and services income, and the high proportion of total distributor revenues typically provided by the sales and servicing of an air compressor line. Ultimately, the switching costs associated with changing prime suppliers—as measured in terms of obsoleted inventory investments as well as forgone vested interests in an installed base—are high.

On their part, producers are highly dependent for maintaining and building market share on resellers, many of which are their sole representatives in important trading areas. Further, field service for existing equipment, an important consideration to buyers of new equipment, is a distributor responsibility on which producers have become increasingly reliant. Producers rely on their primary resellers, as well, to serve as channels to a second distribution level, the mill supply houses and broad line industrial distributors, the great majority of which would not have annual sales of air compressors sufficient to warrant their being called on by a producer's salesforce.

Thus, in this product-market, long-term supplier-reseller relationships preserve the dominance of the market leaders and represent formidable entry barriers for challengers. Nevertheless, although it has taken one 10 years and the other 20 years to do so, Sullair and Atlas-Copco have successfully established themselves as major U.S. suppliers. In each case, the key to market entry was a superior product offering, but the lack of an installed base was a major deterrent to developing distribution. In the meantime, product line superiority, defined in terms of breadth of product offerings and market acceptance, served to protect the established distribution networks of Ingersoll-Rand and Joy from competitive intrusion. It was the networks of weaker producers that became targets of opportunity for the new entrants. Even there, gaining distribution by displacing extant competitors was a slow process.

There were few, if any, other options, as Sullair's experience demonstrated. Starting from scratch to gain market access through a chain of company-owned stores imposed intolerable financial drains on the new company and led to severe distortions in its product policy. Seeking to develop independent distribution by bringing new distributors into the market would not have been a promising strategy either. Local trading areas are limited in revenue potential and in the number of distributors each can support. Those already in place relied heavily on service revenues, an income stream not immediately available to new distributors representing equally new producers. Future change, then, may come only with significant product innovation, or substantial growth in market demand, and/or as existing producers leave the industry for lack of financial resources to ride out its depressions.

Product-Market Study IV

BOOM-AND-BUST IN THE MARKET FOR OIL COUNTRY TUBULAR GOODS

In the first quarter of 1986, U.S. world oil prices fell dramatically. To the domestic oil country tubular goods (OCTG) industry, the lower oil prices represented one more blow in a market already reeling from a collapse in demand, influx of foreign-produced goods, and chaotic pricing. Between 1982 and 1985, OCTG sales in the United States had dropped from an estimated $4.75 billion (7.3 million tons) to $2 billion (3.5 million tons); import shares of the OCTG U.S. market grew from 43.7% to 62.8% over this same period (see Exhibit IV.1).

The demand for oil country tubular goods is driven by the level of oil and gas drilling, a measure of which is the weekly count of oil drilling rigs in operation. As Exhibit IV.2 indicates, drilling activity soared to record heights in 1981–1982 and then abruptly crashed. Through 1983–1984, drilling levels remained volatile before slumping in 1985; in 1986, drilling activity dropped to a 12-year low. The boom-and-bust syndrome was fueled not only by a slump in drilling but also by the huge OCTG field stock inventories amassed during the 1981–1982 boom to protect against having to curtail drilling activities for lack of pipe. According to industry sources, an estimated 40% of the 6.6 million tons of pipe purchased in the United States in 1981–1982 had been stockpiled.

Chaotic OCTG market conditions then led to divestitures, the closing of pipe mills, and a spate of bankruptcies among oil field supply distributors, and distributors that survived had to cope with considerable market changes. Buyer behavior, especially among the large oil companies, had changed markedly. What was a routine convenience purchase at established list prices became the subject of intense competitive negotiation among users, producers, and distributors. The oil field distributors, through which tubular goods had traditionally flowed, adopted survival strategies to min-

355

Exhibit IV.1   OCTG Domestic Market

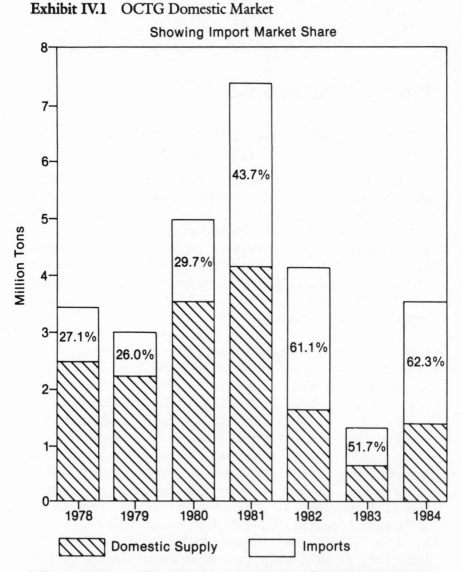

Showing Import Market Share

imize inventory risks and to hedge sources of supply. No longer able to count on distributor brand loyalty and unwilling to leave user price negotiations in the hands of resellers, the major OCTG suppliers became more actively involved in contract negotiations at the user level. A market that was once stable with regard to price, distributor margins, the flow of goods through resellers, market share positions among producers, and the dominance of domestic suppliers producers was, by 1985, unstable in all these dimensions.

Exhibit IV.2   Weekly Rig Count

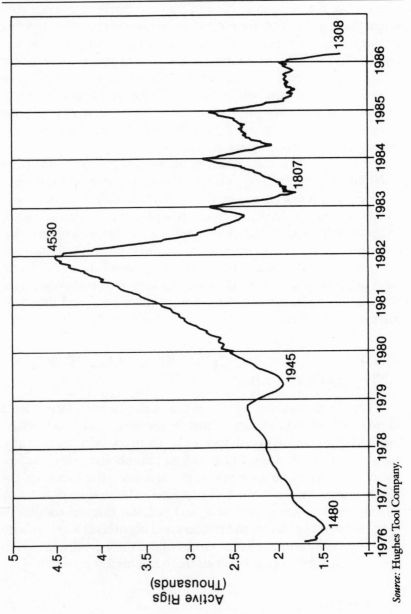

Source: Hughes Tool Company.

While the three preceding product-market studies have focused on distribution in stable and growing markets, this one considers the impact of severe economic decline on channels structure of three major OCTG suppliers, two of which are domestic and one foreign. OCTG distribution in the United States is an example of a system that by long tradition was tightly limited in number of reseller firms and successfully defended against foreign intrusion. However, it was a system that was not able to survive, intact, the extremes of prosperity and depression and the sea change in purchasing modes among major users.

A close look at the OCTG industry also indicates how price instability and the level of inventory risk shape distribution systems as producers, resellers, and users each seek to shift the risk burden to others in the chain or, alternatively, to assume the risk in pursuit of a competitive advantage. This study also confirms an observation made previously: that confronted by sets of common external market conditions, competing firms tend to respond in very different ways depending on their individual business environments—their financial strength, the nature of their product lines, and their past market orientations.

PRODUCTS, PRODUCT SUPPLY, MARKETS, AND DISTRIBUTION

The discussion that follows presents the essential facts related to products, sources of supply, market segments and buyer behavior, and patterns of distribution. We then consider three major competitors—U.S. Steel, Lone Star, and Sumitomo, a leading Japanese producer of tubular products—and how the events of the 1980s shaped their distribution strategies. The study concludes by relating the economic, technical, and political external conditions to the behavior of buyers and resellers and how these, together with the individual business circumstances of the three producers, led to dramatic changes in each firm's strategy and market positioning.

The Product

OCTG encompasses three different types of products: drill pipe, casing, and tubing. These are used respectively to drill oil or gas wells, to line the well hole, and to bring oil and gas to the surface. In an average well, casing accounts for 79% of the OCTG consumed, tubing for 19%, and drill pipe for 2%.

Tubular goods are made in many grades and sizes to meet the requirements of wells drilled to different depths in varied geological environments. There are three OCTG product quality levels, called tiers. Tier 1 products are super-high-strength products that are used in critical applications, such as extremely deep wells under high pressure and temperature conditions and in corrosive environments. Tier 2 OCTG are the high-strength grades used in moderately deep wells. Tier 2 OCTG may also be specified where corrosive conditions are expected or for applications where the consequences of a leak or blowout would be particularly severe, such as in offshore drilling. Tier 3 includes the lowest-grade OCTG and is used in shallow wells. Many of these wells are in existing oil fields where geological conditions are well known.

OCTG products are also defined in terms of the way they are manufactured. Seamless pipe is extruded from a solid steel billet, or round, to form a tube of a certain length, diameter, and wall thickness. ERW (electric-resistance welded) pipe is made by forming flat strips of steel into a tubular form and then welding the seam. Seamless pipe is considered by some to be of higher quality than ERW.

It is standard practice for OCTG users to test each length of pipe for conformance to specifications. Third-party inspectors perform these tests using instruments and techniques that are, by and large, a development of the 1980s. Prior to the beginning of the decade, OCTG could be inspected only on the outer surface and at the ends. New electronic inspection procedures, however, measure internal and external diameters and detect surface irregularities not visible to the naked eye.

Sources of Supply

Before 1980, the largest suppliers in the U.S. OCTG market were the integrated domestic steel producers. By the mid-1980s, several domestic suppliers had quit the OCTG business. Minimills[1] proliferated as OCTG producers. Foreign competition became a dominant force, capturing, at times, more than 60% of the market. As a result of a shortage of OCTG in 1980–1981, many U.S. mills had been forced to put their customers and distributors on alloca-

---

[1] Minimills are small, cost-efficient mills that rely on scrap steel as a primary raw material.

tion. Consequently, users and distributors turned to foreign sources to meet their needs. While foreign OCTG had been imported into the United States since the 1950s, it was not until the tremendous increase in demand in 1980–1981 that offshore mills made significant inroads in the domestic market.

Further, in an effort to increase OCTG production in response to rising demand, domestic producers allegedly relaxed their quality standards. This coincided with the development of new electronic inspection techniques, and it soon became apparent to OCTG users that foreign pipe, particularly from Europe and Japan, was of equal or better quality than domestic.

The 1980–1981 boom in drilling activity in the United States was echoed worldwide. As demand for OCTG rose, new tube mills were built throughout the world, most notably in Europe and Japan but also in developing nations such as Mexico, Brazil, and South Korea. Most of these plants, though, did not come on line until after the 1982 crash, and the result was extreme excess capacity. The United States was and is the largest market for OCTG, and so the output of the new mills had flowed strongly into this market.

New foreign mills, particularly those in developing nations, could produce OCTG at lower cost and could compete on price, despite greater shipping costs. The rising strength of the dollar against most foreign currencies during this period further enhanced the cost advantage of foreign suppliers and made the U.S. market increasingly attractive.

During 1980 and 1981, imports made their greatest gains, rising from less than .8 million tons in 1979 to over 3 million in 1982 (see Exhibit IV.1). By the middle of the decade, the U.S. government intervened to negotiate voluntary restraint agreements (VRAs) with most countries exporting OCTG to the United States. Imports from these countries were then limited to a fixed percentage of U.S. demand. Japan had the highest quota, with 17.5%.

The response to VRAs by foreign suppliers in Japan and Europe was predictable. Those suppliers capable of producing high-quality pipe concentrated on the first tier of the OCTG market, selling high-strength super-alloy products at premium prices. OCTG suppliers in developing nations, on the other hand, focused on Tier 3, using their cost advantage to compete via price. They were joined in that market segment by several new domestic suppliers.

Domestic companies such as Armco, CF&I, LTV, Lone Star Steel, and U.S. Steel positioned themselves primarily in the Tier 2 market segment, as did Algoma, a Canadian producer. Quality foreign mills were also in this market, but their participation was constrained by the VRAs.

Market Segments and Buying Behavior

The buyers of oil country tubular goods are the drilling contractors and the oil companies, including the 20 major oil companies and over 2,000 smaller independents. The independents concentrate on shallow wells and account for 65–85% of all the wells drilled in the United States. However, in terms of footage drilled, the majors represent the larger part of the OCTG market—an estimated two-thirds to three-quarters. The majors are the primary buyers of Tier 1 and 2 products, the independents, of Tier 3 OCTG.

Buying behavior varied between the major oil companies and the independents, but for both, purchasing practices changed markedly following the sharp run-up in OCTG demand in 1979–1981 and again after the precipitous decline starting in 1982.

The decision factors for the independent oil companies, according to an OCTG product manager, had always been "price, price, and price; a lot of independents operate on a shoestring." The financial condition of many independents aside, few had reason to consider factors other than price. Conditions in the shallow wells drilled by the independents were not as critical as those in deep wells. The products they used were standard carbon-grade OCTG (Tier 3) available from a host of domestic and foreign producers.

The thousands of small independents were not called on by supplier sales reps; they relied instead on local distributors for OCTG, MRO supplies, and credit. In many instances, in fact, distributors helped to finance the drilling costs of independents by taking fractional interests in drilling ventures.

As for the major oil companies, prior to 1979 they had allocated their purchases of OCTG primarily to the domestic mills, based on such criteria as mill capacity, new plant investment commitments, and services rendered. Contracts were based typically on list prices that included a traditional 6% distributor margin. Procurement managers then chose distributors based on the services they offered in the field, such as maintaining MRO stocks and providing 24-hour service.

Given the potentially high cost of pipe failure in deep-well and offshore drilling, quality was of primary concern to the majors. Availability of supply was also critical. While the cost of pipe was small relative to the overall costs of completing a well, rig downtime as a result of pipe shortages meant quickly escalating drilling costs and, more important, postponed revenues from oil production. A one-year OCTG supply-on-hand was not unusual.[2] These considerations explain, in large part, the oil companies' reliance on the large domestic mills.

During the 1979–1981 boom in drilling activity, quality, sources of supply, and now price became even more critical concerns for buyers. Domestic mills raised their prices in the face of extreme shortages, opening the door to sales of imported product. The shift to offshore sources was given further impetus by an alleged slippage in domestic mill quality. Some oil companies reportedly experienced OCTG reject rates as high as 25% from domestic mills during this period, as compared with less than 3% from foreign sources. At the same time, offshore tubular goods prices were 10–25% below those of domestic suppliers.

For many oil companies, considerations of price, availability, and quality—of Japanese OCTG, in particular—were sufficient to overcome a strong buy-American preference. In addition, their readiness to return domestically produced pipe that did not meet quality standards became increasingly apparent, especially with the development of new testing techniques. In fact, after 1982, some companies undertook programs for testing their inflated inventories of OCTG as a way of reducing stocks. Below-standard pipe was then returned to the mills for credit based on original purchase prices.

Heightened buyer sensitivity to price and quality considerations had two other effects: First, large buyers began to use competitive bidding in allocating mill purchases. Many oil companies developed lists of approved bidders based on the quality of OCTG their mills could produce.

---

[2] One major oil company reported holding inventories worth close to $1 billion in the field at the end of 1981. While this amounted to an 18-month supply based on 1981 projections of 1982–1985 drilling rates, it represented a 36-month supply based on actual 1982–1983 experience. By the end of 1986, this company had liquidated about three-quarters of the inventory it had on hand at the end of 1981.

In addition, the majors unbundled their OCTG requirements from their MRO requirements in selecting OCTG distributors. Thus, competitive bidding eroded not only mill margins but distributor margins as well.

When the key buying factor had been OCTG availability, the balance of power had been with the domestic mills; after 1980 the balance shifted in favor of user-customers, and the key buying factors became price and quality. What had been referred to by one OCTG manager as "traditional steel oligopolistic pricing" turned into a fiercely competitive market.[3]

Distribution

Prior to 1980, there were only 75–80 OCTG distributors (called oil field supply houses), and the major domestic producers franchised very selectively. At that time, virtually all OCTG moved through the supply houses. Given an assured 6% functional discount, an OCTG distributorship was reputed to be quite lucrative and highly sought after.

During the run-up in demand of 1980–1981, distributors were able to command margins above the 6% discount on OCTG, and as their inventories grew in value, they were encouraged to borrow against their existing OCTG stocks in order to purchase more.

Many established oil field supply houses began to sell foreign pipe. Placed on allocation by the domestic mills, they turned to overseas OCTG sources. At the same time, numerous others clamored to enter the OCTG market. Previously shut out by the selec-

---

[3] In the summer of 1984, for example, LTV Corporation began to offer deep discounts from book prices to select customers. These discounts were 25.5% for high-strength grades and 20.2% for carbon grades. LTV's discount preserved the distributors' traditional 6% functional discount and the 2% discount for prompt payment. This pricing was frequently expressed as 25.5 + 6 + 2 and 20.2 + 6 + 2.

In mid-1985, Lone Star Steel announced a further price discount of 33 + 3 + 2 for both carbon and high-strength grades, and it was reported that Lone Star would offer 36 + 2 if buyers elected not to have their purchases pass through distributors. Lone Star, however, returned to the previous prices of 25.5 + 6 + 2 and 20.2 + 6 + 2 within two weeks.

Six weeks after Lone Star rescinded its price move, U.S. Steel announced a price increase to 10 + 6 + 2 for carbon steel and 15 + 6 + 2 for high-strength steel. The domestic industry, with the exception of Lone Star, followed the U.S. Steel price move. Shortly before the U.S. Steel price was to become effective, Lone Star indicated to key customers that it would not follow suit. Several days later, U.S. Steel returned to 25.5 + 6 + 2 and 20.2 + 6 + 2. Within days, CF&I pushed discounts to 32 + 6 + 2 on high-strength grades and 22 + 6 + 2 on carbon grades. Lone Star returned to its 33 + 3 + 2 offer, and U.S. Steel responded with 33 + 6 + 2, which Lone Star matched.

tive franchising practices of the domestic producers, these individuals and firms found sources of OCTG in the foreign producers, as well. Some of these new entrants did not stock OCTG, but rather sold it "straight off the dock" on consignment from their foreign suppliers under agency arrangements.

In 1982, a shakeout in distribution occurred with the crash in OCTG demand. Many newcomers did not have the financial strength to weather the downturn. Even the established distributors, particularly those who had speculated heavily in inventory, were financially devastated. As prices for OCTG dropped to half their 1982 high, the value of distributors' inventories plummeted. Banks called loans backed by inventory collateral, forcing some distributors into bankruptcy. Others gave up holding any inventory and became OCTG brokers or sales agents.

Following the 1982 crash, there was a significant shift in the respective roles of suppliers and resellers in OCTG distribution. Suppliers had always had salespeople calling on large end-users, but they stepped up their efforts to promote their own brands in the end-market. According to our questionnaire survey responses from six large OCTG producers, approximately 10% of their 1985 sales were being made directly to users with no distributor involvement, and 50% were negotiated with users and serviced and billed through distributors. The remaining 40% were made to distributors, agents, and brokers with no involvement on the part of these suppliers at the user level. The six producers also reported that approximately 60–65% of their revenues in that year were derived from sales, either direct or through distributors, to the 20 major oil companies.

During this time, suppliers often negotiated directly with buyers that then negotiated separately with distributors about services (i.e., inventory, credit, delivery) and about the portion of the 6% functional discount the distributors would retain in return for those services. Some buyers expected minimal services from distributors and allowed a 3% margin or less. Other buyers granted the full 6% but bargained for more services in return.

In addition to shifting roles in the industry, massive shifts in the locus of inventories occurred among oil companies, distributors, and the mills, as they typically have during recurrent periods of shortage and excess supply. At the distributor level, OCTG inventory investments were high relative to sales and slow to turn over. In addition, distributors that did stock full lines of pipe lost their competitive advantage against suppliers that served users directly out of mill stocks, and against other distributors, agents, and brokers that did no stocking at all. This, and the propensity of the majors to make their purchases of pipe and oil field MRO supplies independently of one another in the 1980s, undermined the traditional industry role of the full-line stocking distributor, or oil field supply house.

COMPETITORS

Against this backdrop, we look now at the long-term strategies as of 1985–1986 of three major producers—U.S. Steel, Lone Star Steel, and the leading foreign mill, Sumitomo—and at how they adapted to the events of the 1980s. Each profile begins with a comment from a manager in a fourth OCTG-producing company.

U.S. Steel

> U.S. Steel sets the pricing policy for OCTG in the U.S.A. [It] also produces the full-size range. With the opening of their new Fairfield, Alabama, seamless mill they are now perceived as the highest-quality U.S. OCTG producer. [They] sell through a captive distributor (Oilwell Supply) and through many of our independent distributors.

Although U.S. Steel (USS) was primarily a supplier of seamless pipe for deeper wells, it had the broadest product line of any OCTG producer and a larger share of the U.S. market than any other domestic producer. Its Fairfield plant, started in construction in 1981, was the most modern, lowest-cost, seamless OCTG production facility in the United States.

All sales were made through the USS distribution network, which, as of 1986, included 23 distributor firms, many of which had been USS's OCTG distributors for more than 75 years. Of these, 7 sold only tubular goods; the remaining 16 operated stores as well and served the broad supply needs of drilling rig operators, although OCTG accounted for over half of their sales. U.S. Steel's Oilwell Supply Division (OWD) accounted for the largest distributor share of USS's OCTG tonnage sales. OWD operated more than 100 stores.

Sales activities centered in Houston, where a small, three-person sales team was assigned to sell OCTG as well as standard pipe and line pipe. Six other sales reps—in Tulsa, Dallas, and New Orleans—called both on end-users and distributors; each covered 20–25 key accounts. They had the back-up support of six technical service personnel located at different points in the Southwest as well as a product engineer and two metallurgists.

In 1982, the collapse in demand triggered significant changes in USS's OCTG marketing program. USS sales reps became increasingly active in calling on major user-customers, because these accounts had gone to competitive bidding in sourcing OCTG supplies and were actively soliciting bids from sometimes as many as 50 different mills, distributors, brokers, and agents. Nevertheless, all orders that were negotiated directly by USS sales reps were passed through USS distributors.

By 1983, OCTG inventories were not yet under control, and distributors were not yet reordering pipe for stock. To stimulate sales, USS initiated two programs, one designed to encourage distributors to buy for stock, and the second to stress the quality of USS pipe. Under the first, the so-called 90-day program, USS offered to ship and stock tubular goods in any location for a distributor or a major oil company and to invoice the buyer for the stock as used. As of 1986, more than half of USS's OCTG sales had resulted in their maintaining field stocks in support of individual contracts with stocking distributors and end-users. Under the second program, USS offered to pay all inspection costs if the reject rate exceeded 3%; under 3%, USS would pay the inspection costs on rejected pieces only.

USS managers regarded these programs as successful in meeting their sales objectives and in helping to boost USS's market share among domestic producers. Other factors contributed, as well:

1. The sales base provided through the Fairfield contracts[4]

2. The high quality of the output of the new Fairfield mill as

[4] Eleven of U.S. Steel's large customers, including both distributors and users, had assisted in financing the construction of the Fairfield mill in return for first claim on the mill's production when the mill came on stream. (It will be recalled that when ground was broken for the Fairfield plant in early 1981, there were severe shortages of OCTG.) The contractual arrangement with the 11 participants also provided that they would commit to buy, at prevailing market prices, specified amounts of the mill's output. All 11 had continued to honor contractual commitments.

well as of USS's Lorain works as the result of process improvements

3. Competitive pricing
4. A market shift back to the historical mix of 60% seamless and 40% ERW from the 60/40 balance in favor of the latter, adopted during the 1980–1981 surge in OCTG demand

Nevertheless, U.S. Steel's management feared that a continuing decline in OCTG demand and the growing influx of foreign tubular goods would adversely affect its facilities' operating rates, as well as the financial strength of its stocking distributors. With all of its sales volume going through its distributor network, a network that was generally regarded as the strongest in the industry, USS management was concerned about preserving the system's viability.

Lone Star Steel

Lone Star Steel has been able to convince a lot of people that their welded product is as good as seamless. Most majors, however, won't use welded under severe drilling conditions or in offshore waters. They [LSS] have been offering large discounts for the past year in an attempt to secure business away from U.S. Steel. [They] also sell through a network of independently owned distributors. Their advertising always stresses the "buy-American" label.

An OCTG supplier since 1952, Lone Star Steel (LSS) was second to U.S. Steel in its share of the U.S. OCTG market.[5] It produced only ERW pipe, however, while U.S. Steel concentrated heavily on seamless OCTG. LSS historically had an estimated 20–25% market share among the six domestic producers, but with foreign suppliers accounting for 40–60% of U.S. OCTG consumption, LSS held 8–15% of the market overall.

Although LSS had held a strong position both among the major oil company users of OCTG and the independents, the influx of low-price, lower-grade imported pipe beginning in the late 1970s had weakened it in the latter market segment. In deep-well appli-

---

[5] Incorporated in 1952 as an independent publicly owned company producing seam annealed (ERW) line pipe, casing, and tubing, Lone Star Steel was acquired 16 years later by Northwest Industries. In 1985, LSS was spun off to Northwest Industries stockholders as part of a leveraged buyout refinancing that converted the parent company from a publicly held to a private organization. Lone Star Steel, however, continued as a publicly held corporation.

cations, LSS ERW competed primarily with seamless produced by USS. Lone Star's high-quality ERW pipe was priced competitively with USS's seamless pipe. Both sold at prices above those of other domestic and foreign ERW producers.

LSS had increased its distributor network from 13 in 1983 to 41 in 1986. The expansion was motivated primarily by the desire to tap specific "distributor-controlled" volume, that is, to reach users that were strongly loyal to particular distributors and tended to buy the brands of pipe the distributor stocked. However, according to Byron Dunn, manager of steel pipe sales, the incremental volume gained through adding distribution had not fulfilled his expectations.

> First, the programs they [distributors] "controlled" began to
> dwindle as drilling operations declined. Second, much of the con-
> trolled volume shifted to lower-priced imports. And, actually, many
> of the new distributors didn't try to convert their regular accounts
> to Lone Star; they went after the mill-direct volume that we had
> negotiated with our major oil company customers and tried to get a
> piece of it by cutting prices. They could shave their margins because
> they had low overheads; there were just a few people on the pay-
> rolls; they didn't stock; they'd get a purchase order, go to the bank
> for a loan to finance it, and order the pipe out of one of our ware-
> houses. In the meantime, our old-line stocking distributors from
> whom the majors regularly purchased got really upset because now
> there were more LSS-authorized distributors going after a fixed
> amount of mill-direct business and at bargain-basement prices.

> Distributors may be getting nervous anyhow. Not long ago
> both Hughes Tool and Smith International eliminated distributors
> in selling drilling bits and now go direct. Some think that could
> happen with OCTG. Without the volume that bits and tubular
> goods bring in, there isn't much left.

With a 55-person marketing organization, 7 field warehouses for stocking pipe, a trucking subsidiary, and a network of 41 inde-pendent distributors, LSS operated as a full-service company. Its advertising slogan, "We've got you covered," emphasized that LSS field stocks were available for overnight delivery anywhere in the "oil patch," and LSS field service personnel were on call 24 hours a day to handle any problems at drill sites across the United States.

A primary purpose of the sales organization was to develop and maintain strong personal relationships among purchasing per-sonnel and drilling operators in the oil companies. According to

Dunn, "Personal relationships are key although somewhat less important now with a new generation of purchasing managers; but still very, very important."

A field salesforce of 20 technically trained reps, including 6 branch managers, called on OCTG users as well as on LSS distributors. Although the LSS salesforce was larger than that of U.S. Steel, LSS sales reps sold only line pipe, casing, and tubing. They and top officers of LSS called at all levels of management in customer companies. They negotiated sales of OCTG with users and arranged to have contracts filled through LSS distributors.

Less than 5% of total sales were made directly with no distributor involvement. About 80–85% were negotiated with users by LSS field sales personnel and processed through distributors, with the distributor receiving a margin based on performing functions such as stocking and credit. These were called "mill-direct" sales. Another 10–15% of sales were made by distributors, with no involvement by LSS reps with the distributors' customers. The amount of tonnage moving in this last category varied considerably, depending on how price competitive LSS was at any point in time. When it was pricing aggressively, sales to distributors not in the mill-direct category increased significantly as a percentage of total sales.

In retrospect, Dunn saw very little contribution to the LSS sales program from the added distributor coverage: "Except for providing credit, they aren't doing anything we're not doing ourselves every day. In fact, we really function as a distributor except that we don't finance sales."

After more than doubling the number of LSS distributors between 1983–1986, Lone Star's management announced in June 1986 that it had given notice to its 41 distributors that all existing distributor arrangements were terminated.

Sumitomo

> Sumitomo supplies the best pipe in the world. All other OCTG producers are measured against their quality standards. [They] sell through distributors and trading companies. I believe they underprice their product, which isn't necessary.

Having received American Petroleum Institute certification of its OCTG products in 1952, Sumitomo Metals, the world's largest producer of OCTG, was the first Japanese company to sell oil country tubular goods in the U.S. market. It was represented in the United States by the Sumitomo Corporation of America (SCA), a trading company in the Sumitomo group of companies. As of 1986, SCA was estimated by industry sources to have 5–7% of the U.S. market.

SCA's market entry was facilitated in 1955 by a steel strike in the United States, which created OCTG shortages for the next two years. During that time, SCA's sole distributor was a company in Tulsa. In 1957, Sumitomo established its own captive distribution operation, Union Pipe, which was formed as a joint venture among Sumitomo Metals, SCA, and an American business person with previous experience as an importer. Union Pipe established a market presence for Sumitomo at a time when other distributors were reluctant to represent Japanese mills.

In 1973–1975, during the OPEC oil embargo when oil prices went from $1.25 to $11 a barrel, SCA realized substantial gains in market share. It also gained substantial market share in 1979–1981 when OPEC raised prices from $11 to $28, and oil sold on the black market for $40 a barrel. During these two shortage periods, the total volume of OCTG increased greatly and so, too, did Sumitomo's share of the market. Imports from all sources had gone to 25% of the domestic market by 1973–1975. Although the imports' share of the U.S. market dropped after the oil embargoes, in the 1980–1981 boom, foreign-sourced product went to 50% of the U.S. market, and after that to 65%.

With the run-up in demand, Sumitomo expanded to 16 distributors. Under long-term supply contracts to major oil companies, it also made commitments that preempted a large portion of its mill capacity in Japan. In 1982, when the market crashed, many of these orders, as well as many distributor orders, were canceled. The cancellations shocked Sumitomo personnel, whose grounding in Japanese business norms had created expectations that such contractual commitments were binding.

May 1982 was remembered by SCA's management as a frantic time. SCA attempted to stop orders placed on the mills but was able to cancel only about 20% because some shipments were already in transit, and reducing mill output would have resulted in drastic layoffs. Instead SCA accepted in-transit and in-process product and stocked it in Houston. This amounted to over 100,000 tons. Pipe that had been transferred to SCA at $1,200 per ton was liquidated over the next three years at about $700 per ton. Most of this tonnage moved into the spot market through SCA distributors and was eventually purchased by other resellers and by small independent drillers.

By March 1985, SCA's excess inventories had been sold. During the fall in the market, some of Sumitomo's distributors went out of business, and Sumitomo dropped other distributors who had canceled orders. The boom-and-bust experience of 1981–1983 had underscored for SCA's management the financial vulnerability of selling through stocking distributors. By comparison, the major oil companies had demonstrated greater financial stability, most of them accepting deliveries on orders and taking OCTG shipments into stock to be worked off gradually in their own drilling programs.

As of 1986, Sumitomo's distributors functioned primarily as finders of sales opportunities, receiving a 3% commission on sales subsequently negotiated by SCA. They had, in fact, been advised by SCA's management not to purchase for stock, because prices had been falling steadily for 20 months. To satisfy current purchase orders, distributors then resorted to buying their OCTG supplies in the spot market at going rates.

In late 1985, SCA had sold about 30% of its tonnage directly to users under noncancelable orders at negotiated prices for delivery in six months. By March 1986, direct business had increased to 60%; 30% more was negotiated directly with user-customers and serviced through distributors; 10% went to distributors for sale to their own customers. Like those of other Japanese importers, SCA prices were about 6% below domestic mill prices for comparable grades and quality.

In 1986, SCA had 15 direct sales reps selling OCTG as well as standard pipe and line pipe. Approximately 80% of the Houston SCA group's time was spent in selling pipe made from special alloys and/or specially threaded. This special pipe was intended for use in drilling in high-stress environments and might sell for $10,000–40,000 a ton as compared with $1,000 for carbon-grade steel or $3,000 for stainless. Since negotiations with the major oil company customers for special pipe were highly technical and carried on over extended time periods, SCA managers saw no role for distributors at this stage in the development of the market. Further, special pipe was produced and shipped only to order.

In reflecting on his experience through the late 1970s and 1980s, one SCA vice president commented,

> The oil patch is a very closed society, difficult to break into. The domestic suppliers will always take a certain percentage of the U.S. market because of the support of domestic users based on feelings of patriotism and a sense of obligation to the steel industry which nurtured the oil and gas industry in its early years. The traditional relationship between the oil and steel industries in the United States has been a very strong factor in this market. Some major oil companies have a rule that they will limit their purchases of foreign OCTG to 20% of their total needs. Japanese technology, however, is appreciated by the oil companies.

Commentary

The profound changes in U.S. market conditions after 1981 affected these three OCTG suppliers differently.

U.S. Steel. Early in the development of an OCTG marketing strategy, U.S. Steel managers must have made two key choices: first, to depend heavily on distributors for selling, stocking, customer credit, and customer service; and second, to limit the number of such distributors. Relying on a distributor network would reduce the commitment of resources needed to sell direct to user-customers, resources such as a large field sales and service organization, warehousing facilities, and capital to finance inventories and receivables. In addition, the oil field supply houses, with their broad lines, would be able to reach the smaller independents.

Selective distribution was a logical option, as well, for several reasons: Customers were concentrated mostly in the southwestern states, limited distribution could reduce intrabrand price competition, and selective distribution maximized each reseller's market share—making the USS line an important income generator for its resellers and reducing any incentives resellers might have for carrying competing lines. Basically, such considerations tend to support the case for selective franchising in any product-market environment.

The demand increase of 1979–1981 and the ensuing market collapse, however, forced significant changes. By 1986, more than half of USS OCTG sales resulted from USS's maintaining field stocks in support of individual contracts with stocking distributors and end-users. At the same time, in responding to competitive pricing moves that eliminated the distributor margin, USS managers provided for the traditional 6% distributor discount. While the move showed distributor support, it was not the profit guarantee it once was for distributors, given severe price competition and the current propensity of the oil companies to claim a portion of this margin in negotiating selectively for distributor services.

Among its possible weaknesses, USS might include the relatively small size of its salesforce. As of 1985, six sales reps in all covered the oil fields in the Southwest, and all carried standard pipe and line pipe in addition to OCTG. We could speculate that any move to expand its direct selling activities, however, might be perceived as undermining the distributor's role and reversing USS's long history of distributor support. Similarly, history and relationships might be a deterrent in USS's making any change in the traditional 6% distributor margin in response to competitive price moves. The reluctance to make moves that might adversely affect its distributors could be explained by the company's continuing reliance on this network to service large accounts and to reach the many small independents.

Lone Star Steel.   Unlike U.S. Steel, Lone Star had traditionally relied heavily on its direct field salesforce, which included 20 technically trained reps calling on oil field operators. It also had seven field warehouses for stocking pipe and a trucking company subsidiary. Because of its extensive direct sales and customer service operations, Lone Star was sometimes described in the industry as a "distributor with a captive mill." Nevertheless, LSS had traditionally maintained an independent distributor network to reach accounts smaller than those which could be economically served through its direct salesforce.

LSS's strategy in the 1980s, however, was ambivalent regarding distribution. On the theory that some distributors "controlled" certain "loyal" pieces of business, LSS marketing managers franchised 25 new distributors in 1983–1986. For reasons described above by Byron Dunn, this move failed to improve sales volume, and LSS terminated all 41 distributors. Ambivalence was reflected as well in Lone Star's pricing actions; its price announcements sometimes provided for a 6% distributor margin, sometimes 3%, and sometimes none.

Under the high cost/price/quality pressures of the OCTG markets in the 1980s, LSS management would seem to be facing a difficult choice: whether to continue to carry its extensive direct sales and service operation or to rely more on resellers to reach user-customers. LSS's opting to rely on its direct salesforce may be interpreted as choosing to focus on the larger accounts that could be economically served on a direct basis. Given the weakened condition of OCTG distribution generally, and LSS's recent moves, which seemed certain to alienate its distributors, it was unlikely that the latter option—relying on resellers—was a viable one.

Sumitomo.   One SCA manager's comment, "The U.S. oil patch is a closed society," succinctly sums up three decades of the world's largest producer of OCTG trying to establish a position in the world's largest market. Sumitomo, the first Japanese company to enter the U.S. market, had significant strengths: a high-quality, low-cost product and technical skills in developing new products for use in hostile oil well drilling environments. It was at a significant disadvantage, however, in a market environment with a strong buy-American preference and where relationships between the domestic suppliers and their distribution networks were of long standing.

Sumitomo's initial entry strategy conformed to well-established patterns: its sales reps called on major oil companies, and it relied on what distributors it could enlist to stock pipe, extend credit, and provide customer service. The market collapse in 1982, however, left SCA managers with the bitter memories of having had to accept and dispose of large stocks of OCTG, the orders for which had been canceled. It left them with a lack of confidence in distributors' financial viability in times of stress and a concern for their failure to honor commitments.

Sumitomo's U.S. strategy as of the late 1980s may be seen as a response to this experience, and as a recognition of its strengths, weaknesses, and import limitations. In going for the high end of the market, it was building on technical strength and on its position as a cost-efficient producer of quality product. It was also seeking to optimize unit profits by positioning itself in a product-market segment that gave high priority to performance and exhibited less sensitivity to price. Further, it was a segment that could be served directly with little or no need for distributor involvement. Finally, long lead times were not likely to be a significant deterrent for Tier 1 customers, since prices were individually negotiated and not subject to short-run fluctuations.

Exhibit IV.3 Determinants of Competitive Distribution Strategies for Oil Country Tubular Goods

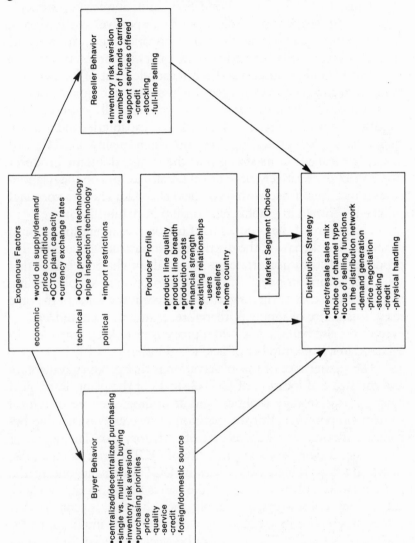

CONCLUSIONS

Thus, for U.S. Steel, Lone Star, and Sumitomo, strategic choices were shaped internally by their respective cost positions, their product quality, and their past histories, and externally by changing buyer behavior and by U.S. government intervention in the form of trade restrictions. Each responded out of its own particular circumstances. U.S. Steel built on its traditionally strong market position, its financial resources, and its well-established distribution; Sumitomo relied on its technical capabilities and modern mill capacity for producing low-cost, high-quality OCTG; and Lone Star leaned on its strength as the largest domestic producer of ERW, on its relatively large direct salesforce, and on the relationships it had established with its oil field clientele. These factors, and others, are shown in graphic relationship in Exhibit IV.3.

The nature of the product itself has been a significant factor in shaping buyer behavior, the distribution infrastructure, and OCTG producers' distribution strategies. One salient characteristic is that OCTG is a "line stopper": If there is disruption in supply, drilling stops, and the cost of rig downtime far exceeds the value of the pipe itself. Ensurance of supply is, therefore, critical to users and leads— everything else being equal—to a propensity to keep ample stocks on hand in a large number of field locations.

The significance of this observation is that product availability and the size and location of OCTG stocks at the plant, in oil field supply house storage facilities, and at drilling sites are of critical strategic importance. But the question of inventory ownership becomes compounded by the degree of inventory risk. The degree of risk tends to be a function primarily of OCTG price fluctuations, which are exacerbated over cycles in OCTG demand due to the lack, or surfeit, of inventories in the hands of producers, resellers, and drilling contractors. Inevitably, then, holding inventories becomes an exercise in speculation, with important implications for OCTG distribution. The interplay between the need for product availability and ensurance of supply, on the one hand, and the risks of inventory ownership, on the other, have thus been central in shaping OCTG distribution throughout the history of the industry. It is likely that those two factors in combination have been the single most important factor behind the dramatic changes that have taken place in the boom-and-bust environment of the past decade.

• Appendix B •

Glossary of Terms

Agents. Commission agents: individuals and organizations that represent the producer in selling its products; also called manufacturers' representatives, or MRs; the agent does not take title to the goods and is usually compensated by a commission calculated as a percentage of the selling price established by the producer.

Backward integration. Undertaking upstream operations in the manufacturing/distribution chain, for example, a producer integrating backward into components or raw materials production or an intermediary developing its own manufacturing operation, as alternatives to the external sourcing of these products.

Brand. A product group identified by producer name or by the trademark of the reseller (see *private brand*).

Brokers. Individuals and organizations that negotiate sales of goods—usually on a fee basis; may represent either buyers or sellers but tend not to establish long-term relations with particular principals, as do agents.

Bundled purchasing. Buying multiple products, generally related in use and of different brands, in a single transaction.

Buying cycle. The stages in the process of making a purchasing decision and receiving the product.

Captive distributor. A distributor owned by a manufacturer, which sells goods produced by other departments in the corporate structure as well as goods sourced from outside suppliers.

Channels of distribution. The product distribution system, including the producer's direct salesforce, independent distributors, agents, brokers, and captive distributors.

Competitors, first-tier, second-tier. From the producer's perspective, first-tier competitors are major, direct competitors; second-tier competitors are composed of smaller firms often spe-

377

cializing in a particular product line and/or a specific market niche.

Direct distribution. The producer selling directly to end-user customers through its own salesforce.

Direct sales branches. Field sales locations staffed by the producer's direct sales and sales administrative personnel.

Distribution costs. The full costs of negotiating a sale, fulfilling the contract, and providing product-related service, including the salaries of sales, service, and staff personnel; warehousing and delivery costs; sales office expenses; allocated overheads; and resale margins (see Chapter 4).

Distribution intensity. The relative density of intermediary organizations authorized to carry a producer brand in a trading area; *exclusive distribution* refers to a condition in which the producer has only one distributor in a market area; *selective distribution* refers to a policy of limited franchising, with only certain resellers being selected; and *intensive distribution* is the practice of making franchises widely available, usually to conform to customer preferences for purchasing convenience. These terms are used only in a general sense and have never been quantifiable, since degree of intensity is relative to the size of the product-market and to the sales volume and number of potential customers for a specific brand.

Distributor margin. The difference between the distributor's selling price and cost of goods sold expressed as a percentage of selling price.

DIY. Do-it-yourself, a market segment for many industrial products, consisting of consumers who buy supplies, components, materials, and machinery for use in home maintenance, hobbies and crafts, and the building, maintenance, and repair of equipment.

End-users. Purchasers that consume the product or use it as materials and components in the goods they make and sell; user-customers.

Exclusivity, brand or territorial. Conditions specified in some franchise agreements; brand exclusivity requires that the distributor carry only the supplier's brand and not any directly competing products; in granting territorial exclusivity, the producer agrees that it will appoint no other distributor in the defined geographic area that may carry the brand.

External intermediaries. See *intermediaries*.

Forward integration. The undertaking of operations that bring the producer closer to end-user customers by engaging in downstream manufacturing operations and/or distribution functions previously performed by outside organizations in the manufacturing/distribution chain.

Independent distributors. Individuals and firms that buy from producers and other intermediaries and sell to other resellers and to end-users.

Indirect distribution. The producer selling to end-user customers through independent resellers.

Installed base. The volume of products of a specific product category and of a given brand currently in use in a market area.

Intermediaries. All agents and resellers that sell the producer's product line, including the producer's salesforce. The term *external intermediaries* refers to outside distributors and agents and excludes the direct salesforce and any captive distribution arm.

Intrabrand competition. Competition, usually among resellers or between resellers and the direct salesforce, for sales of same-brand products.

List price. The price of the product or service listed in the supplier's price book, not net of any discounts or rebates.

Mail-order retailer. A reseller that enters and fulfills customer orders through the mail.

Manufacturers' reps. See *agents*.

Market niche. A product-market segment defined usually in terms of the product's application, a class of customer, and/or a geographic market area.

MRO. Maintenance, repair, and operating supply items; consumable supplies used in the production of goods and services.

Multichannel distribution. Distribution systems that include the producer's salesforce as well as other types and classes of intermediaries.

OEM. Original equipment manufacturer; a producer of goods for end-use markets or for use in the assembly of end-products.

Private brand. A reseller brand used to identify goods manufactured to the reseller's specifications by its own and/or outside manufacturing sources, as opposed to a manufacturer's or national brand.

Producer. A manufacturer of goods; a supplier of services.

Product-market. The total market for a product class or group.

Product-market segment. That portion of a product-market defined in terms of geographic area, end-use application, customer demographic characteristics, type of customer institution, and/or buyer behavior.

Product-specific investments. Capital investments made for facilities, equipment, and personnel resources for use in stocking, selling, and servicing a specific product line.

Push/pull. Terms referring to marketing strategies that focus the producer's pricing, advertising, promotion, and personal selling programs on the resellers (push) or, alternatively, on end-user (pull) levels. In a push strategy, the producer's objective is to motivate resellers to take on the burden of demand generation at the end-user level. In a pull strategy, the producer concentrates its marketing resources on generating end-market demand, relying primarily on intermediaries for demand fulfillment, thus "pulling" the product through channels of distribution.

Resellers. See *independent distributors*.

SIC code. The Standard Industrial Classification code of the U.S. Commerce Department, classifying product categories and subcategories identified by SIC numbers.

Spiff. A cash payment or an award of value made by the producer to a salesperson representing an intermediary for selling a unit of the producer's product; a form of incentive compensation.

Spot market. A market in which transactions between buyers and sellers take place at "spot" prices, which fluctuate daily, and, sometimes, hourly to reflect supply-demand conditions. In spot markets, intermediaries are often brokers that have knowledge of sources of available supplies and of the needs of customers in the market. Spot markets exist for a wide range of commodities and materials (e.g., petroleum products, grain, minerals, and fabricated items such as oil country tubular goods). Spot markets may often be used to move product supplies in excess of those which can be channeled through direct sales and resellers, which compose producers' established distribution systems.

Supplier. In this context, *supplier, vendor, producer,* and *manufacturer* are synonymous terms.

Trading area. A geographically defined market area; major centers of trade are known as primary markets, smaller centers as secondary markets, and sparsely populated territories as tertiary markets.

VAR. Value-added reseller; a reseller that purchases goods from a primary producer and adds value through product assembly, modification, and/or customization.

INDEX

A

Abrams, Lee N., 265n
Account potential, distribution boundaries based on, 156–158
Account restrictions, 36–37
Account type specialization, 114
Addison Electrical Supply (AES), 191
Advanced Micro Devices, 258n, 315
Adversarial collaboration, in industrial systems, 281–282
Advertising costs, 63, 66–67
Advertising-to-sales (A/S) ratios, 66–67
After-sale service, 32–33, 34–35
Agents, 212
  defined, 256
  in history of industrial distribution, 241–242
  sales by, 207, 212, 213
  size of, 213
Airco electrode plant, 90
Air compressors. *See* Stationary air compressors
Aldcroft, Derek H., 239n
Algoma, 361
Allegheny Ludlum Industries (ALI), 86
Allen-Bradley (A&B), 79, 96–97
Alloy Rods (AR)
  distribution channels of, 86–87
  resource availability and, 86–89, 92–94
Alster, Norman, 328n
American Hospital Supply, 221
American Mining Congress International Coal Show, 67
Ames Manufacturing Company, 236
Anheuser-Busch, Inc., 265
Apple Computer, 13
Armco, 361
Arrow Electronics, 12, 96, 170, 320
  channel blocking and, 258n
  on distributors, 99

interpersonal relationships at, 140
stock of, 314
suppliers of, 315
AT&T, 12
Atchinson, Cynthia, 264n
Atlas-Copco of North America (ACNA), 4
  background of, 15–16, 351
  distribution strategy of, 344
  entry into U.S. market, 97, 351–352
  market segments served by, 343
  market share of, 344
  product-market study of, 351–352

B

Baker, H. J., & Bros., 242
Balance of power, in industrial systems, 281–282
Barry Controls, 29
Bechtel, 340
Becton Dickinson & Company, distribution negotiations at, 43, 52–56, 115
Beddia, Paul, 90, 92
Berg, Andrew, 264n
Bergman Wire & Cable Company, 126n
Block franchising, 178
Blue Chip distributors, 50, 295–296, 297
Bonoma, Thomas V., 67n
Boston Manufacturing Company, 233–234
Boundaries, distribution, 152–158
  based on account potential or size of order, 156–158
  based on customers, 153–155
  based on product classification, 155–156
Brand exclusivity, 78–79
Brand preference, value of, 136–137
Brands, private, 250, 282
Brand switching costs, 140–143

383